Sustainability Sciences in Asia and Africa

Sustainable Agriculture and Food Security

Series Editor

Rajeev K. Varshney, WA State Agricultural Biotechnology Centre, Centre for Crop and Food Innovation, Food Futures Institute, Murdoch University, Murdoch, WA, Australia

Sustainability as a scientific endeavor requires broad understanding of the interconnections in our global environment. It is therefore positioned at the crossroads of research in natural sciences, technology and human sciences. This book series supports the global efforts by providing timely coverage of the progress, opportunities, and challenges towards achieving sustainable development goals (SDGs) in Asia and Africa. It fosters the research in transdisciplinary academic fields spanning across agriculture and global food security, health and diseases, access to clean water and affordable energy, climate resilience and effects of global climate change on biodiversity, water and land. This series is planned to narrate the stories of scientific development and progression towards SDGs from the regions of Asia and Africa. The overall aim of the series is to provide a comprehensive publication platform and act as a knowledge engine in the growth of sustainability sciences with a special focus on developing nations. The series constitute of four sub-series each focusing on a specific theme closely linked to a set of SDGs. Each sub-series is leaders and experts in agricultural sciences, environmental sciences, biomedicine and medical sciences respectively. Books in this series are useful for environmentalists, biotechnologists, plant scientists, medical practitioners, conservation biologists, policy makers, NGOs and scientific bodies working in the areas of sustainability sciences in the developing regions.

This book series support the global efforts towards sustainability by providing timely coverage of the progress, opportunities, and challenges of sustainable food production and consumption in Asia and Africa. The series narrates the success stories and research endeavors from the regions of Africa and Asia on issues relating to SDG 2: Zero hunger. It fosters the research in transdisciplinary academic fields spanning across sustainable agriculture systems and practices, post- harvest and food supply chains. It will also focus on breeding programs for resilient crops, efficiency in crop cycle, various factors of food security, as well as improving nutrition and curbing hunger and malnutrition. The focus of the series is to provide a comprehensive publication platform and act as a knowledge engine in the growth of sustainability sciences with a special focus on developing nations. The series will publish mainly edited volumes but some authored volumes. These volumes will have chapters from eminent personalities in their area of research from different parts of the world.

Vanika Garg • Robert J. Henry
Rajeev K. Varshney
Editors

DNA of Sustainability

Genomic Insights into Food Security Challenges

Springer

Editors
Vanika Garg
WA State Agricultural Biotechnology Centre, Centre for Crop and Food Innovation, Food Futures Institute
Murdoch University
Murdoch, WA, Australia

Robert J. Henry
Centre for Crop Science
Queensland Alliance for Agriculture and Food Innovation
University of Queensland
St Lucia, QLD, Australia

Rajeev K. Varshney
WA State Agricultural Biotechnology Centre, Centre for Crop and Food Innovation, Food Futures Institute
Murdoch University
Murdoch, WA, Australia

ISSN 2730-6771 ISBN 2730-678X (electronic)
Sustainability Sciences in Asia and Africa
ISSN 2730-6798 ISSN 2730-6801 (electronic)
Sustainable Agriculture and Food Security
ISBN 978-981-95-5108-8 ISBN 978-981-95-5109-5 (eBook)
https://doi.org/10.1007/978-981-95-5109-5

This Springer imprint is published by the registered company Springer Nature Singapore Pte Ltd.
The registered company address is: 152 Beach Road, #21-01/04 Gateway East, Singapore 189721, Singapore

Series Preface

Global food production systems are under significant pressure. Based on current estimations, food production systems will need to double production levels in order to keep up with the needs of the expanding global population. While the UN has an ambitious goal of ending world hunger, achieving food security, improving nutrition, and promoting sustainable agriculture by 2030, the stark reality is that 1 in 12 people go hungry every day. According to data from the UN, the number of people facing moderate-to-severe food insecurity has increased by a staggering 383 million people in just four years, demonstrating the scale of this global crisis.

The challenge is further exacerbated by climate change, which has had a profoundly negative impact on agricultural productivity and farm incomes worldwide, particularly in developing nations in Asia and Africa, which house the largest proportion of people suffering from malnutrition and other nutrition-related issues.

The agri-food systems in these regions are severely constrained by the subsistence nature of farming, declining land and other agricultural resources and increasing environmental pollution, soil and biodiversity degradation, and climate change. Therefore, this book series, "Sustainable Agriculture and Food Security," has been planned to support the global efforts toward sustainability by providing timely coverage of the progress, opportunities, and challenges of sustainable food production and consumption in Asia and Africa. The series narrates the success stories and research endeavors from the regions of Africa and Asia on issues relating to SDG 2: Zero Hunger. It fosters research in transdisciplinary academic fields spanning sustainable agriculture systems and practices from pre-seed to post-harvest. At its core, the Series aims to provide a comprehensive publication platform and serve as a knowledge hub that fosters the growth of sustainability sciences with a special focus on developing nations.

This book, *DNA of Sustainability*, edited by my co-editors Dr Vanika Garg and Professor Robert J Henry, provides a comprehensive overview of genetics, genomics, molecular breeding, and bioinformatics approaches that can support the development of crops tailored for the future. From the construction of pangenomes that uncover desirable traits and guide pre-breeding programs, to practical insights on access and the implementation of low-cost genotyping platforms, this 12-chapter book contains an abundance of timely resources for future crop production. Importantly, this book offers a global perspective, ensuring relevance in those regions that most require these approaches to adapt, enhance, and future-proof their agricultural systems. I would like to thank all contributing authors for their valuable insights and for their continued efforts to deliver a more equitable, sustainable, and nutritious food future.

I am sure this book is a great resource for students, researchers, farmers, educators and extension agents, policymakers, and other stakeholders engaged in transforming agriculture around the world.

I wish to extend my sincere thanks and gratitude to the Springer staff, particularly Sanchi Bhimrajika, Associate Editor (Books, Medicine and Life Sciences), and Nick Melchior, Editorial Director (Books, Medicine and Life Sciences, Asia), for their constant support for the accomplishment of this compendium. The cooperation received from my supervisors and research leaders, such as Prof Andrew Deeks (Vice Chancellor), Prof Peter Eastwood (Deputy Vice Chancellor–Research), Prof Peter Davies (Pro Vice Chancellor), and my laboratory colleagues—Annapurna Chitikineni, and Nicholas Ingram from Murdoch University (Australia), is also gratefully acknowledged. I would like to thank my family members—Monika Varshney, Prakhar Varshney, and Preksha Varshney—for their love and support in discharging my duties as Series Editor.

Murdoch, WA, Australia Rajeev K. Varshney

Preface

The agricultural systems that support global food and nutritional security are under immense pressure. With the global population expected to exceed 9 billion by 2050, food production will need to increase dramatically while using fewer resources and producing less environmental harm. The dual challenge of increasing productivity and sustainability is further compounded by the unpredictable impacts of climate change, including erratic rainfall patterns, extreme temperatures, shifting pest and pathogen pressures, and land degradation. These challenges are particularly severe in low- and middle-income countries across Asia, Africa, and Latin America, where farming is often subsistence-based, resources are limited, and access to advanced technologies remains restricted. To meet these challenges head-on, the future of agriculture must be underpinned by innovation, equity, and a deep understanding of plant biology driven by the power of genomics.

In this context, *DNA of Sustainability* brings together a timely and comprehensive collection of twelve chapters showcasing how genomics is reshaping the way we breed, conserve, and cultivate crops. Contributions from researchers across Australia, China, India, the USA, Kenya, Brazil, and Spain explore topics ranging from reference genomes and integrative omics to big data, gene editing, and AI-driven breeding. The volume highlights that sustainability is not merely about producing more food but about producing it responsibly, equitably, and resiliently with genomics as the foundation for transformation.

Recent advances, from telomere-to-telomere genome assemblies to pangenomes, are revealing unprecedented levels of genetic diversity and variation within and across species. Harnessing this diversity, particularly from underutilized crops and wild relatives offers immense potential for resilience and adaptability. Equally critical are efforts to strengthen bioinformatics infrastructure, open-access data platforms, and capacity building, especially in the Global South, to ensure equitable participation in the genomic revolution.

Bridging discovery and application, the book examines how affordable genotyping, synthetic genomics, and data-driven breeding are enabling precision crop improvement. The integration of AI, machine learning, and multi-omics data is accelerating genetic gains and helping decode complex traits such as drought tolerance and nutrient efficiency. Together, these innovations are driving a paradigm shift toward more efficient, informed, and inclusive agricultural systems.

Beyond technologies, this book underscores the importance of vision, collaboration, and inclusivity. Achieving sustainability in agriculture requires that genomic tools be made accessible to all across continents, disciplines, and sectors. It calls for stronger public-private partnerships, open science, gender-inclusive capacity building, and policy environments that promote responsible innovation. The future of agriculture will not be shaped solely by genomics but by how effectively we integrate these tools into broader socioeconomic and ecological systems.

This volume brings together a constellation of leading voices in the field, offering a timely and forward-looking perspective on how genomics can serve as the backbone of future agricultural systems. The editors hope that this book serves not only as a compendium of the latest advances but also as a source of inspiration, dialogue, and action. Whether you are a researcher, breeder, student, policymaker, or interested reader, you are invited to explore the chapters ahead and join in building a more resilient, equitable, and sustainable future.

The editors also express their gratitude to friends and collaborators from the wider research community for their encouragement, cooperation, and valuable support in numerous ways. Rajeev K. Varshney sincerely acknowledges the unwavering support of his wife, *Monika*, son, *Prakhar*, and daughter, *Preksha*, whose patience and understanding enabled him to fulfill his editorial responsibilities. Vanika Garg extends her heartfelt gratitude to her husband, *Saurabh*, and daughter, *Veda*, for their constant encouragement and moral support during the preparation of this book, alongside her research commitments at Murdoch University. Robert J. Henry deeply appreciates the support and patience of his family throughout the development of this work.

The editors extend their deepest thanks to all contributors, colleagues at Murdoch University and the University of Queensland, for sharing their expertise and to the broader scientific community for advancing the frontiers of plant genomics. They also express sincere appreciation to the Springer team, particularly *Sanchi Bhimrajka* (Associate Editor) and *Nandhini Viswanathan* (Production Editor, Books), for their continuous support and dedication in bringing this compendium to completion.

Murdoch, WA, Australia — Vanika Garg
St Lucia, QLD, Australia — Robert J. Henry
Murdoch, WA, Australia — Rajeev K. Varshney

Contents

Part III Advanced Applications in Crop Improvement

Series Editor and Editorial Board Members

About the Series Editor

Rajeev K. Varshney
WA State Agricultural Biotechnology Centre, Centre for Crop and Food Innovation, Food Futures Institute, Murdoch University, Murdoch, WA, Australia

Rajeev K Varshney FRS, is a renowned agricultural scientist with over two decades of experience in genomics and molecular breeding. Currently, he holds multiple prestigious positions at Murdoch University in Australia. He has made a significant impact on global food security by creating genomic resources for major "orphan" tropical crops, using these resources to identify genetic loci and candidate genes for drought and pest tolerance in staple crops in sub-Saharan Africa and India, and spearheading international programs that deliver superior crop varieties to some of the world's most impoverished farmers. He is an accomplished author, with more than 600 publications to his name, and has been recognized as a Clarivate Highly Cited Researcher for eleven consecutive years (2014–2024).

Editorial Board Members

About the Editors

Vanika Garg is a Senior Research Fellow in Genomics and Bioinformatics at the Centre for Crop and Food Innovation, Food Futures Institute, Murdoch University, Australia. She earned her PhD in 2019 from the International Crops Research Institute for the Semi-Arid Tropics (ICRISAT), India, where she investigated stress responses in legumes using multi-omics approaches. With over a decade of experience in high-throughput genomic data analysis, she has developed impactful genomic resources, including reference genomes, pangenomes, genetic markers, superior haplotypes, and candidate genes, contributing significantly to both fundamental research and applied crop improvement. She has published more than 65 research papers in leading journals such as *Nature*, *Nature Biotechnology*, *Nature Genetics*, *PNAS*, *Trends in Plant Science*, and *Plant Biotechnology Journal*. Her work has garnered over 5,000 citations, with an h-index of 33, and she continues to play a pivotal role in integrating genomics with crop improvement programs, fostering innovations that enhance agricultural sustainability and global food security.

Robert J. Henry is currently working as a Professor at Innovation in Agriculture, The University of Queensland, Australia. His research area is focused on the study of agricultural crops using molecular tools. His current research targets plant genome sequencing for the capture of novel genetic resources for the diversification of food crops to deliver improved food products. He is a fellow of the Australian Academy of Technological Sciences and Engineering. He is a highly cited scientist (ISI) in agriculture.

Rajeev K. Varshney FRS is an agricultural scientist specializing in genomics, genetics, molecular breeding, and capacity building in developing countries. As the Director of the Centre for Crop & Food Innovation, the Director of the WA State Agricultural Biotechnology Centre, and the International Chair of Agriculture and Food Security at the Food Futures Institute (FFI) at Murdoch University, Australia, he is at the forefront of pioneering agricultural research, delivering translational scientific outcomes aimed at solving the most significant challenges facing global food production. Throughout his illustrious 20-plus-year career, he has made outstanding contributions to improving food and nutrition security, and has popularized genomics in crop breeding programs at research institutes across the globe. He has been honored with elected and honorary fellowships from a dozen academies/societies around the world—including the Royal Society—and received over 20 prestigious awards for his research.

Contributors

Nikita Aggarwal Division of Genetics & Plant Breeding, Faculty of Agriculture (FoA), SKUAST-Kashmir, Sopore, Jammu and Kashmir, India

Mahpara Bashir Division of Genetics & Plant Breeding, Faculty of Agriculture (FoA), SKUAST-Kashmir, Sopore, Jammu and Kashmir, India

Amanda Bullones Instituto de Hortofruticultura Subtropical y Mediterránea "La Mayora", Consejo Superior de Investigaciones Científicas-Universidad de Málaga (IHSM, CSIC-UMA), Málaga, Spain

Departamento de Biología Molecular y Bioquímica, Universidad de Málaga (UMA), Málaga, Spain

Quinn R. Campbell Department of Tropical Plant and Soil Science, University of Hawai'i at Mānoa, Honolulu, HI, USA

Annapurna Chitikineni WA State Agricultural Biotechnology Centre, Centre for Crop and Food Innovation, Food Futures Institute, Murdoch University, Murdoch, WA, Australia

Reshmi Rani Das IRRI-South Asia Hub, Hyderabad, India

Noe Fernandez-Pozo Instituto de Hortofruticultura Subtropical y Mediterránea "La Mayora", Consejo Superior de Investigaciones Científicas-Universidad de Málaga (IHSM, CSIC-UMA), Málaga, Spain

Khushboo Fulara WA State Agricultural Biotechnology Centre, Centre for Crop and Food Innovation, Food Futures Institute, Murdoch University, Murdoch, WA, Australia

Vanika Garg WA State Agricultural Biotechnology Centre, Centre for Crop and Food Innovation, Food Futures Institute, Murdoch University, Murdoch, WA, Australia

Chunli Guo Guangdong Key Laboratory of Plant Epigenetics, College of Life Sciences and Oceanography, Shenzhen University, Shenzhen, China

Shenzhen Engineering Laboratory for Marine Algal Biotechnology, Guangdong Technology Research Center for Marine Algal Biotechnology, Longhua Innovation Institute for Biotechnology, College of Life Sciences and Oceanography, Shenzhen University, Shenzhen, China

Saurabh Gupta Curtin Medical Research Institute, Curtin Medical School, Curtin University, Perth, WA, Australia

UWA Centre for Medical Research, University of Western Australia, Perth, WA, Australia

Robert J. Henry Queensland Alliance for Agriculture and Food Innovation, University of Queensland, Brisbane, QLD, Australia

ARC Centre of Excellence for Plant Success in Nature and Agriculture, University of Queensland, Brisbane, QLD, Australia

Zhangli Hu Guangdong Key Laboratory of Plant Epigenetics, College of Life Sciences and Oceanography, Shenzhen University, Shenzhen, China

Shenzhen Engineering Laboratory for Marine Algal Biotechnology, Guangdong Technology Research Center for Marine Algal Biotechnology, Longhua Innovation Institute for Biotechnology, College of Life Sciences and Oceanography, Shenzhen University, Shenzhen, China

Guangdong Provincial Key Laboratory of Functional Substances in Medicinal Resources and Healthcare Products, School of Life Sciences and Food Engineering, Hanshan Normal University, Chaozhou, China

Farkhandah Jan Division of Genetics & Plant Breeding, Faculty of Agriculture (FoA), SKUAST-Kashmir, Sopore, Jammu and Kashmir, India

Diego Jarquin Agronomy Department, University of Florida, Gainesville, FL, USA

Sehijpreet Kaur Agronomy Department, University of Florida, Gainesville, FL, USA

Aamir W. Khan Division of Plant Science and Technology, University of Missouri, Columbia, MO, USA

Mohd Anwar Khan Division of Genetics & Plant Breeding, Faculty of Agriculture (FoA), SKUAST-Kashmir, Sopore, Jammu and Kashmir, India

Yiran Li Guangdong Key Laboratory of Plant Epigenetics, College of Life Sciences and Oceanography, Shenzhen University, Shenzhen, China

Shenzhen Engineering Laboratory for Marine Algal Biotechnology, Guangdong Technology Research Center for Marine Algal Biotechnology, Longhua Innovation Institute for Biotechnology, College of Life Sciences and Oceanography, Shenzhen University, Shenzhen, China

Enrique Lopez-Gomez Instituto de Hortofruticultura Subtropical y Mediterránea "La Mayora", Consejo Superior de Investigaciones Científicas-Universidad de Málaga (IHSM, CSIC-UMA), Málaga, Spain

Jie Luo Yazouwan National Laboratory, Sanya, China

School of Breeding and Multiplication (Sanya Institute of Breeding and Multiplication), Hainan Yazhou-Bay Seed Laboratory, Hainan University, Sanya, China

Rishitha Magunta The Millet Lab, Department of Plant Sciences, School of Life Sciences, University of Hyderabad, Hyderabad, Telangana, India

Reyazul Rouf Mir Division of Genetics & Plant Breeding, Faculty of Agriculture (FoA), SKUAST-Kashmir, Sopore, Jammu and Kashmir, India

WA State Agricultural Biotechnology Centre, Centre for Crop and Food Innovation, Food Futures Institute, Murdoch University, Murdoch, WA, Australia

Mehanathan Muthamilarasan The Millet Lab, Department of Plant Sciences, School of Life Sciences, University of Hyderabad, Hyderabad, Telangana, India

Ana Carolina Campana Nascimento Agronomy Department, University of Florida, Gainesville, FL, USA

Laboratory of Intelligence Computational and Statistical Learning (LICAE), Department of Statistics, Federal University of Viçosa, Viçosa, MG, Brazil

Moysés Nascimento Agronomy Department, University of Florida, Gainesville, FL, USA

Laboratory of Intelligence Computational and Statistical Learning (LICAE), Department of Statistics, Federal University of Viçosa, Viçosa, MG, Brazil

Muthiah Naveenkumar The Millet Lab, Department of Plant Sciences, School of Life Sciences, University of Hyderabad, Hyderabad, Telangana, India

Marie-Noelle Ndjiondjop M'bé Research Station, Africa Rice Center (AfricaRice), Bouaké, Côte d'Ivoire

Henry T. Nguyen Division of Plant Science and Technology, University of Missouri, Columbia, MO, USA

Shri Hari Prasad WA State Agricultural Biotechnology Centre, Centre for Crop and Food Innovation, Food Futures Institute, Murdoch University, Murdoch, WA, Australia

Shatabdi Deb Proma Agronomy Department, University of Florida, Gainesville, FL, USA

Mukesh Rathore Division of Genetics & Plant Breeding, Faculty of Agriculture (FoA), SKUAST-Kashmir, Sopore, Jammu and Kashmir, India

Ali Raza Guangdong Key Laboratory of Plant Epigenetics, College of Life Sciences and Oceanography, Shenzhen University, Shenzhen, China

Shenzhen Engineering Laboratory for Marine Algal Biotechnology, Guangdong Technology Research Center for Marine Algal Biotechnology, Longhua Innovation Institute for Biotechnology, College of Life Sciences and Oceanography, Shenzhen University, Shenzhen, China

Javier Santos-del Rio Instituto de Hortofruticultura Subtropical y Mediterránea "La Mayora", Consejo Superior de Investigaciones Científicas-Universidad de Málaga (IHSM, CSIC-UMA), Málaga, Spain

Departamento de Biología Molecular y Bioquímica, Universidad de Málaga (UMA), Málaga, Spain

Vitor Seiti Sagae Agronomy Department, University of Florida, Gainesville, FL, USA

Laboratory of Intelligence Computational and Statistical Learning (LICAE), Department of Statistics, Federal University of Viçosa, Viçosa, MG, Brazil

Dinesh Kumar Saini Department of Plant and Soil Science, Texas Tech University, Lubbock, TX, USA

Valeria Serrano-García Instituto de Hortofruticultura Subtropical y Mediterránea "La Mayora", Consejo Superior de Investigaciones Científicas-Universidad de Málaga (IHSM, CSIC-UMA), Málaga, Spain

Shaina Sharma IRRI-South Asia Hub, Hyderabad, India

Shen Shuangqian Yazouwan National Laboratory, Sanya, China

Pooja Shukla The Millet Lab, Department of Plant Sciences, School of Life Sciences, University of Hyderabad, Hyderabad, Telangana, India

Vikas Kumar Singh IRRI-South Asia Hub, Hyderabad, India

Pallavi Sinha IRRI-South Asia Hub, Hyderabad, India

Parvaze Ahmad Sofi Division of Genetics & Plant Breeding, Faculty of Agriculture (FoA), SKUAST-Kashmir, Sopore, Jammu and Kashmir, India

Vasileia Spyridaki WA State Agricultural Biotechnology Centre, Centre for Crop and Food Innovation, Food Futures Institute, Murdoch University, Murdoch, WA, Australia

Laha Supriya The Millet Lab, Department of Plant Sciences, School of Life Sciences, University of Hyderabad, Hyderabad, Telangana, India

Cassandria Geraldine Tay Fernandez WA State Agricultural Biotechnology Centre, Centre for Crop and Food Innovation, Food Futures Institute, Murdoch University, Murdoch, WA, Australia

Rajeev K. Varshney WA State Agricultural Biotechnology Centre, Centre for Crop and Food Innovation, Food Futures Institute, Murdoch University, Murdoch, WA, Australia

Mohit Verma Department of Biomedical Informatics, Biostatistics, and Medical Epidemiology (BBME), University of Missouri, Columbia, MO, USA

Peterson W. Wambugu Kenya Agricultural and Livestock Research Organization, Genetic Resources Research Institute, Nairobi, Kenya

Qamar U. Zaman Yazouwan National Laboratory, Sanya, China

Mengkai Zhao Guangdong Key Laboratory of Plant Epigenetics, College of Life Sciences and Oceanography, Shenzhen University, Shenzhen, China

Shenzhen Engineering Laboratory for Marine Algal Biotechnology, Guangdong Technology Research Center for Marine Algal Biotechnology, Longhua Innovation Institute for Biotechnology, College of Life Sciences and Oceanography, Shenzhen University, Shenzhen, China

Chapter 1
Introduction: DNA of Sustainability—Genomic Insights into Food Security Challenges

Vanika Garg, Robert J. Henry, and Rajeev K. Varshney

Abstract The accelerating challenges of global food security, driven by climate change, population growth, and limited natural resources, demand transformative approaches to crop improvement. DNA of Sustainability: Genomic Insights into Food Security Challenges explores how plant genomics, computational biology, and integrated multi-omics are reshaping agricultural science to meet these urgent needs. The chapters in this volume trace the journey from the generation of complete genome assemblies and comprehensive pangenomes, through the characterization of genetic diversity and management of large-scale genomic data, to the deployment of enabling technologies such as low-cost genotyping platforms, synthetic genomics, and artificial intelligence. Building on these foundations, the book examines cutting-edge applications including genomic selection, integrated omics strategies, single-cell and spatial technologies, trait discovery, and genomics-informed gene editing. Together, these contributions provide a detailed blueprint for transforming genome-scale insights into resilient, high-yielding, and climate-adapted crops, offering practical and innovative solutions for the future of sustainable agriculture.

Keywords Food security · Climate resilience · Sustainable agriculture · Crop improvement · Crop genetics · Genomics

---

V. Garg (✉) · R. K. Varshney
WA State Agricultural Biotechnology Centre, Centre for Crop and Food Innovation, Food Futures Institute, Murdoch University, Murdoch, WA, Australia
e-mail: Vanika.Garg@murdoch.edu.au

R. J. Henry
Queensland Alliance for Agriculture and Food Innovation, University of Queensland, Brisbane, QLD, Australia

V. Garg et al. (eds.), *DNA of Sustainability*, Sustainability Sciences in Asia and Africa,
https://doi.org/10.1007/978-981-95-5109-5_1

1.1 Introduction

1.1.1 DNA of Sustainability: Genomic Insights into Food Security Challenges

Ensuring global food security in the face of rapid population growth, climate change, environmental degradation, and finite natural resources is one of the defining scientific and societal challenges of our time. Agriculture must produce more food, of higher nutritional value, with fewer inputs, and in more challenging environments than ever before. Traditional plant breeding, while remarkably successful over the past century, is no longer sufficient on its own to meet these urgent demands. The genomics revolution, fuelled by advances in sequencing technologies, computational biology, and integrated multi-omics, is transforming the landscape of crop improvement. The ability to generate complete, highly accurate genome assemblies, explore genetic diversity at unprecedented depth, and harness computational power to guide breeding decisions offers an unprecedented opportunity to develop climate-resilient, high-yielding, and nutritionally superior crops.

This volume, *DNA of Sustainability*, brings together leading experts to explore the full arc of this transformation, from foundational genomic resources to enabling technologies to advanced applications in breeding and crop improvement. The book is structured in three interconnected parts. *Part I* focuses on the genomic resources that underpin modern plant science, highlighting the assembly, characterization, and organization of plant genomes. *Part II* presents the enabling tools and technologies, both laboratory-based and computational, that allow researchers and breeders to translate genomic data into practical solutions. *Part III* showcases advanced applications, where these resources and tools are integrated to address complex breeding challenges, accelerate innovation, and deliver impact in the field. Together, these chapters offer a comprehensive reference for researchers, breeders, policymakers, and students working towards sustainable agricultural futures. A brief overview of the chapters is provided below.

1.2 Part I: Foundations of Genomic Resources

1.2.1 Chapter 2: Telomere-to-Telomere Genomes as a Foundation for Crop Improvement and Food Security

The book opens with a discussion of one of the most significant technical achievements in plant genomics: telomere-to-telomere (T2T) genome assemblies. These complete, gap-free representations of crop genomes provide the ultimate reference for understanding gene content, structural variation, and regulatory elements. This chapter by Garg et al. details the unique challenges posed by complex plant genomes, often polyploid, rich in repetitive elements, and highly heterozygous, and how advances in long-read sequencing, optical mapping, and scaffolding algorithms have made T2T assemblies possible. Case studies across multiple species illustrate how these assemblies are enabling pangenomics, precise functional gene identification, and improved breeding strategies. The authors position T2T genomes as a transformative foundation for tackling food security by providing breeders with the most accurate possible blueprint for crop improvement.

1.2.2 Chapter 3: Pangenomes, Super-Pangenomes, and Beyond

Building upon the concept of a single reference genome, this chapter introduces the pangenome, a representation of the full genetic repertoire of a species, comprising both core and dispensable genes. By incorporating multiple individuals, pangenomes capture genetic variation underlying adaptation, disease resistance, and other agriculturally important traits. This chapter by Khan et al. traces the rapid evolution of pangenomics into the era of "super-pangenomes" that span entire genera, integrating data from wild relatives and domesticated lines. Such broad datasets illuminate evolutionary trajectories, domestication events, and reservoirs of untapped diversity. These insights have profound implications for breeding strategies, enabling the targeted use of wild variation to create crops resilient to biotic and abiotic stresses.

1.2.3 Chapter 4: Exploring Genetic Diversity Using Genomic Approaches

This chapter surveys the diverse genomic tools now available for assessing and exploiting genetic variation in plants. From high-coverage genome assemblies to targeted amplicon sequencing and transcriptomics, these approaches are uncovering alleles associated with yield, quality, and stress tolerance. Wambugu et al. emphasize the value of wild germplasm as a reservoir for traits not present in elite cultivars and describe strategies for incorporating this diversity into breeding programmes through genome editing and precision introgression. Importantly, the chapter considers how climate change and anthropogenic pressures may reshape plant genetic diversity over time and how genomics can guide conservation and adaptive breeding strategies.

1.2.4 Chapter 5: Challenges and Opportunities for Genomic Data Resources

The proliferation of genomic data has brought with it new challenges: data storage, management, standardization, and accessibility. This chapter by Lopez-Gomez et al. examines these issues in detail, noting the complexities of harmonizing nomenclature, managing multiple genome versions, and integrating dispersed databases. The authors also highlight opportunities for innovation, such as AI-driven mining of genomic datasets and the construction of pan-genomic reference frameworks. By addressing the technical and infrastructural challenges of data stewardship, this chapter lays out a road map for ensuring that the explosion of genomic data can be fully leveraged for agricultural innovation.

1.3 Part II: Enabling Technologies and Tools

1.3.1 Chapter 6: Development of Low-Cost Genotyping Platforms

Cost-effective genotyping is essential for ensuring that genomic-assisted breeding reaches resource-limited settings. This chapter by Das et al. reviews the spectrum of genotyping platforms, from low-density SNP assays to high-throughput sequencing, and their applications in genomic selection, marker-assisted selection, and trait mapping. It discusses collaborative initiatives such as CGIAR's shared genotyping resources and explores innovative bioinformatics solutions for maximizing data utility. The emphasis is on democratizing access to advanced molecular tools, enabling breeding programs worldwide to accelerate genetic gains and respond to local agricultural challenges.

1.3.2 Chapter 7: Advances in Synthetic Genomics (SynGen) for Crop Improvement and Beyond

Synthetic genomics represents a bold frontier in plant science, allowing the design and construction of customized genomes for enhanced performance. This chapter by Raza et al. provides a conceptual and technical overview of SynGen, with applications ranging from yield and stress tolerance to metabolic engineering and minimal-genome design. The author addresses the integration of SynGen with multi-omics, biofoundries, and emerging gene-drive technologies, while also engaging with the ethical, regulatory, and societal considerations surrounding synthetic crops. They argue that, with transparent governance and responsible innovation, SynGen could be a game-changer for sustainable agriculture.

1.3.3 Chapter 8: Big Data, Machine Learning, and Artificial Intelligence for Crop Improvement

This chapter explores how AI and machine learning can transform the vast, complex datasets of modern plant breeding into actionable knowledge. Spyridaki et al. define big data in the agricultural context, outline the challenges of data integration, and illustrate how ML models can detect patterns across genomic, phenotypic, and environmental datasets. Case studies show how AI can optimize breeding decisions, predict trait performance, and accelerate variety release. The chapter envisions a future where AI-driven decision support becomes an integral component of breeding pipelines.

1.4 Part III: Advanced Applications in Crop Improvement

1.4.1 Chapter 9: Genomic Selection for Crop Improvement in the Post-NGS Era

Genomic selection (GS) has emerged as a powerful approach for predicting the genetic value of breeding lines, using genome-wide markers to model quantitative traits. This chapter details how GS can improve selection accuracy, reduce breeding cycles, and optimize resource allocation. Sagae et al. also discuss multi-trait models, genotype-by-environment interactions, and the integration of GS with other omics data. In doing so, they highlight how GS is reshaping breeding strategies for complex traits in major crops.

1.4.2 Chapter 10: Integrated Omics Approaches to Accelerate New Strategies for Crop Improvement

Modern crop improvement increasingly depends on integrating genomics with transcriptomics, proteomics, metabolomics, and epigenomics. This chapter examines how such integration can elucidate the molecular basis of stress tolerance, yield, and quality traits and how it can guide targeted interventions. Zaman et al. highlight emerging areas such as 3D genome architecture and its role in gene regulation, underscoring the potential of holistic, systems-level approaches to deliver resilient and productive crops.

1.4.3 Chapter 11: Advances and Applications of Single-Cell and Spatial Omics for Crop Improvement

Single-cell and spatial omics technologies are offering unprecedented resolution into plant cell types, states, and interactions. This chapter by Gupta et al. describes their application in identifying rare cell populations, mapping developmental processes, and dissecting stress responses at cellular resolution. Technical challenges, data analysis workflows, and integration with AI and gene editing are also discussed. The authors present a compelling case for the transformative potential of these tools in understanding and manipulating plant traits.

1.4.4 Chapter 12: Trait Discovery in the Post-NGS Era

The identification of genes underlying agronomically important traits remains a cornerstone of crop improvement. This chapter by Rathore et al. outlines the advances in QTL mapping, association studies, and fine mapping enabled by NGS and pangenomics. It also considers how integration with phenomics and high-throughput phenotyping can accelerate trait discovery. Future directions include the use of single-cell genomics, synthetic biology, and AI to resolve complex traits more efficiently.

1.4.5 Chapter 13: Tailoring Future Crops with Genomics-Informed Gene Editing

The final chapter brings together the themes of the book, showcasing how genomics can inform precise, efficient gene-editing strategies using CRISPR, TALENs, ZFNs, and emerging tools. Shukla et al. describe how combining gene editing with genomic selection, speed breeding, and novel delivery systems can accelerate the development of climate-resilient, nutrient-rich crops. They also address challenges such as off-target effects and regulatory hurdles, concluding with a vision for the role of gene editing in sustainable agriculture.

1.5 Summary and Outlook

In summary, this book, *DNA of Sustainability*, charts the trajectory of plant genomics from foundational resources to transformative applications. It highlights how technological innovations, integrated datasets, and cross-disciplinary collaborations are driving a new era of precision agriculture, one capable of addressing the intertwined challenges of food security, environmental resilience, and economic sustainability. By linking genome-scale knowledge with practical breeding objectives, the tools and approaches described in this volume offer a clear blueprint for developing the crops of the future.

Part I
Foundations of Genomic Resources

Chapter 2
Telomere-to-Telomere Genomes as a Foundation for Crop Improvement and Food Security

Vanika Garg, Khushboo Fulara, and Rajeev K. Varshney

Abstract Food security is a pressing global challenge exacerbated by rapid population growth, climate change, and diminishing natural resources. To meet the increasing demand for food, it is crucial to develop resilient, high-yielding crops capable of thriving in diverse and challenging environments. Plant genomics has become a pivotal tool in addressing food security, providing transformative insights into crop yield enhancement, disease resistance, and stress tolerance. Among the most significant advancements in this field is the advent of telomere-to-telomere (T2T) genome assemblies, which offer complete, gap-free representations of crop genomes. This chapter examines the challenges associated with assembling complex crop genomes, which often contain extensive repetitive sequences, polyploid structures, and high heterozygosity. It discusses recent technological advancements that have facilitated the generation of high-resolution reference genomes and provides an overview of their progress across various plant species. Additionally, the chapter explores the broad applications of T2T genome assemblies, including pangenomics, functional gene identification, and precision breeding. By enabling more accurate and comprehensive genomic insights, these assemblies are poised to drive advancements in crop research and accelerate the development of resilient, high-yielding crop varieties.

Keywords Reference genomes · Next-generation sequencing · Long-read technologies · Plant genomics · Crop improvement · Pangenomics · Precision breeding

V. Garg · K. Fulara · R. K. Varshney (✉)
WA State Agricultural Biotechnology Centre, Centre for Crop and Food Innovation, Food Futures Institute, Murdoch University, Murdoch, WA, Australia
e-mail: Rajeev.Varshney@murdoch.edu.au

V. Garg et al. (eds.), *DNA of Sustainability*, Sustainability Sciences in Asia and Africa,
https://doi.org/10.1007/978-981-95-5109-5_2

2.1 Introduction

Food security remains one of the most critical global challenges of the twenty-first century, as the world's population is projected to reach nearly 10 billion by 2050 (Hickey et al. 2019). This rapid population growth, compounded by the adverse effects of climate change, land degradation, and diminishing natural resources, places unprecedented strain on agricultural systems. To meet the increasing demand for food, there is an urgent need to develop resilient, high-yielding crops that can thrive in diverse and challenging environmental conditions. Achieving this goal necessitates a deep understanding of plant biology and innovative tools to accelerate crop improvement.

Plant genomics has emerged as a cornerstone in addressing global food security, providing transformative insights into crop yield enhancement, disease resistance, and stress tolerance (Tuberosa et al. 2017). Among the most significant advancements in this field is crop genome sequencing, which deciphers the genetic blueprints of plants, offering a comprehensive view of their biology. Since the publication of the Arabidopsis genome in 2000 (Arabidopsis Genome Initiative 2000), plant genome sequencing has expanded rapidly, with thousands of nuclear plant genomes now sequenced, covering a diverse range of species, predominantly angiosperms (Fig. 2.1; Bernal-Gallardo and de Folter 2024). This foundational knowledge underpins modern breeding strategies, enabling precise improvement of critical traits such as drought resilience, nutritional quality, and pest resistance. High-quality reference genomes have become indispensable for unraveling molecular mechanisms, identifying genetic variations and superior haplotypes, facilitating molecular breeding and genomic selection, and paving the way for the development of pangenomes to capture species-wide genetic diversity (Kong et al. 2023; Gupta et al. 2024).

Despite these advances, significant gaps remain in our understanding of crop genomes, particularly in regions characterized by high complexity, such as telomeres, centromeres, and repetitive sequences. Historically, technological limitations in sequencing and assembly have resulted in incomplete genome assemblies, impeding the ability to fully explore genome organization, structural variations, and evolutionary history. Recent breakthroughs in sequencing technologies and assembly methods, however, have ushered in a new era of plant genomics, marked by the advent of telomere-to-telomere (T2T), gap-free assemblies. These complete genome assemblies provide unprecedented access to previously inaccessible genomic regions, collectively referred to as "dark matter," and offer transformative potential for crop research and improvement (Garg et al. 2024).

T2T assemblies have proven invaluable across multiple plant species, including *Arabidopsis thaliana*, *Oryza sativa* (rice), and *Zea mays* (maize), among others (Li et al. 2021; Naish et al. 2021; Song et al. 2021; Hou et al. 2022; Chen et al. 2023). These assemblies enable the detailed exploration of high-complexity regions, facilitating the discovery of genetic variations, repetitive elements, and duplication events that were previously obscured. Moreover, T2T genomes provide critical insights into genome evolution, centromere biology, and telomeric structures while delivering practical benefits such as enhancing disease resistance and optimizing

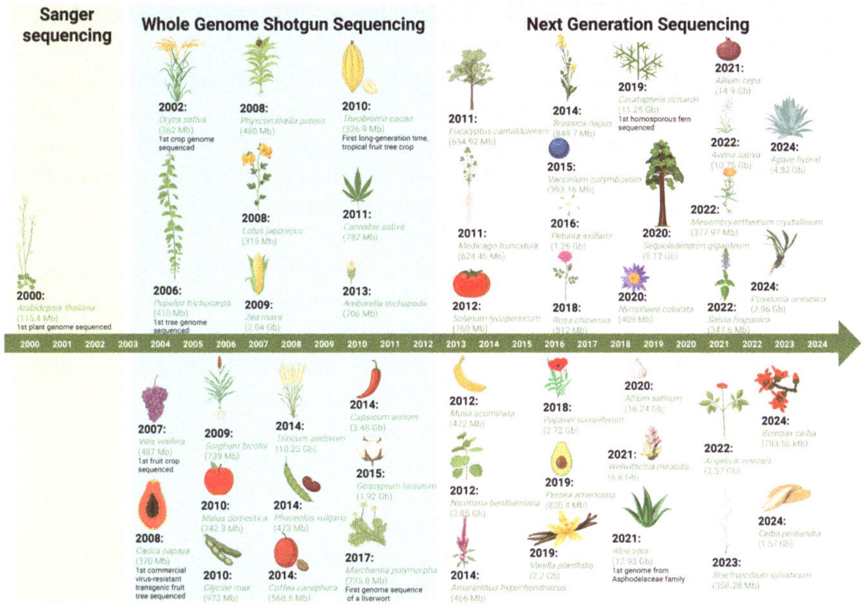

Fig. 2.1 Timeline of plant genome sequencing advancements using different sequencing technologies. The figure illustrates the progression of plant genome sequencing from 2000 to 2024, categorized into three major sequencing approaches: Sanger sequencing (green panel), Whole Genome Shotgun (WGS) sequencing (blue panel), and Next-Generation Sequencing (NGS) (purple panel). Each entry represents a sequenced plant genome, including its species name, genome size (Mb/Gb), and notable firsts, such as the first crop genome (*Oryza sativa*, 2002). The timeline highlights how advancements in sequencing technologies have enabled the assembly of increasingly complex and larger plant genomes, facilitating breakthroughs in plant genomics, evolutionary biology, and crop improvement. (Created with Biorender.com (https://BioRender.com))

breeding programs. By providing comprehensive and high-accuracy representations of crop genomes, high-quality reference genome assemblies have become a cornerstone for modern genomics, with wide-ranging implications for food security and sustainability.

Here, we examine the challenges of assembling complex crop genomes and highlight recent technological advances enabling high-quality T2T reference genomes. We provide an overview of their current status across plant species and discuss their applications for advancing genomics-assisted crop improvement.

2.2 Challenges in Genome Assembly

2.2.1 Repetitive Regions

Repetitive regions are sequences of variable length that occur in multiple copies throughout the genome and include transposable elements and tandem repeat arrays. Repetitive regions play a key role in gene regulation, chromatin organization, genomic variation, and genome size (Anderson et al. 2019; Domínguez et al. 2020). These sequences act as drivers of evolution by facilitating adaptation and diversification in plants. Repetitive elements are integral to key developmental processes, stress responses, and environmental adaptation. Acting as cis-regulatory elements, such as enhancers or repressors, repeats can modulate the expression of nearby genes (Makarevitch et al. 2015; Yokosho et al. 2016). For instance, a recent study demonstrated that inverted repeats of maize transposons regulate drought tolerance and yield-related traits, illustrating the functional importance of repetitive sequences (Sun et al. 2023b). However, these regions, including telomeres, sub-telomeres, nucleolar organizing regions (NORs), and some portions of centromere kinetochores, represent some of the most challenging genomic regions to assemble. These regions are often misassembled or entirely unassembled in genome projects due to the inherent limitations of sequencing technologies, particularly those relying on short-read platforms. Since many draft genomes were constructed using short-read technologies, repetitive regions were frequently overlooked, resulting in incomplete or erroneous assemblies. Understanding and characterizing repetitive sequences is a critical aspect of genome biology.

Centromeres are chromosomal regions composed predominantly of satellite DNA and retrotransposons. They ensure sister chromatid cohesion and proper segregation during cell division (Oliveira and Torres 2018). However, centromeres' highly repetitive and structurally complex nature makes them particularly challenging to assemble. Centromeric regions remain unresolved in most plant genome assemblies, limiting our understanding of their organization, evolution, and function. This lack of resolution hampers insights into chromosome segregation and the genetic and epigenetic processes occurring within these regions. Millions of bases within pericentromeric and centromeric regions remain poorly characterized, excluding them from many genetic and functional studies.

Telomeres, the protective caps at the ends of chromosomes, play a vital role in maintaining genome stability. These structures prevent the loss of genetic information during DNA replication and protect against chromosomal degradation. Plant telomeres typically consist of tandemly repeated DNA sequences spanning several hundred base pairs. Similar to centromeres, telomeres can influence the expression of adjacent genes, serving as regulatory elements (Shakirov et al. 2022). Understanding telomere structure and function is essential for deciphering the mechanisms that ensure genome integrity and for identifying genes regulated by these sequences.

Further, nucleolar organizing regions (NORs) are specialized chromosomal segments that house rRNA gene clusters. These regions play a pivotal role in ribosome biogenesis, cellular metabolism, and overall genome organization. NORs often exhibit extensive repeat sequences, complicating their assembly and characterization. However, resolving NOR structure can reveal mechanisms underlying nucleolar dynamics and their impact on gene expression and cellular function (Navrátilová et al. 2022). Addressing the challenges posed by repetitive regions will advance our understanding of genome architecture and pave the way for more complete and accurate genome assemblies.

2.2.2 *Polyploidy*

Polyploidy, the presence of multiple sets of chromosomes, is another significant challenge in genome assembly, particularly in plants. Polyploid genomes, such as those of wheat, cotton, and sugarcane, contain multiple closely related sub-genomes, making it difficult to distinguish between homeologous loci. In recently formed polyploids, homeologous sequences can be nearly identical, often leading to the collapse of these regions into a single contig or the generation of chimeric sequences that erroneously combine different homeologs (Kong et al. 2023). This complexity is compounded by the amplification of repetitive elements in polyploid genomes, further complicating assembly efforts. Specialized bioinformatics approaches, such as subgenome-aware assembly pipelines, are often required to reconstruct each genome copy accurately. A tiered strategy, which involves sequencing diploid progenitors to establish a foundation for subsequent polyploid genome assemblies, has proven effective in some cases (Zhuang et al. 2019).

2.2.3 *Heterozygosity*

Heterozygosity, the presence of different alleles at the same locus within an individual's genome, poses a significant challenge in plant genome assembly, particularly for outcrossing species (VanBuren et al. 2018). This allelic variation disrupts the assumption of a single, consensus sequence, leading to potential ambiguities in contig construction. Consequently, assembly algorithms often struggle to accurately resolve these variations, resulting in chimeric or fragmented contigs that fail to represent the true genetic landscape. In polyploid plants, this challenge is compounded by the need to distinguish between allelic and homoeologous differences, further increasing the complexity of assembly. Accurate phasing and representation of haplotypes are therefore crucial for generating high-quality plant genome assemblies, requiring advanced computational strategies and sequencing technologies.

2.3 Strategies for Achieving High-Quality Genome Assemblies

2.3.1 *Sequencing Platform*

2.3.1.1 Long Read Sequencing

The rapid expansion of plant genome sequencing has been driven by remarkable advancements in sequencing technologies. Over the past few decades, sequencing methods have become faster, more accurate, and cost-effective, revolutionizing the ability to determine DNA sequences. Traditionally, short-read sequencing technologies, dominated by Illumina, were used to generate millions of short reads (typically 100–300 base pairs in length) that were then used to assemble the genome. However, these short reads were error-prone and insufficient for accurately resolving complex genomic regions, particularly those rich in repetitive sequences or exhibiting high heterozygosity. The advent of long-read sequencing, often referred to as third-generation sequencing, has revolutionized plant genomics by enabling the assembly of high-quality, near-complete genomes. Long-read technologies produce substantially longer reads, reducing assembly ambiguity and improving the resolution of repetitive and structurally complex regions. Early advancements included Pyrosequencing by 454 Life Sciences in the late 1990s and early 2000s, generating reads of several thousand base pairs. However, it was Pacific Biosciences (PacBio) and Oxford Nanopore Technologies (ONT) that drove major advancements in long-read sequencing. PacBio's Single Molecule Real-Time (SMRT) sequencing platform, introduced in the 2010s, provided reads tens of thousands of base pairs in length. Subsequent improvements led to the Sequel II system, followed by Revio, which generates high-fidelity (HiFi) reads with >99% accuracy using circular consensus sequencing (CCS). These HiFi reads have significantly improved genome assembly accuracy, particularly in resolving complex regions. Oxford Nanopore Technologies (ONT) employs a distinct approach, measuring electrical conductivity changes as DNA passes through nanopores. ONT can generate ultra-long (UL) reads up to 4 Mb, making it particularly useful for assembling large and highly repetitive plant genomes (Xu et al. 2022). Long-read sequencing has become indispensable for various genomic applications, earning recognition as Method of the Year in 2022 (Marx 2023).

A critical prerequisite for long-read sequencing is efficient DNA extraction, as high-quality, high-molecular-weight (HMW) DNA is essential for generating long and accurate reads. Unlike short-read sequencing, which tolerates fragmented DNA, long-read technologies require intact, high-purity DNA with minimal shearing. Extraction protocols typically involve gentle cell lysis, careful handling to prevent fragmentation, and purification steps to remove contaminants such as polysaccharides and phenolics, which are common in plant tissues. Optimized methods, including nuclei isolation and ultra-HMW DNA extraction using CTAB-based or commercial kits, have been developed to recover long DNA fragments suitable for sequencing.

2.3.1.2 Long-Range Scaffolding

Long-read sequencing technologies produce highly contiguous draft genome assemblies. However, chromosome-scale assemblies are essential for different applications such as marker-assisted breeding, quantitative genetics, and comparative genomics. Traditionally, high-density genetic maps were used to anchor contigs and scaffolds into chromosomes. However, these maps often suffer from errors in ordering and orientation, fail to anchor sequences in regions with low recombination rates or sparse marker density, and require segregating populations, which may not always be available (Mascher and Stein 2014). Modern long-range scaffolding approaches address these limitations by leveraging long-range chromatin interactions or optical maps for contig anchoring. One of the most widely adopted techniques is Hi-C, a high-throughput method originally developed to study the three-dimensional organization of genomes. Hi-C involves cross-linking chromatin, fragmenting DNA, and ligating interacting fragments before sequencing them. By analyzing the interaction patterns between genomic regions, Hi-C data enable the accurate scaffolding of contigs into chromosome-length assemblies. This approach has been instrumental in constructing chromosome-scale genome assemblies for several plant species, including barley, chickpea, pigeonpea, and soybean (Mascher et al. 2017; Garg et al. 2022). Another powerful long-range scaffolding technique is optical mapping, a non-sequencing approach that generates high-resolution genome maps using fluorescently labeled DNA molecules. In this method, single DNA molecules are passed through nanochannels, where they are imaged to create physical maps of the genome. Optical mapping provides a resolution of up to 10 kilobases, significantly higher than that achieved with traditional sequencing technologies, which rely on short-read data. These high-resolution maps help improve genome assemblies by resolving complex structural variations, ordering contigs, and validating scaffolds (Garg et al. 2023). By integrating Hi-C and optical mapping data with long-read sequencing data, chromosome-scale genome assemblies with enhanced accuracy and completeness can be achieved.

2.3.2 Assembly Algorithms

Assembly algorithms are computational methods used to assemble DNA sequences into a contiguous genome representation. The most commonly used algorithms for plant genome assemblies include the de Bruijn graph (DBG) and the Overlap-Layout-Consensus (OLC) methods. The de Bruijn graph method breaks down the genome into k-mers (short overlapping sequences of length k) and constructs a graph where each k-mer represents a node, and edges connect nodes that overlap by $k-1$ nucleotides. The graph is then traversed to generate contigs, which are then merged into scaffolds. The DBG method is efficient and can handle large and complex genomes, but may suffer from errors due to the inherent fragmentation of the genome into k-mers (Turner et al. 2018). On the other hand, the OLC method relies on overlapping reads to assemble the genome. Reads are aligned to one another based on the degree of sequence similarity, and overlapping regions are identified to construct contigs. These contigs are then oriented and ordered based on the read overlaps to generate scaffolds. The OLC method is known to produce high-quality assemblies, particularly for small and less complex genomes, but it may be computationally intensive and require significant computational resources. Both the DBG and OLC methods have been successfully used to assemble plant genomes and have their advantages and limitations. The choice of method depends on factors such as genome size, complexity, and available resources. Recent advancements in sequencing technologies and assembly algorithms have led to the development of hybrid assembly approaches that combine the strengths of both the DBG and OLC methods, resulting in high-quality genome assemblies for a wide range of plant species, such as pea, *Aegilops tauschii* (a progenitor of bread wheat), sweet potato, and loblolly pine (Yang et al. 2017, 2022; Zimin et al. 2017a, b).

2.3.3 Experimental Evidence and Manual Curation

Experimental evidence and manual curation play crucial roles in ensuring the accuracy and completeness of genome assemblies, especially in complex regions. Techniques like Fluorescence In Situ Hybridization (FISH) and Chromatin Immunoprecipitation Sequencing (ChIP-Seq) provide valuable insights into genomic architecture and validate challenging regions such as centromeres, NORs, and telomeric sequences. For instance, FISH utilizes fluorescently labeled DNA probes to visualize genomic regions, aiding in the confirmation of structural features like NORs and telomeric sequences, as demonstrated in barley to exclude chloroplast DNA from nuclear genomes (Navrátilová et al. 2022). Similarly, ChIP-Seq, targeting the centromere-specific histone H3 variant (CENH3), validates centromeric regions and has been extensively applied to plants like Arabidopsis, maize, and wheat. Alongside these techniques, manual curation complements computational methods by resolving misassemblies and filling gaps through high-density genetic linkage maps and Bacterial Artificial Chromosome (BAC) clones. Although time-consuming, manual curation enhances the confidence and precision of genome assemblies, integrating experimental validations to produce high-quality, reliable genomic assemblies.

2.3.4 Quality Assessment of Genome Assembly

It is crucial to validate the quality of genome assemblies to ensure their accuracy and completeness. Several computational and experimental methods can be used to validate the quality of genome assemblies, and the choice of method depends on the specific needs of the analysis and the resources available. These tools can be used individually or in combination.

(a) *Benchmarking Universal Single-Copy Orthologs (BUSCO)* is a widely used tool for the assessment of genome assembly and annotation completeness by assessing the presence and intactness of highly conserved and essential genes that are expected to be present as single-copy orthologs across diverse species (Manni et al. 2021).

(b) *Quality Assessment Tool for Genome Assemblies (QUAST)* is a user-friendly tool that computes a range of metrics (such as contig length, scaffold lengths, the number of gaps, N50, L50, and the number of misassemblies) needed for gauging and comparing the quality of assemblies. QUAST can evaluate assembly quality even in the absence of a reference genome and, thus, is especially advantageous when working with new species that lack a fully sequenced and annotated reference genome (Gurevich et al. 2013).
(c) *Merqury* is a reference-free, robust method that evaluates the quality, accuracy and completeness of genome assemblies by comparing k-mers in a *de novo* assembly to those found in unassembled high-accuracy reads (Rhie et al. 2020).
(d) *GenomeScope* is used to estimate genome size, heterozygosity, and repeat content from sequencing reads using kmer frequency distributions (Ranallo-Benavidez et al. 2020).
(e) *LTR Assembly Index (LAI)* is a reference-free genome metric that evaluates assembly continuity using LTR retrotransposons. A higher LAI suggests a more accurate and complete assembly of the LTR retrotransposons, indicating a high-quality genome assembly (Ou et al. 2018).
(f) *GenomeQC* is an easy-to-use and interactive web framework for calculating contiguity and completeness metrics for genome assemblies and annotations. It takes genome assembly and genome annotation files as input and provides different statistics, such as N50, L50, gene and repeats space completeness, and the number and size of gene models, exons, and transcripts (Manchanda et al. 2020).
(g) *Purge Haplotigs* is a pipeline designed to automate the reassignment of allelic contigs and to assist in the manual curation of genome assemblies (Roach et al. 2018).

2.3.5 Accurate Gene Prediction

Once the reference genome sequences are assembled, the next step is the identification of functional elements, particularly the protein-coding genes. There are several approaches used for gene prediction in plant genomes, which can be broadly categorized into *ab initio*, homology-based, and evidence-based methods. The *ab initio* methods predict genes solely based on the properties of a gene sequence, such as the presence of start and stop codons, splice sites, and open reading frames. *Ab initio* strategies tend to miss small introns and small genes. Homology-based methods employ sequence similarity to known genes or proteins from closely related organisms, so they are useful for identifying conserved genes but may miss novel or highly divergent genes. Evidence-based methods use additional sources of evidence, such as *de novo* transcriptome assembly and/or transcriptome data, to support gene predictions, help identify novel or low-abundance genes, and refine the structure and boundaries of predicted genes, including the addition of UTRs. Traditionally, short-read RNA-seq data have been used for transcriptome description. However, advancements now allow for long-read RNA sequencing using platforms like PacBio's IsoSeq and Oxford's cDNA sequencing technology. Long-read RNA sequencing can provide complete and accurate reads spanning full-length transcripts, allowing for the identification of novel isoforms, alternative splicing events, and complex gene structures. Long-read data provides more accurate and comprehensive transcript annotation, enhancing the quality of gene models and reducing annotation errors that can arise from short-read data gaps or misalignments. At present, the common deficiencies in genome annotation are incorrectly annotated genes, pseudogenes, chimeras, and delineating genes (Xie et al. 2024). To annotate a comprehensive set of genes that accurately reflects the functional elements within a plant genome, a hybrid gene prediction pipeline that combines *ab initio*, homology-based, and evidence-based methods along with manual curation is required.

2.4 Status of Complete T2T Reference Genomes

Assembling complete, gapless genomes is challenging, but the right combination of sequencing and assembly strategies has been successful in a few cases (Table 2.1, Figure 2.1). Recently, the first complete T2T assembly for the human genome was reported (Nurk et al. 2022). While the first genome assembly from the Human Genome Project, completed in 2003, was an unprecedented achievement in the field of genetics, it did not provide a complete picture of the human genome. Since then, the human reference genome sequence has been continually improved, and 20 years later, the T2T Consortium presented a complete 3.055 billion base pair human genome sequence (T2T–CHM13), which included previously uncharted regions and corrected earlier inaccuracies by employing multiple sequencing technologies, including PacBio HiFi, ONT-UL, Hi-C, and optical maps. The T2T human genome encapsulated highly repetitive centromeric satellite arrays, recent segmental duplications, and the short arms of all five acrocentric chromosomes, thus paving the way for more comprehensive studies on these and other important genomic regions (Nurk et al. 2022).

The genome sequence of *Arabidopsis*, the model plant species, was first reported in the year 2000, and for 20 years, the centromeres, telomeres, and ribosomal DNA repeats remained unassembled. In 2021, Naish et al. broke this barrier by developing a genome sequence of the *A. thaliana* accession Columbia (Col–CEN) that resolved all five centromeres. This assembly was generated using long-read ONT sequencing and polishing with high-accuracy PacBio HiFi reads and Bionano optical mapping data. The Col–CEN genome sequence provided an in-depth understanding of the structure and evolution of the *Arabidopsis* centromeres. In one of the similar efforts, a high-quality, near-complete Col-0 genome sequence (Col-XJTU) was developed by integrating the ONT ultra-long reads, PacBio HiFi reads, and Hi-C data. Out of five chromosomes, Chr3 and Chr5 were resolved entirely in the T2T manner, and Chr4 was completely resolved except for the NOR, which comprised long repetitive DNA fragments (Wang et al. 2022).

Table 2.1 Status of telomere-to-telomere genome assemblies in different plant species

Species	Assembly size (Mb)	Technologies used	Centromeres	Telomeres	Reference
Arabidopsis (*Arabidopsis thaliana*)	133.73	PacBio HiFi, Oxford Nanopore, Hi-C	5/5	8/10	Wang et al. (2022)
	132.08	PacBio HiFi, Oxford Nanopore, Hi-C, Bionano	5/5	10/10	Naish et al. (2021)
	133.92	PacBio HiFi, Oxford Nanopore, Illumina	5/5	9/10	Hou et al. (2022)
Banana (*Musa acuminata*)	484.06	Oxford Nanopore, Illumina, Bionano	11/11	22/22	Belser et al. (2021)
Bitter melon (*Momordica charantia*)	295.60	PacBio HiFi, Illumina, Hi-C	11/11	18/22	Fu et al. (2023)
Brachypodium distachyon	274.75	PacBio HiFi, Hi-C	5/5	7/10	Chen et al. (2024a)
Brachypodium hybridum	535.10	PacBio HiFi, Hi-C	15/15	27/30	Chen et al. (2024a)
Brachypodium stacei	258.94	PacBio HiFi, Hi-C	10/10	19/20	Chen et al. (2024a)
Cherry (*Prunus campanulata*)	266.23	PacBio HiFi, Oxford Nanopore, Hi-C, Illumina	8/8	13/16	Jiang et al. (2025)
Chili pepper (*Capsicum annuum*)	3103.12	PacBio HiFi, Oxford Nanopore, Hi-C, Illumina	12/12	24/24	Chen et al. (2024b)
Chinese bayberry (*Myrica rubra*)	292.60	PacBio HiFi, Hi-C	8/8	15/16	Zhang et al. (2024a)
Grapevine (*Vitis vinifera*)	504.70	PacBio HiFi, Oxford Nanopore, Hi-C	19/19	38/38	Wang et al. (2024)
Maize (*Zea mays*)	2178.60	PacBio HiFi, Oxford Nanopore, Illumina, Bionano	10/10	20/20	Chen et al. (2023)
Pear (*Pyrus pyrifolia*)	501.20	PacBio HiFi, Oxford Nanopore, Hi-C, Illumina	17/17	34/34	Sun et al. (2023a)
Potato (*Solanum tuberosum*)	773.47	PacBio HiFi, Oxford Nanopore, Hi-C	12/12	24/24	Yang et al. (2023)
Rice (*Oryza sativa*)	385.70	PacBio HiFi, Oxford Nanopore, Illumina, Hi-C	12/12	24/24	Shang et al. (2023)
	397.71	PacBio HiFi, Illumina	12/12	NA	Li et al. (2021)

(continued)

Table 2.1 (continued)

Species	Assembly size (Mb)	Technologies used	Centromeres	Telomeres	Reference
	395.77	PacBio HiFi, Illumina, Bionano	12/12	22/24	Song et al. (2021)
	391.56	PacBio HiFi, Illumina, Bionano	12/12	19/24	Song et al. (2021)
Sorghum (*Sorghum bicolor*)	724.85	PacBio HiFi, Oxford Nanopore, Hi-C	10/10	20/20	Li et al. (2024)
	722.95	PacBio HiFi, Oxford Nanopore, Hi-C, Illumina	10/10	20/20	Ding et al. (2024)
Soybean (*Glycine max*)	1016.38	PacBio HiFi, Bionano	20/20	40/40	Garg et al. (2023)
	1013.74	PacBio HiFi, Bionano	20/20	39/40	Garg et al. (2023)
	1007.24	PacBio HiFi, Oxford Nanopore, MGI, Bionano	20/20	39/40	Zhang et al. (2023)
Strawberry (*Fragaria vesca*)	220.79	PacBio HiFi, Oxford Nanopore, Hi-C	7/7	14/14	Zhou et al. (2023)
Watermelon (*Citrullus lanatus*)	369.32	PacBio HiFi, Oxford Nanopore, Hi-C, Bionano	11/11	22/22	Deng et al. (2022)
Wild rice (*Oryza rufipogon*)	418.80	PacBio HiFi, Oxford Nanopore, Hi-C	12/12	22/24	Huang et al. (2024)

NA not available

In the case of maize, the T2T genome of Mol17 was reported with each chromosome assembled as a single contig. The 2178.6 Mb genome, with over 99.99% base accuracy, unveiled the structural features of repetitive regions, including ultra-long TAG repeats and a fully assembled 26.8 Mb NOR with 2974 rDNA copies. Additionally, the assembly of all ten centromeres provided detailed insights into CentC-rich and CentC-poor centromere structures, advancing the understanding of complex repetitive regions in plant genomes (Chen et al. 2023). Similarly, the complete genome assembly of bittermelon (*Momordica charantia* L. var. *abbreviata* Ser.) with a size of 295.6 Mb was reported. The genome sequence was assembled into 11 chromosomes, of which six were gap-free. The seven-base telomeric repeat (CCCTAAA at the 5′ end or TTTAGGG at the 3′ end) was used as a sequence query to identify 18 telomeres (Fu et al. 2023). Furthermore, a T2T gap-free genome assembly of the red-skinned pear (*Pyrus pyrifolia*) was developed, significantly improving upon previous pear genomes. The 501.20 Mb genome, with a contig N50 of 29.26 Mb, assembled all 17 chromosomes to the T2T level, identifying

34 telomeres and centromeres enriched with CEN198 and LTR Gypsy elements. The genome revealed that segmental duplications played a key role in expanding disease resistance genes and stress response pathways, while gene duplication events in *MYB10* and *MYB114* contributed to anthocyanin biosynthesis, influencing fruit color (Sun et al. 2023a). Furthermore, the first T2T and gap-free genome assembly of sorghum (CHBZ) was reported using PacBio HiFi reads, ONT, and Hi-C data. The assembled genome comprises 724.85 Mb, effectively resolving all 3913 gaps present in the previous sorghum BTx623 reference genome and effectively annotating all 10 centromeres and all 20 telomeres (Li et al. 2024). Very recently, the first complete assembly of a flowering cherry was generated using PacBio HiFi, UL-ONT, Illumina, and Hi-C sequencing. The assembly spanned 266.23 Mb with a contig N50 of 31.6 Mb, achieving 98.9% BUSCO completeness and a quality value of 48.75. It identified 13 telomeres and putative centromere regions across 8 pseudochromosomes. Comparative analysis with draft assembly revealed over 336,000 SNPs, 107,000 indels, 1400 SVs, and 1402 newly annotated genes (Jiang et al. 2025).

2.5 Applications of Complete Reference Genomes

The transition from fragmented to T2T genome assemblies in plants represents a paradigm shift in our ability to decipher the complexities of plant genomes. These complete, gapless sequences, extending from one telomere to the other, provide an unprecedented level of resolution, enabling researchers to explore genomic landscapes that were previously obscured. Below are some of the applications of these assemblies for crop improvement (Fig. 2.2).

2.5.1 Understanding Genome Architecture and Evolutionary Dynamics

The complete assemblies illuminate the intricate architecture of plant genomes, revealing the complete context of repetitive elements, which constitute a substantial portion of plant DNA. For example, in *Arabidopsis*, T2T assemblies have provided a comprehensive view of centromeric regions, revealing the organization of centromeric repeats and their influence on chromosome segregation (Naish et al. 2021). These complete sequences are instrumental in studying chromosome evolution, polyploidization events, and the dynamic interplay between genome structure and function. For instance, in polyploid crops like wheat, high-quality assemblies are crucial for understanding subgenome interactions and the evolution of homoeologous gene families (International Wheat Genome Sequencing Consortium (IWGSC) 2018; Jiao et al. 2025; Wang et al. 2025). Furthermore, complete genomes

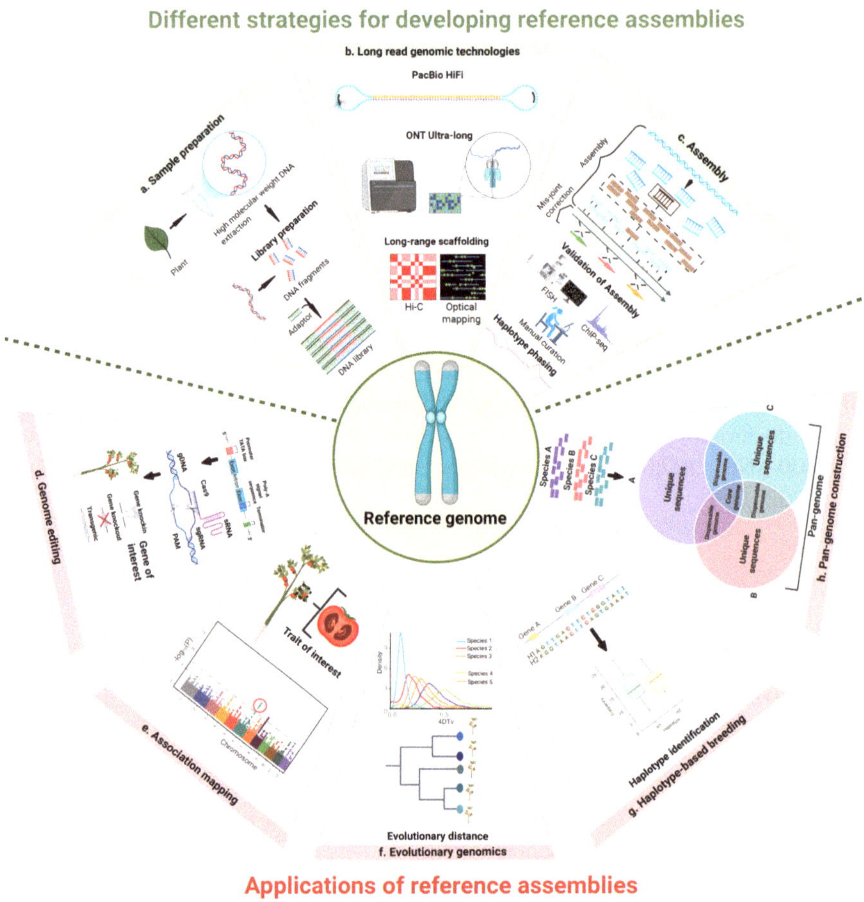

Fig. 2.2 Strategies for developing T2T reference genomes and their applications. The figure illustrates the process of generating high-quality reference genomes and their downstream applications. The top half (green section) outlines different strategies for developing T2T reference assemblies, including (a) sample preparation, (b) long-read sequencing technologies (PacBio HiFi and Oxford nanopore ultra-long reads), and (c) genome assembly (scaffolding, haplotype phasing, and validation). The bottom half (red section) highlights key applications of reference genomes for biological research and crop improvement, including (d) genome editing, (e) association mapping, (f) evolutionary genomics, (g) haplotype-based breeding, and (h) pangenomics. (Created with Biorender.com (https://BioRender.com))

facilitate the precise tracing of gene family evolution, revealing the mechanisms underlying gene duplication, divergence, and functional specialization. This enhanced understanding of evolutionary dynamics is critical for identifying structural variations associated with phenotypic diversity and for unraveling the genetic basis of agronomically important traits.

2.5.2 Making Centromeres Accessible for Biological Research and Crop Improvement

T2T genomes are essential for understanding the complex biology of plant centromeres and telomeres. These regions play critical roles in chromosome stability, cell division, and genome evolution. T2T assemblies provide a complete view of their structure and function, enabling researchers to investigate their dynamics and identify key regulatory elements. For instance, T2T assemblies have revealed the organization of centromeric repeats and their influence on chromosome segregation in Arabidopsis (Naish et al. 2021). Similarly, the T2T assembly of cotton (*Gossypium hirsutum*) uncovered a recent centromere repositioning event specific to chromosome D08, involving the inactivation of an ancestral centromere and the formation of a novel, satellite repeat-based centromere. This discovery provides insights into centromere evolution in polyploid crops and its potential role in genome stability and trait development (Hu et al. 2025). In another study, the near gapless assemblies of a tetraploid *Brachypodium hybridum* and its two diploid ancestors, *Brachypodium distachyon* and *Brachypodium stacei* revealed extensive sequence, structural, and epigenetic variations of centromeres at base-pair resolution between closely related *Brachypodium* genomes. The analysis highlighted that centromeres exhibit species-specific repeats, with homogenization accompanied by satellite repeat bursts and retrotransposon purging (Chen et al. 2024a). Additionally, these complete sequences can facilitate the study of telomere dynamics and their role in plant aging and stress responses, potentially leading to strategies for extending plant lifespan and improving crop resilience.

2.5.3 Enhancing Gene Annotation and Accelerating Functional Genomics Research

Complete genome sequences significantly improve the accuracy and completeness of gene annotation, particularly for genes located in repetitive regions or near telomeres and centromeres, which are often missed or mis-annotated in fragmented assemblies. T2T genomes enable the comprehensive mapping of regulatory elements, including promoters, enhancers, and long non-coding RNAs, leading to a refined understanding of gene regulation. For instance, in wild rice, T2T assemblies have facilitated the accurate annotation of genes involved in grain development and stress response, enhancing the power of functional genomics studies (Huang et al. 2024). Furthermore, a gap-free genome of a doubled monoploid potato (*Solanum tuberosum*) developed by combining ONT and Hi-C data, coupled with multiple gap-closing strategies, helped to identify large tandem clusters of genes (patatin, terpene synthase, and cupin) with agronomical importance (Yang et al. 2023). Similarly, in Chinese bayberry (*Myrica rubra*), a T2T genome assembly significantly improved gene annotation and facilitated the identification of allelic variations associated with fruit quality traits. By integrating GWAS with a high-quality reference genome, researchers identified genetic variants linked to key fruit characteristics, such as color, which were regulated by MYB transcription factors and an MLP-like protein influencing anthocyanin synthesis (Zhang et al. 2024a). By providing a complete and accurate reference for gene location and structure, T2T genomes enhance the power of genome-wide association studies (GWAS), transcriptomics, and proteomics analyses, enabling researchers to identify causal genes and regulatory networks with greater precision. This enhanced understanding of gene function and regulation is crucial for developing targeted breeding strategies and for engineering plants with improved traits.

2.5.4 Transforming Plant Breeding and Crop Improvement Strategies

T2T genomes offer significant advantages for plant breeding by enabling the precise identification and characterization of agronomically important genes and alleles. Breeders can now accurately map and track desirable traits, such as disease resistance, yield potential, and stress tolerance, across complete chromosomes, eliminating the ambiguity associated with fragmented assemblies. This precision facilitates marker-assisted selection (MAS) and genomic selection (GS), accelerating the development of improved crop varieties. For example, in maize, T2T assemblies have aided in the identification of genes associated with kernel size and composition, enabling the development of high-yielding varieties (Hufford et al. 2021). Furthermore, T2T genomes are invaluable for understanding and manipulating complex traits controlled by genes located in previously inaccessible regions of the genome, such as those within repetitive regions or near telomeres and centromeres. T2T assemblies can reveal the complete sequence of disease-resistance gene clusters, enabling the development of more effective breeding strategies for disease-resistant crops. For instance, the availability of a reference-quality genome for *Aegilops sharonensis*, a diploid wild relative of wheat, was instrumental in identifying the *Sr62* gene conferring stem rust resistance, enabling its subsequent molecular cloning and application in enhancing wheat's disease resistance (Yu et al. 2022). The completeness of T2T genomes provides a valuable resource for advancing breeding strategies, offering new opportunities for the precise improvement of key agronomic traits.

An emerging strategy empowered by high-quality genome assemblies is haplotype-based breeding, which leverages the complete and phased structure of genomes to identify and select favorable haplotypes associated with important traits. T2T genomes provide haplotype-resolved assemblies that enable the dissection of allele combinations across loci, facilitating the identification of superior haplotypes for traits such as yield, stress tolerance, and quality. This approach goes beyond individual SNPs or genes, considering the collective impact of linked alleles and their interactions, which is particularly relevant for polygenic traits. For example, in legumes and cereals, haplotype-based selection has been used to track combinations of beneficial alleles associated with drought tolerance (Sinha et al. 2020; Singh et al. 2024). With T2T genomes, haplotypes can be more accurately defined, including those spanning centromeric and pericentromeric regions, which are often rich in important gene clusters but were previously inaccessible. This approach enables more effective genomic prediction, targeted introgression of elite haplotypes, and the design of next-generation breeding populations tailored to specific environments or market needs.

2.5.5 *Advancing Comparative Genomics and Phylogenomics Studies*

Complete genome sequences enable more accurate and comprehensive comparative genomics studies, allowing researchers to explore the evolutionary relationships between plant species with greater precision. T2T genomes provide a robust framework for phylogenomic analyses, revealing the conserved and divergent features of plant genomes across diverse lineages. They have been used to study the evolutionary history of several crops, such as cassava, and chili pepper, shedding light on evolutionary events, gene family expansions, and adaptive traits (Xu et al. 2023; Chen et al. 2024b). Similarly, in one of the most important medicinal plants, neem (*Azadirachta indica)*, the T2T genome assembly has provided insights into karyotype evolution, highlighting structural rearrangements, chromosomal fusions, and the expansion of gene families linked to bioactive compound biosynthesis. Comparative genomics analyses have traced the evolution of metabolic pathways responsible for azadirachtin production, revealing lineage-specific gene duplications and rearrangements that contribute to neem's medicinal properties (Liu et al. 2025). The enhanced understanding of genome evolution is crucial for identifying key adaptations and developing strategies for crop improvement. Furthermore, comparative genomics studies using T2T genomes can reveal the mechanisms underlying genome size variation and polyploidization, providing insights into the evolutionary forces that have shaped plant genomes.

2.5.6 *Empowering Synthetic Biology and Genome Editing Applications*

T2T genomes provide a complete blueprint for synthetic biology and genome editing applications. Researchers can now design and engineer complex genetic constructs with greater precision, targeting specific genes and regulatory elements across complete chromosomes. This capability is essential for developing novel plant varieties with enhanced traits, such as improved nutritional content, increased biomass production, and enhanced stress tolerance. For instance, T2T assemblies can facilitate the design of synthetic gene circuits that regulate plant development or metabolism. Additionally, these complete sequences enable the precise targeting of gene editing tools, such as CRISPR–Cas, to modify specific genes or regulatory elements, leading to enhanced precision and reduced off-target effects. For example, a T2T allele-aware, chromosome-level assembly of cultivated alfalfa, covering 32 allelic chromosomes, was generated using HiFi and Hi-C data. This high-quality assembly facilitated a genome editing protocol that simultaneously disrupted all target gene alleles to create null mutants in one generation, thus demonstrating the need for high-quality, haplotype-resolved assemblies for allele-specific gene editing systems (Chen et al. 2020). As genome editing and synthetic biology continue to advance, the completeness of T2T genomes will play a crucial role in enabling precise, allele-specific modifications, accelerating the development of next-generation crops with improved traits.

2.5.7 Enabling Pangenomics for Unraveling Genomic Diversity

Pangenomes are crucial for capturing the full genomic diversity within a species, as they represent the complete set of genes and structural variations (Khan et al. 2024; Garg et al. 2025). Constructing pangenomes from T2T genomes offers a more accurate and comprehensive view, especially in capturing repetitive regions and structural variants often missed in fragmented assemblies. This is particularly important for understanding the genetic basis of phenotypic variation and for identifying genes associated with local adaptation and crop improvement. Pangenomes constructed from T2T genomes allow for the identification of core genes that are present in all varieties, and dispensable genes that are present in some varieties but not others. This is essential for understanding gene evolution and breeding efforts. Recently, T2T assemblies for 27 *Citrullus* genotypes (including seven watermelon species and *C. lanatus* subsp. *cordophanus*) were reported, which were used to construct a T2T super-pangenome that expanded the reference genome by 399.2 Mb and 11,225 genes. Comparative analyses unveiled the genomic landscape of the genus, shedding light on its domestication history and the distribution of structural variations (SVs). These findings also facilitated the identification of genes in wild *Citrullus* species that could enhance biotic and abiotic resistance and resilience in cultivated watermelon (Zhang et al. 2024b).

2.6 Future Perspectives

The emergence of T2T genome assemblies represents a transformative step in plant genomics, offering an unprecedented level of completeness and accuracy. These assemblies are poised to accelerate discoveries in plant biology and drive innovations in crop improvement. As sequencing technologies continue to evolve, the routine generation of T2T genomes across diverse and complex plant species will become more feasible. However, one of the key challenges in this space is the requirement for large amounts of high-quality DNA, which can be difficult to obtain, particularly from species with challenging DNA extraction processes or cases where sample availability is limited. Advancements that allow high-fidelity sequencing from minimal DNA quantities would be instrumental in expanding the applicability of T2T genomics. Additionally, the cost of producing and analyzing high-quality genome assemblies remains a limiting factor. Reducing financial and technical barriers through improved sequencing workflows and accessible computational tools will be critical for broadening participation in genomics research. Future efforts will focus on integrating multi-omics data with these complete genomes, enabling a deeper understanding of gene regulation and genotype-phenotype relationships. Furthermore, pangenomics, powered by T2T assemblies, will revolutionize our comprehension of species-wide diversity, facilitating the identification of novel genes for crop improvement. The precise targeting of gene editing tools, guided by complete genomic blueprints, will enhance the development of designer crops with improved traits, while sophisticated bioinformatics and AI tools will aid in data analysis and interpretation. Ultimately, the widespread implementation of T2T genome assemblies will play a key role in addressing global challenges such as food security and climate resilience. By providing a complete genetic blueprint, these assemblies will support sustainable agricultural practices and the development of climate-adapted, high-yielding crop varieties.

References

Anderson SN et al (2019) Transposable elements contribute to dynamic genome content in maize. Plant J 100(5):1052–1065

Arabidopsis Genome Initiative (2000) Analysis of the genome sequence of the flowering plant *Arabidopsis thaliana*. Nature 408:796–815

Belser C et al (2021) Telomere-to-telomere gapless chromosomes of banana using nanopore sequencing. Commun Biol 4(1):1047

Bernal-Gallardo JJ, de Folter S (2024) Plant genome information facilitates plant functional genomics. Planta 259:117

Chen H et al (2020) Allele-aware chromosome-level genome assembly and efficient transgene-free genome editing for the autotetraploid cultivated alfalfa. Nat Commun 11(1):2494

Chen J et al (2023) A complete telomere-to-telomere assembly of the maize genome. Nat Genet 55(7):1221–1231

Chen C et al (2024a) Three near-complete genome assemblies reveal substantial centromere dynamics from diploid to tetraploid in Brachypodium genus. Genome Biol 25(1):63

Chen W et al (2024b) Two telomere-to-telomere gapless genomes reveal insights into Capsicum evolution and capsaicinoid biosynthesis. Nat Commun 15(1):4295

Deng Y et al (2022) A telomere-to-telomere gap-free reference genome of watermelon and its mutation library provide important resources for gene discovery and breeding. Mol Plant 15(8): 1268–1284

Ding Y et al (2024) A telomere-to-telomere genome assembly of Hongyingzi, a sorghum cultivar used for Chinese Baijiu production. Crop J 12(2):635–640

Domínguez M et al (2020) The impact of transposable elements on tomato diversity. Nat Commun 11(1):4058

Fu A et al (2023) Telomere-to-telomere genome assembly of bitter melon (Momordica charantia L. var. abbreviata Ser.) reveals fruit development, composition and ripening genetic characteristics. Hortic Res 10(1):uhac228

Garg V et al (2022) Chromosome-length genome assemblies of six legume species provide insights into genome organization, evolution, and agronomic traits for crop improvement. J Adv Res 42: 315–329

Garg V et al (2023) Near-gapless genome assemblies of Williams 82 and Lee cultivars for accelerating global soybean research. Plant Genome 16(4):e20382

Garg V et al (2024) Unlocking plant genetics with telomere-to-telomere genome assemblies. Nat Genet 56(9):1788–1799

Garg V et al (2025) An Australian chickpea pan-genome provides insights into genome organization and offers opportunities for enhancing drought adaptation for crop improvement. Plant Biotechnol J. https://doi.org/10.1111/pbi.70192

Gupta S et al (2024) The genome of Haberlea rhodopensis provides insights into the mechanisms for tolerance to multiple extreme environments. Cell Mol Life Sci 81(1):117

Gurevich A et al (2013) QUAST: quality assessment tool for genome assemblies. Bioinformatics (Oxford, England) 29(8):1072–1075

Hickey LT et al (2019) Breeding crops to feed 10 billion. Nat Biotechnol 37(7):744–754

Hou X et al (2022) A near-complete assembly of an *Arabidopsis thaliana* genome. Mol Plant 15(8): 1247–1250

Hu G et al (2025) A telomere-to-telomere genome assembly of cotton provides insights into centromere evolution and short-season adaptation. Nat Genet. https://doi.org/10.1038/s41588-025-02130-4

Huang J et al (2024) Haplotype-resolved gapless genome and chromosome segment substitution lines facilitate gene identification in wild rice. Nat Commun 15(1):4573

Hufford MB et al (2021) De novo assembly, annotation, and comparative analysis of 26 diverse maize genomes. Science (New York, NY) 373(6555):655–662

International Wheat Genome Sequencing Consortium (IWGSC) (2018) Shifting the limits in wheat research and breeding using a fully annotated reference genome. Science (New York, NY) 361(6403):eaar7191

Jiang D et al (2025) The telomere-to-telomere genome of flowering cherry (*Prunus campanulata*) reveals genomic evolution of the subgenus Cerasus. GigaScience 14. https://doi.org/10.1093/gigascience/giaf009

Jiao C et al (2025) Pan-genome bridges wheat structural variations with habitat and breeding. Nature 637(8045):384–393

Kong W et al (2023) Recent advances in assembly of complex plant genomes. Genomics Proteomics Bioinformatics 21(3):427–439

Li K et al (2021) Gapless indica rice genome reveals synergistic contributions of active transposable elements and segmental duplications to rice genome evolution. Mol Plant 14(10):1745–1756

Li M et al (2024) Telomere-to-telomere genome assembly of sorghum. Sci Data 11(1):835

Liu J et al (2025) Comparative genomics provides insights into the biogeographic and biochemical diversity of meliaceous species. Nat Commun 16(1):2603

Makarevitch I et al (2015) Transposable elements contribute to activation of maize genes in response to abiotic stress. PLoS Genet 11(1):e1004915
Manchanda N et al (2020) GenomeQC: a quality assessment tool for genome assemblies and gene structure annotations. BMC Genomics 21(1):193
Manni M et al (2021) BUSCO update: novel and streamlined workflows along with broader and deeper phylogenetic coverage for scoring of eukaryotic, prokaryotic, and viral genomes. Mol Biol Evol 38(10):4647–4654
Marx V (2023) Method of the year: long-read sequencing. Nat Methods 20(1):6–11
Mascher M, Stein N (2014) Genetic anchoring of whole-genome shotgun assemblies. Front Genet 5:208
Mascher M et al (2017) A chromosome conformation capture ordered sequence of the barley genome. Nature 544(7651):427–433
Naish M et al (2021) The genetic and epigenetic landscape of the Arabidopsis centromeres. Science 374(6569):eabi7489
Navrátilová P et al (2022) Prospects of telomere-to-telomere assembly in barley: analysis of sequence gaps in the MorexV3 reference genome. Plant Biotechnol J 20(7):1373–1386
Nurk S et al (2022) The complete sequence of a human genome. Science 376(6588):44–53
Oliveira LC, Torres GA (2018) Plant centromeres: genetics, epigenetics and evolution. Mol Biol Rep 45(5):1491–1497
Ou S, Chen J, Jiang N (2018) Assessing genome assembly quality using the LTR Assembly Index (LAI). Nucleic Acids Res 46(21):e126
Ranallo-Benavidez TR, Jaron KS, Schatz MC (2020) GenomeScope 2.0 and Smudgeplot for reference-free profiling of polyploid genomes. Nat Commun 11(1):1432
Rhie A et al (2020) Merqury: reference-free quality, completeness, and phasing assessment for genome assemblies. Genome Biol 21(1):245
Roach MJ, Schmidt SA, Borneman AR (2018) Purge Haplotigs: allelic contig reassignment for third-gen diploid genome assemblies. BMC Bioinformatics 19(1):460
Shakirov EV, Chen JJ-L, Shippen DE (2022) Plant telomere biology: the green solution to the end-replication problem. Plant Cell 34(7):2492–2504
Shang L et al (2023) A complete assembly of the rice Nipponbare reference genome. Mol Plant 16(8):1232–1236
Singh P et al (2024) Superior haplotypes of key drought-responsive genes reveal opportunities for the development of climate-resilient rice varieties. Commun Biol 7:89
Sinha P et al (2020) Superior haplotypes for haplotype-based breeding for drought tolerance in pigeonpea (*Cajanus cajan* L.). Plant Biotechnol J 18(12):2482–2490
Song J-M et al (2021) Two gap-free reference genomes and a global view of the centromere architecture in rice. Mol Plant 14(10):1757–1767
Sun M et al (2023a) Telomere-to-telomere pear (*Pyrus pyrifolia*) reference genome reveals segmental and whole genome duplication driving genome evolution. Hortic Res 10(11):uhad201
Sun X et al (2023b) The role of transposon inverted repeats in balancing drought tolerance and yield-related traits in maize. Nat Biotechnol 41(1):120–127
Tuberosa R, Frascaroli E, Salvi S (2017) Leveraging plant genomics for better and healthier food. Curr Opin Food Sci 16:100–105
Turner I et al (2018) Integrating long-range connectivity information into de Bruijn graphs. Bioinformatics (Oxford, England) 34(15):2556–2565
VanBuren R et al (2018) Extreme haplotype variation in the desiccation-tolerant clubmoss Selaginella lepidophylla. Nat Commun 9(1):13
Wang B et al (2022) High-quality *Arabidopsis thaliana* genome assembly with nanopore and HiFi long reads. Genomics Proteomics Bioinformatics 20(1):4–13
Wang X et al (2024) Telomere-to-telomere and gap-free genome assembly of a susceptible grapevine species (Thompson Seedless) to facilitate grape functional genomics. Hortic Res 11(1):uhad260

Wang Z et al (2025) Near-complete assembly and comprehensive annotation of the wheat Chinese Spring genome. Mol Plant. https://doi.org/10.1016/j.molp.2025.02.002

Xie L et al (2024) Technology-enabled great leap in deciphering plant genomes. Nat Plants 10(4): 551–566

Xu Y et al (2022) Oxford nanopore technology: revolutionizing genomics research in plants. Trends Plant Sci 27(5):510–511

Xu X-D et al (2023) Telomere-to-telomere assembly of cassava genome reveals the evolution of cassava and divergence of allelic expression. Hortic Res 10(11):uhad200

Yang J et al (2017) Haplotype-resolved sweet potato genome traces back its hexaploidization history. Nat Plants 3(9):696–703

Yang T et al (2022) Improved pea reference genome and pan-genome highlight genomic features and evolutionary characteristics. Nat Genet 54(10):1553–1563

Yang X et al (2023) The gap-free potato genome assembly reveals large tandem gene clusters of agronomical importance in highly repeated genomic regions. Mol Plant 16(2):314–317

Yokosho K et al (2016) Retrotransposon-mediated aluminum tolerance through enhanced expression of the citrate transporter OsFRDL4. Plant Physiol 172(4):2327–2336

Yu G et al (2022) *Aegilops sharonensis* genome-assisted identification of stem rust resistance gene Sr62. Nat Commun 13(1):1607

Zhang C et al (2023) The T2T genome assembly of soybean cultivar ZH13 and its epigenetic landscapes. Mol Plant 16(11):1715–1718

Zhang S et al (2024a) T2T reference genome assembly and genome-wide association study reveal the genetic basis of Chinese bayberry fruit quality. Hortic Res 11(3):uhae033

Zhang Y et al (2024b) Telomere-to-telomere Citrullus super-pangenome provides direction for watermelon breeding. Nat Genet 56(8):1750–1761

Zhou Y et al (2023) The telomere-to-telomere genome of Fragaria vesca reveals the genomic evolution of Fragaria and the origin of cultivated octoploid strawberry. Hortic Res 10(4): uhad027

Zhuang W et al (2019) The genome of cultivated peanut provides insight into legume karyotypes, polyploid evolution and crop domestication. Nat Genet 51(5):865–876

Zimin AV, Stevens KA et al (2017a) An improved assembly of the loblolly pine mega-genome using long-read single-molecule sequencing. GigaScience 6(1):1–4

Zimin AV, Puiu D et al (2017b) Hybrid assembly of the large and highly repetitive genome of *Aegilops tauschii*, a progenitor of bread wheat, with the MaSuRCA mega-reads algorithm. Genome Res 27(5):787–792

Chapter 3
Pangenomes, Super-Pangenomes and Beyond

Aamir W. Khan, Saurabh Gupta, and Henry T. Nguyen

Abstract The conventional reliance on a single reference genome falls short in capturing the full spectrum of genetic diversity within a population. Pangenomes overcome this limitation by integrating genetic variations from multiple individuals, offering a more comprehensive view of species-wide genomic diversity. A pangenome typically comprises the core genome, shared by all individuals, and the dispensable or accessory genome, which contains genes present in some but not all individuals. These variable regions often include genes related to adaptation, stress resistance, and other traits shaped by environmental and evolutionary pressures. By revealing both conserved and unique genomic elements, pangenomes provide critical insights into intra-species variation, structural rearrangements, and functional genomics. With the rapid advancements in sequencing technologies and the exponential growth of genomic data, plant pangenomics has progressed beyond single-species frameworks. This progression has given rise to *super-pangenomes*, which span all genomic sequences across an entire genus. These super-pangenomes are particularly valuable as they incorporate both domesticated and wild species, enabling deeper insights into evolutionary trajectories, domestication events, and haplotype patterns unique to wild lineages. Such insights are essential for understanding genetic diversity, tracing evolutionary dynamics, studying domestication, and formulating advanced breeding strategies. This chapter explores the development and significance of pangenomes and super-pangenomes, emphasizing their critical role in harnessing wild genomic variation for crop improvement.

Keywords Pangenome · Super-pangenome · SVs · CNVs · Graph

A. W. Khan · H. T. Nguyen (✉)
Division of Plant Science and Technology, University of Missouri, Columbia, MO, USA
e-mail: nguyenhenry@missouri.edu

S. Gupta
Curtin Medical Research Institute, Curtin Medical School, Curtin University, Perth, WA, Australia

UWA Centre for Medical Research, University of Western Australia, Perth, WA, Australia

V. Garg et al. (eds.), *DNA of Sustainability*, Sustainability Sciences in Asia and Africa,
https://doi.org/10.1007/978-981-95-5109-5_3

3.1 Introduction

Over the past decade, the field of genomics has transitioned from focusing on individual reference genomes to exploring the full spectrum of genomic diversity within species. While reference genomes remain foundational to genetic research, it has become increasingly clear that a single reference genome is insufficient to represent the complete gene content and structural variability present across populations. This recognition has catalyzed the rise of pangenomics, a transformative approach that captures both the conserved (core genome) and variable (dispensable or accessory genome) regions across multiple individuals of a species (Golicz et al. 2016a, b; Della Coletta et al. 2021).

Advances in long-read sequencing technologies have played a pivotal role in this shift. These technologies have made it possible to assemble complete and contiguous genomes with high accuracy, revealing large structural variations (SVs) that were previously inaccessible through short-read assemblies (Rhie et al. 2021; Marx 2023; Garg et al. 2024). SVs, including insertions, deletions, duplications, inversions, and translocations, represent a major source of genetic diversity and have been shown to influence important agronomic traits (Saxena et al. 2014; Yuan et al. 2021). Unlike SNPs and small InDels, SVs often affect gene presence-absence, regulatory elements, and genome architecture, making their inclusion essential for breeding and trait discovery.

The pangenome framework addresses this need by constructing a collective genomic resource that integrates gene content from multiple accessions, thereby enabling a population-scale view of genomic diversity (Golicz et al. 2020). This approach has proven especially valuable in crop species where domestication and selective breeding have narrowed genetic diversity. By including both cultivated and diverse wild relatives, researchers can uncover novel genes associated with stress tolerance, adaptation, and yield improvement.

Most early pangenomes were built using cultivated varieties, which limited the extent of genetic diversity captured. To overcome this, the concept of a super-pangenome was introduced. Unlike species-level pangenomes, super-pangenomes encompass genomic information from multiple species within a genus, particularly wild species with deep evolutionary divergence (Khan et al. 2020). These wild genomes often contain unique haplotypes and alleles that have been lost in modern cultivars but are critical for resilience to biotic and abiotic stresses (Bohra et al. 2022).

In this chapter, we explore the emerging landscape of pangenomics and super-pangenomics, beginning with the various strategies for constructing pangenomes, including de novo and reference-based approaches. We discuss the critical role of SVs and transposable elements (TEs) in shaping genomic diversity and highlight how pangenomic frameworks have been applied in crop improvement, evolutionary biology, and stress-resilience research. We also consider future directions in the field, including pan-transcriptomics, graph-based genome representations, and their implications for next-generation breeding strategies.

3.2 Emergence of Plant Pangenomics

The concept of the pangenome was first introduced by Tettelin et al. (2005) in a landmark study on *Streptococcus agalactiae* (Group B Streptococcus, GBS), a bacterial species known for causing neonatal infections. By sequencing the genomes of eight GBS strains, they discovered that bacterial genomes consist of two distinct components: a *core genome* shared by all strains, and a *dispensable genome* comprising genes present in only some strains. Remarkably, approximately 80% of the genes in any given strain were part of the core genome. Pangenome modeling revealed that GBS had an *open pangenome*, meaning that sequencing additional strains would likely reveal new genes indefinitely. This foundational work demonstrated how pangenomics could be used to investigate inter-strain variation, population structure, and evolutionary dynamics—principles that would later influence broader applications across many species.

The pangenome concept was later extended to humans. In 2010, Li et al. reported approximately 5 Mb of novel DNA sequences that were absent from the human reference genome but present in individuals of Asian and African ancestry. These sequences showed population-specific distribution patterns consistent with known human migration routes, highlighting the insufficiency of a single reference genome to capture species-wide diversity (Li et al. 2010). This study reinforced the need for a population-scale genomic framework in higher organisms, including humans and plants.

In plants, the first reported application of the pangenome concept was in soybean (*Glycine soja*), the wild progenitor of cultivated soybean (*Glycine max*) (Li et al. 2014). This study used whole-genome sequencing and de novo assembly of seven phylogenetically diverse *G. soja* accessions to construct a species-level pangenome. The results showed a gene content distribution of ~80% core and ~20% dispensable genes. The dispensable genome exhibited higher variability and evolutionary rates and was enriched in genes related to disease resistance (R-genes), flowering time, oil and fatty acid biosynthesis, and overall plant growth traits critical for environmental adaptation and domestication.

A parallel effort in rice (*Oryza sativa*) further solidified the value of plant pangenomics. Schatz et al. (2014) constructed a rice pangenome using de novo genome assemblies of three genetically divergent cultivars: Nipponbare (temperate *japonica*), IR64 (*indica*), and DJ123 (*aus*). These cultivars represented major subpopulations of cultivated rice. Comparative analyses revealed extensive SVs, including megabase-scale genomic segments unique to each accession, many of which were not captured in the existing reference genome. Most early plant pangenome studies have focused on identifying SVs and TEs, as they are major contributors to genome size variation, gene content differences, and adaptive traits among individuals.

3.3 Role of Structural Variants and Transposable Elements in Genome Evolution

SVs are major contributors to genome evolution, playing a crucial role in shaping the genomic architecture and adaptability of plants. These variations, which have evolved over millions of years, enable plants to survive and thrive in diverse and challenging ecological environments (Zhou et al. 2019). SVs include a wide range of chromosomal rearrangements such as insertions, deletions, duplications, inversions, copy number variations (CNVs), and translocations (Saxena et al. 2014). Among these, CNVs of specific genes are often directly linked to key agronomic traits like disease resistance, stress tolerance, and flowering time (Cook et al. 2012; Zhou et al. 2019). For example, the expansion of resistance (R) gene families frequently correlates with improved pathogen resistance in many plant species.

SVs can range in size from small InDels (~50 base pairs) to megabase-scale rearrangements and may impact both coding sequences and regulatory regions. Such alterations can influence gene expression, protein structure, or even result in gene loss or gain. These changes often lead to significant phenotypic effects and are especially relevant in crop improvement efforts targeting yield, stress adaptation, and trait durability.

TEs are another major force driving genome evolution, particularly in plants, where they can occupy majority of the genome. TEs contribute to genome size expansion, introduce genetic variation, and play critical roles in gene regulation and evolutionary innovation (Fedoroff 2012). Due to their repetitive nature and widespread distribution, TEs are often concentrated in non-coding and intergenic regions, but they can also insert into or near genes, impacting their expression.

TEs are broadly classified into two main classes: DNA transposons and retroelements. DNA transposons include miniature inverted-repeat TEs (MITEs), Helitrons, and several superfamilies such as DTC, DTA, DTT (Tc1-Mariner), DTM (Mutator), and DTH (PIF-Harbinger). Retroelements, on the other hand, include long terminal repeat (LTR) elements such as Copia and Gypsy as well as non-LTR elements like LINEs and SINEs.

TEs have been shown to regulate important traits in several crop species. For instance, in wheat (*Triticum aestivum*), RLG_famc7.3 TEs activate the expression of enhancer-like RNAs (ELE-RNAs) in genomic regions associated with spike development and rust resistance (Xie et al. 2023). In apple (*Malus domestica*), the gene MdRFNR1 (root-type ferredoxin-NADP$^+$ oxidoreductase), which enhances drought tolerance by regulating the redox system, exists in two allelic forms: MdRFNR1-1 and MdRFNR1-2. Only MdRFNR1-1, which harbors a MITE insertion in its promoter (MITE–MdRF1), shows strong induction under drought stress (Niu et al. 2022).

Despite their biological significance, the high proportion of repeat elements in plant genomes poses major analytical challenges. Repetitive sequences can complicate genome assembly and annotation, necessitating the use of computational strategies such as soft masking, which masks tandem repeats during sequence alignment and assembly to improve bioinformatics efficiency. On the experimental front, selecting the right approach for pangenome construction depends heavily on genome complexity and the quality of available datasets. High-resolution identification of SVs and TEs requires carefully designed sequencing strategies, robust reference genomes, and effective computational pipelines.

3.4 Different Approaches for Pangenome Construction

The construction of plant pangenomes has evolved significantly over the past decade, transitioning from reference-dependent, linear methods to dynamic, graph-based frameworks. Broadly, two main strategies have been employed: iterative mapping and assembly-based approaches, and the more recent graph-based pangenome methods.

3.4.1 Iterative Mapping and Assembly-Based

This approach represents one of the earliest strategies in plant pangenome development. It involves aligning short-read sequencing data from multiple accessions to a reference genome, followed by de novo assembly of unmapped reads. The resulting pangenome comprises the reference genome and novel genomic regions absent from the reference, capturing gene presence-absence variations (PAVs). While cost-effective and scalable, this method is inherently limited by its reliance on a single reference genome and short-read data, which reduces the resolution for SVs and repetitive regions. Notable iterative mapping and assembly approach-based plant pangenomes were reported in *Brassica*, rice, wheat, tomato (*Solanum lycopersicum*), and chickpea (*Cicer arietinum* L.), where short-read sequencing data were mapped to the reference genome, and unmapped reads were de novo assembled to identify the genes missing in the reference genome. This constituted the dispensable genome and was largely limited to a smaller portion of the pangenome, as most of these studies developed the pangenome with cultivated accessions. The first Brassica pangenome was reported by sequencing nine Brassica lines, including one wild-type *B. macrocarpa* (Golicz et al. 2016a). The reference genome Chinese kale rapid cycling line (TO1000) was used as the base genome for mapping the sequencing data from nine lines with an iterative mapping and assembly approach. The pangenome assembly was 587 Mb, with a total of 61,379 genes. The study shed light on PAVs in genes associated with flowering time, glucosinolate, and auxin pathways. The first instance of a comprehensive rice pangenome was reported by Wang et al. 2018, where the authors investigated genetic variations in ~3000 rice genomes in a large-scale study. The study estimated 268 Mb novel sequences missing from the Nipponbare reference genome and identified ~12,000 full-length novel genes using an iterative mapping and assembly method. The GO enrichment and pangenome analyses revealed dispensable GO terms enriched for regulating immune and defense responses and ethylene metabolism. The study shed insights into the evolutionary history and domestication of rice using this large genomic resource and identified unique haplotypes in various subpopulations.

Similarly, large-scale germplasm sequencing of 725 phylogenetically and geographically diverse lines was performed to develop tomato pangenome (Gao et al. 2019). The study identified substantial gene loss during tomato domestication and breeding, particularly affecting disease resistance genes. The pangenome analysis revealed a rare version of the TomLoxC gene that influences tomato flavor, which was selected against during early domestication but has been recently reintroduced in modern breeding. Another major study presented a comprehensive genomic survey of chickpea diversity based on sequencing 3366 chickpea genomes, including 3171 cultivated and 195 wild accessions. One of the key aspects was the construction of the first chickpea pangenome (592.58 Mb) using the cultivated accessions, immediate wild progenitor *C. reticulatum* and *C. echinospermum* genomes (Varshney et al. 2021). The study developed a comprehensive map of genetic variations across cultivated chickpea and its wild progenitors. This study reported the largest number of individuals sequenced for a plant species and identified superior haplotypes for important traits. While iterative mapping enabled the early establishment of plant pangenomes, the limitations of two-dimensional representation and short-read sequencing led to the development of more comprehensive models based on high-quality genome sequences of multiple individuals.

3.4.2 Graph-Based Pangenome

Graph-based pangenomes represent the next generation of pangenome construction. Rather than relying on a single reference, this approach uses high-quality de novo genome assemblies from diverse individuals to construct a nonlinear representation of the genome. In a graph structure, nodes represent genomic variants and edges define the relationships between them. This allows for precise modeling of complex genomic features such as SVs, gene duplications, and rearrangements. The current approach encompasses developing high-quality de novo genome assemblies using a long-read sequencing approach and leveraging them for developing the pangenome. High-quality reference genomes are essential for trait-based mapping (Garg et al. 2022, 2024). Long-read sequencing has produced high-quality complete genomes for several plant species, such as *Arabidopsis*, maize (*Zea mays*), soybean, rice, haberlea, and watermelon (*Citrullus lanatus*) (Hou et al. 2022; Chen et al. 2023; Garg et al. 2023; Zhang et al. 2023; Shang et al. 2023; Gupta et al. 2024; Deng et al. 2022). Further, the short-read sequencing data is mapped to the reference pangenome to accurately genotype the SVs across the population. Initial efforts in graph-based pangenomics focused on crops like tomato and soybean. In tomato, Alonge et al. (2020) sequenced 100 accessions using long-read sequencing, identifying over 238,000 SVs and generating one of the most comprehensive "pan-SV" genomes in any crop to date. In soybean, Liu et al. (2020) assembled genomes from 26 diverse accessions and revealed 723,862 PAVs and 27,531 CNVs affecting genes involved in seed lustre, flowering time, and iron deficiency chlorosis, laying the groundwork for SV-guided trait mapping. These two studies laid the foundation for long-read-based pangenome development in plants. Further, rice became a landmark case for graph-based pangenomics with the study by Qin et al. (2021), which assembled high-quality genomes from 33 diverse rice accessions, including 32 Asian and one African rice genome, using PacBio SMRT sequencing. The study identified 171,072 SVs and 25,549 gene CNVs. Further analysis revealed that TE insertions and non-homologous end joining were the dominant mechanisms of SV formation. A graph pangenome incorporating these variations facilitated genome-wide association studies (GWAS), uncovering phenotype-linked variants not detectable through conventional reference-based methods.

In pearl millet (*Pennisetum glaucum*), Yan et al. (2023) constructed a graph-based pangenome using ten new chromosome-scale assemblies alongside one existing reference. The analysis captured 424,085 SVs and revealed an expansion of the RWP–RK transcription factor family, which contributes to heat tolerance, a key adaptive trait in this climate-resilient cereal. Similarly, a cucumber (*Cucumis sativus*) graph pangenome was developed by integrating 12 chromosome-level assemblies, which included the 9930 reference genome and 11 diverse accessions (Li et al. 2022). The study uncovered seven large chromosomal inversions distinguishing wild and cultivated cucumbers and reported ~4.3 million variants, including 56,214 SVs. In another effort, Hufford et al. (2021) sequenced 26 maize founder lines of the Nested Association Mapping (NAM) population and revealed a pangenome comprising over 103,000 genes, of which only one-third were core. SVs, including centromere repositioning and chromosomal rearrangements, were widespread. DNA methylation differences associated with gene expression and regulatory variation were also detected. Later, Gui et al. (2022) developed an even more expansive pangenome referred to as the pan-Zea pangenome by analyzing 721 individuals, including modern cultivars, landraces, and wild relatives (teosintes). This resource spanned 6.71 Gb and contained approximately 4.57 Gb of sequences absent from the B73 reference, identifying 255,821 SVs and showing that 44% of genes were dispensable.

A pangenome study in *Arabidopsis thaliana* by Lian et al. (2024) demonstrated an unusually conserved genome structure despite global distribution. Among 69 accessions, only seven major inversions were identified, and variations in genome size were mainly attributed to differences in centromeric repeat content, not TEs. Meanwhile, in potato (*Solanum tuberosum*), Cheng et al. (2025) constructed a phased graph pangenome of 60 haplotypes representing both cultivated diploids and wild relatives. SVs were primarily driven by TEs, and cultivated potatoes exhibited greater heterozygosity due to hybridization. Deleterious SNPs were found to cluster near deleterious SVs, and ideal haplotypes were computationally designed to assist breeding.

Several graph-based pangenomes have been developed across diverse plant species, leveraging long-read sequencing to reveal SVs and genotype–phenotype associations with unprecedented accuracy (Yang et al. 2022; Guo et al. 2024; Li et al. 2024; Liu et al. 2024; Yano et al. 2025; Zhu et al. 2025). Together, these efforts demonstrate that graph-based approaches represent the future of pangenomics, offering a scalable and comprehensive platform for plant genomic research and breeding.

3.5 Methods for Developing Graph-Based Genomes

The development of graph-based pangenomes has introduced a new paradigm in genomics by enabling the representation of complex variation across populations in a nonlinear, multi-dimensional format. Graph pangenomes provide a flexible framework for integrating SNPs, small InDels, SVs, and PAVs, thereby capturing the full scope of genomic diversity. Several computational methods have been developed to construct these graph-based genomes, and they fall into three primary categories: reference-based variation graph construction, reference-based graph assembly, and reference-unbiased graph construction.

3.5.1 Reference-Based Variation Graph Pangenome

The reference-based genomic variation graph development is the most frequently deployed approach that embeds the genome-wide variations, including the SNPs, small InDels, and large SVs, in the reference genome (Tao et al. 2020). Researchers have used two different methods in the form of de novo assembly-based comparison and read alignments. In the first approach, the genome assemblies developed from long read sequences using assemblers such as hifiasm (Cheng et al. 2021) and Verkko (Rautiainen et al. 2023) are aligned to the reference genome using Minimap2 (Li 2018) and MUMmer (Marçais et al. 2018). The resulting alignments are then fed to tools such as SVIM-asm (Heller and Vingron 2021), SyRi (Goel et al. 2019), SVMU (Chakraborty et al. 2019), and Assemblytics (Nattestad and Schatz 2016) to identify the genomic variations. The reads alignment approach, on the other hand, uses the long reads sequence alignment to reference genome using tools such as Minimap2 (Li 2018) and NGMLR (Sedlazeck et al. 2018) to detect SVs using tools SVIM (Heller and Vingron 2019), pbsv (https://github.com/PacificBiosciences/pbsv), Sniffles (Smolka et al. 2024), CuteSV (Jiang et al. 2020), etc. The genomic variations are subsequently integrated with the reference genome to develop a 3D graph pangenome using vgtoolkit (Hickey et al. 2020). This approach has been applied in several plant pangenomes such as tomato, chickpea, rice, cucumber, etc. (Li et al. 2023; Khan et al. 2024; Shang et al. 2022; Li et al. 2022).

3.5.2 Reference-Based Graph Pangenome

The reference-based graph pangenome approach is conceptually similar to iterative mapping but extends it to graph structures. Here, multiple genome assemblies are incrementally aligned to a single reference genome to generate a sequence graph that represents the collective diversity. This approach has been implemented in tools like Minigraph (Li et al. 2020), which enables rapid sequence-to-graph mapping and SVs identification across individuals.

While Minigraph is computationally efficient and effective for detecting large-scale variations, it is primarily limited to pairwise comparisons between the reference and other genomes. To overcome these limitations, Minigraph-Cactus (Hickey et al. 2024) was introduced as an extension, combining Minigraph's speed with the multi-genome graph alignment capabilities of *Cactus*. This method has demonstrated superior performance in human pangenome studies and is increasingly being considered for complex, multi-species plant pangenomes.

3.5.3 Reference-Unbiased Graph Pangenome

One of the most recent approaches is the development of plant graph pangenomes with a reference unbiased method (Garrison and Guarracino 2023). This approach is implemented in the pangenome graph builder (PGGB) pipeline and aims to reduce the reference bias inherent in other pangenome construction approaches (Garrison et al. 2024). PGGB deploys an all vs all genome alignment blueprint using tools like wfmash for whole genome alignment, seqwish for graph development, and smoothxg for smoothing the graph. This method uses each genome equally, and any of the genomes can be used as a reference. The reference unbiased technique comprehensively encapsulates the complete genetic variations within the pangenome and overpowers the reference bias. This approach has been used in the recently developed grapevine (*Vitis*) pangenome (Liu et al. 2024). PGGB is a computationally demanding workflow as it involves whole-genome alignment of different individuals in the pangenome, and alignment of diverse individuals is a daunting task.

3.6 Super-Pangenomes

The emergence of super-pangenomes marks a pivotal advancement in plant genomics, aiming to capture the full spectrum of genetic diversity within an entire genus rather than within a single species. These expansive resources are constructed by sequencing and analyzing both cultivated species and their wild relatives, offering unprecedented insight into genome evolution, SV, and adaptive potential.

Super-pangenomes have been established for plant species such as rice, tomato, chickpea, grape (*Vitis vinifera*), citrullus, and populus (Shang et al. 2022; Li et al. 2023; Khan et al. 2024; Cochetel et al. 2023; Wu et al. 2023; Shi et al. 2024) by sequencing genus-wide accessions and focusing on the crop wild relatives (Table 3.1).

Super-pangenome construction begins with the genome sequencing of genus-wide accessions, especially focusing on crop wild relatives (CWRs), which are known reservoirs of untapped genetic diversity. These genomes reveal both the gene content and chromosomal arrangement unique to each species. Comparative analyses between cultivated and wild accessions consistently reveal extensive SV, PAV, and gene copy number differences. Importantly, the core genome of a super-pangenome tends to be much smaller than that of a species-level pangenome, reflecting the greater diversity captured across species boundaries.

One of the earliest and most comprehensive plant super-pangenomes was developed for rice, incorporating 251 accessions spanning cultivated and wild species from both Asia and Africa (Shang et al. 2022). Using Oxford Nanopore long-read sequencing, high-quality genome assemblies were generated, uncovering widespread SVs and gene PAVs. The study identified genetic variants linked to critical traits such as submergence tolerance, seed shattering, and plant architecture. Intriguingly, it also found evidence for parallel selection of similar gene sets during the independent domestication of Asian and African rice species. Further, in tomato, a super-pangenome was constructed from chromosome-scale assemblies of nine wild species and two cultivated varieties (Li et al. 2023). A graph-based framework was used to incorporate these genomes, enabling SV-based GWAS. The analysis uncovered genetic signals associated with fruit metabolites and flavor-related traits and resolved phylogenetic ambiguities by clustering the species into four major clades. Similarly, in chickpea, a genus-wide super-pangenome comprising 495 accessions from all annual wild *Cicer* species and cultivated chickpea was constructed (Khan et al. 2024). This comprehensive resource enabled a reassessment of the divergence timeline between wild and cultivated *Cicer* species and revealed major structural differences that had shaped domestication.

The Citrullus super-pangenome represented another milestone, using telomere-to-telomere (T2T) assemblies of 27 accessions spanning all seven species of the genus (Zhang et al. 2024b). The pangenome analysis identified three large chromosomal rearrangements, which appeared to impact gene expression and duplication events, potentially contributing to domestication traits in watermelon. In grapevine, Cochetel et al. (2023) sequenced and assembled nine wild North American *Vitis* species to develop a genus-level super-pangenome. Using a reference-unbiased, whole-genome alignment-based approach, they captured extensive genomic diversity from sequence to gene-level resolution. Species-specific genome segments were enriched in TEs particularly Gypsy retrotransposons while core genes were linked to essential developmental functions and dispensable genes to species-specific environmental adaptations. Further, a Populus super-pangenome was constructed from 19 genomes spanning multiple *Populus* species and subspecies (Shi et al. 2024). The study identified over 142,000 SVs, many of which played critical roles in local

Table 3.1 Key Graph based pangenomes and super-pangenomes in plants

Crop	Number of species	Super-pangenome	Number of accessions	Sequencing technology	Tool used for pangenome construction	References
Tomato	10	Yes	433	PacBio/ Illumina/ Bionano/Hi-C	VG	Li et al. (2023)
Rice	4	Yes	856	ONT/ Illumina/Hi-C	Minigraph and VG	Shang et al. (2022)
Lettuce	4	Yes	474	Illumina	PanTools	van Workum et al. (2024)
Maize	2	No	732	Illumina	VG	Gui et al. (2022)
Chickpea	8	Yes	495	Illumina/Hi-C	VG	Khan et al. (2024)
Citrullus	7	Yes	27	PacBio HiFi/ ONT/ Illumina/ Bionano/Hi-C	Minigraph	Zhang et al. (2024b)
Grapevine	18	Yes	185	PacBio HiFi/ ONT/Hi-C/ Illumina	VG	Guo et al. (2024)
	6	Yes	484	PacBio HiFi/ ONT/ Illumina/Hi-C	Minigraph-cactus and PGGB	Liu et al. (2024)
Potato	26	No	298	PacBio/ Illumina/Hi-C	VG	Zhu et al. (2025)
Cucumber	1	No	12	PacBio/ Illumina/10X/ Hi-C	VG	Li et al. (2022)
Soybean	1	No	462	ONT/Bionano optical map/Illumina/ Hi-C	Minigraph-cactus and VG	Yano et al. (2025)
	2	No	2927	PacBio/ Illumina/Hi-C	VG	Liu et al. (2020)
Barley	2	No	1391	PacBio HiFi/ Illumina/Hi-C	Minigraph	Jayakodi et al. (2024)
Pearl Millet	1	No	11	PacBio HiFi/ Bionano/ Illumina/Hi-C	VG	Yan et al. (2023)
Brassica	1	No	403	PacBio/ Illumina	VG	Guo et al. (2024)

(continued)

Table 3.1 (continued)

Crop	Number of species	Super-pangenome	Number of accessions	Sequencing technology	Tool used for pangenome construction	References
	1	No	731	PacBio/ONT/Illumina/Bionano/Hi-C	VG	Li et al. (2024)
Pea	3	No	118	PacBio/Bionano/Illumina/10X/Hi-C	Minigraph	Yang et al. (2022)

ONT Oxford Nanopore Technologies

environmental and climatic adaptation. Private genes were found to be especially relevant for niche-specific phenotypic traits and ecological specialization.

More recently, a lettuce (Lactuca) super-pangenome was developed using 474 accessions from four species: *L. sativa* (cultivated lettuce), *L. serriola* (primary gene pool), *L. saligna* (secondary), and *L. virosa* (tertiary) (van Workum et al. 2024). The study identified between 10 and 18 Mbp of novel, non-reference sequence per species, comprising 2000–3000 genes absent in the reference genome. Approximately 50% of the genes were shared across the gene pools, with core genes enriched in transcription factors and dispensable genes in disease-resistance genes, underscoring the genetic divergence across domestication layers.

Most super-pangenomes to date have been constructed using graph-based approaches. These typically involve aligning high-quality wild and cultivated genome assemblies, detecting SVs, and integrating them into a unified graph genome. The resulting 3D pangenomic framework allows for high-resolution genotyping using short-read data mapped to the graph using tools like vg giraffe. This approach facilitates trait-association studies and diversity analyses at a level of resolution not possible with linear genomes.

3.6.1 SVs Genotyping

Accurate genotyping of SVs and small variants from short-read sequencing data remains a challenge due to alignment bias and highly repetitive genomic regions in plant genomes. Several graph-based and machine learning tools have been developed to address these limitations (Table 3.2). General-purpose graph-based tools include VG (variation graph), whose main features are genotyping SNPs, InDels, and all SV types (deletions, insertions, inversions, etc.; Hickey et al. 2020). GraphTyper2, on the other hand, focuses on SNPs and short InDels genotyping with limited SV support (Eggertsson et al. 2019). PanGenie is a k-mer-based tool used to genotype SNPs/InDels and SVs (Ebler et al. 2022). Specialized SV

Table 3.2 SVs genotyping tools for pangenome analysis

Tool	Type of variation			Specialization	Tool link	References
	SNPs	InDels	Structural Variants (SVs)			
vg	✔	✔	✔ (all types)	General-purpose	https://github.com/vgteam/vg	Hickey et al. (2020)
GraphTyper2	✔	✔	✘ (limited)	Small variants	https://github.com/DecodeGenetics/graphtyper	Eggertsson et al. (2019)
PanGenie	✔	✔	✘	K-mer-based	https://github.com/eblerjana/pangenie	Ebler et al. (2022)
BayesTyper	✔	✔	✔ (complex SVs)	Haplotype-aware	https://github.com/bioinformatics-centre/bayestyper	Sibbesen et al. (2018)
Paragraph	✘	✘	✔ (focused on SVs)	SV genotyping	https://github.com/Illumina/paragraph	Chen et al. (2019)
PanDelos	✘	✘	✔ (deletions only)	Deletion detection	https://github.com/InfOmics/PanDelos	Bonnici et al. (2018)
SVDSS	✘	✘	✔ (large/complex SVs)	SV assembly + genotyping	https://github.com/Parsoa/SVDSS	Denti et al. (2023)
Seven Bridges' GRAF	✔	✔	✘	Population-scale	https://www.sevenbridges.com/graf/	–

genotyping tools include a paragraph that is optimized for SVs (deletions, insertions, and duplications) genotyping (Chen et al. 2019). It works with linear or graph references and is a widely used tool. BaycsTyper is used to genotype SNPs, InDels, and complex SVs using a k-mer counting approach (Sibbesen et al. 2018). GGTyper is specialized for genotyping breakpoint-resolved complex SVs (no insertions with novel sequence; Mirus et al. 2024). SVLearn, on the other hand, is a machine learning-based genotyping tool for genotyping biallelic insertions/deletions (Yang et al. 2025b). Some other notable genotyping tools are PanDelos, which focuses only on deletion detection, SVDSS, which combines assembly and graph-based genotyping for large/complex SVs, and Seven Bridges' GRAF, which is used for population-scale genotyping but is limited to small variants (Bonnici et al. 2018; Denti et al. 2023).

3.7 Tools for Plant Pangenome Analysis

Most pangenome tools have focused on the analysis of bacterial genomes, which are comparatively smaller in size and are less complex than eukaryotic genomes. There are fewer tools and pipelines available for plant pangenome analysis, such as EUPAN, ppsPCP, and PSVCP. EUPAN (EUkaryotic PAN-genome analysis toolkit), a software package designed to enable large-scale pangenome studies of eukaryotic organisms with complex genomes (Hu et al. 2017). EUPAN employs a "map-to-pan" strategy as opposed to traditional assembly-based methods and enables detection of gene PAVs at relatively low sequencing depth. EUPAN successfully demonstrated its utility with 453 rice genomes in a previous study. It enabled gene-PAV-based GWAS, complementary to SNP-based GWAS. Further, ppsPCP is a pipeline designed for eukaryotic genome analysis, specifically targeting plants, that addresses a key challenge in pangenome analysis of large genomes (Tahir Ul Qamar et al. 2019). It scans for PAVs across different plant genomes and constructs fully annotated pangenomes. ppsPCP identifies sequences present in query genomes but absent in reference genomes and provides a comprehensive genetic variation analysis interface. Further, a novel pangenome analysis pipeline, PSVCP (Pangenome SV Calling Pipeline), was presented by Wang et al. 2023 that provided insights into functional gene identification in rice. The main pipeline features include preserving position information of novel sequences in a linear format while maintaining genomic context and enabling detection of PAVs, translocations, and inversions. PSVCP was deployed for PAV–GWAS to discover a novel locus (qPH8-1) controlling plant height that was undetectable by SNP–GWAS, highlighting the utility of the pipeline for plant pangenomes, however, PSVCP is limited in representing all genomic variations usually managed by graphical pangenomes. Graph-based pangenome tools, developed mainly for analyzing human genomes, have facilitated the analysis of larger plant genomes.

3.8 Visualization of Pangenome

Visualization of a pangenome is a critical step in the analysis, including focusing on SVs affecting traits/gene expression and visualizing specific candidate genes for further characterization and validations. Visualization of regions such as SVs enriched regions, telomeres, centromeres, and regions syntenic between the individuals makes the visualization an important aspect in pangenome analysis. There are several tools and web browsers available for visualizing the pangenomes. Visualizations could be achieved in two different ways:

3.8.1 *Linear Pangenome Visualization*

Investigating the quality and accuracy of variants detected using pangenome analysis is crucial for downstream applications and validations, as these variations are breeding targets and help improve plant phenotypes. Visualization of genomic variations and assessing the quality of read alignments are important aspects for further pursuing plant improvements. Multiple-genome browsers such as JBrowse (Buels et al. 2016), GBrowse2 (Stein 2013), and UCSC genome browser (Karolchik et al. 2011) have been deployed for genome visualizations and integrated genomic analysis across plant species. These genome browsers serve as a crucial interface for browsing the features of linear pangenomes and provide features such as comparative genomics to analyze the similarity and differences between the pangenome individuals. Users can zoom and focus on specific genomic regions with variations and browse the read alignment to manually inspect the quality and accuracy of genomic variations. These genome browsers have evolved with advances in genomics, and improvised versions such as JBrowse2 (Diesh et al. 2023) have been developed with better features and a focus on SVs and synteny as more pangenomes are established. With the developments in graph pangenomes, efforts to furnish specific tools to visualize these pangenomes are in progress.

3.8.2 *Graph Pangenome Visualization*

Linear pangenome visualizations have limitations in applications, and specific features are needed for graph pangenome visualizations since linear pangenome tools cannot handle such a data structure and manage coordinates in two-dimensional graph representations. There are two classes of graph pangenome visualization tools, viz., whole-genome level and nucleotide level visualizations. Pangenome visualization tools like bandage (Wick et al. 2015), panache (Durant et al. 2021), and gfaviz (Gonnella et al. 2019) exist. Some notable features of bandage include interactive visualization of graph pangenome and graph genome structures, such as de Bruijn graphs, string graphs, and node-based visualization where sequences are represented as nodes and connections as edges. Zooming into and navigation capabilities for exploring large and complex genome graphs, color-coding of nodes based on various properties such as coverage, GC content, etc., are some other features. Bandage deploys BLAST search integration for identifying specific sequences within the graph and supports multiple graph file formats (GFA, FASTG, LastGraph). Some of the unique features in gfaviz are support for hierarchical graph representations through bubbles that can expand/collapse for simplified viewing, advanced path visualization with dedicated functionality for haplotype-resolved assemblies, and on-the-fly graph filtering and transformation without modifying source files. Further, panache is a pangenome visualization tool that displays pangenomes as linear representations with blocks that can be shared across genomes using a layout similar to genome browsers with multiple information tracks (Durant et al. 2021). Panache shows PAV blocks across different genomes and includes navigation tools like pan-chromosome selection, zoom options, and a hollow area finder. It allows interactive exploration with on-the-fly modifications to the core/variable threshold. Panache graph pangenomes have been configured for wheat (Bayer et al. 2022) and brassica (MacNish et al. 2025). Sequence tube maps present an intuitive graphical representation for graph genomes inspired by public transport network maps (Beyer et al. 2019). Bi-directed nodes are represented as rounded rectangles, and representative paths, viz., haplotypes or sequencing reads, are shown as colored lines in sequence tube maps. It can display vg files and show all types of SVs, including deletions, duplications, and inversions.

3.9 Applications of Pangenomes

Plant pangenomes have become essential tools in modern genomics, enabling a deeper understanding of genetic diversity, domestication, and adaptation. They provide a robust framework for exploring evolutionary relationships, identifying deleterious alleles, constructing resistance gene atlases, and mapping genetic variations associated with important traits. These applications extend to breeding, gene editing, and predictive crop improvement.

3.9.1 Evolutionary Genomics

Understanding the relationship between individuals and species can have profound effects on advancing the plant genetics field. Pangenome studies have often focused on the relationship between different subpopulations, such as cultivars, landraces, and breeding lines from the cultivated gene pool and their comparison with wild species to identify their phylogenetic relationship and how the breeders have selected traits for improved trait performance (Varshney et al. 2021). The studies have also focused on revisiting the divergence events between species due to differences in methods based on time differences-based estimation with whole genome SNPs or genes in the individuals/species (Zhuang et al. 2019; Khan et al. 2024). This has resulted in better and accurate estimation of divergence time between the individuals.

3.9.2 Deleterious Alleles in Germplasm

Deleterious alleles are genetic variants that reduce fitness or agricultural performance, often accumulating in crop populations due to genetic bottlenecks, domestication, and breeding practices (Dwivedi et al. 2023). Modern breeding approaches have unintentionally increased the genetic load in many plant species by reducing effective population sizes. Genomic tools like whole-genome sequencing and GWAS have enhanced our ability to identify and characterize deleterious alleles in germplasm collections (Wang et al. 2017). Pangenomes serve as a crucial source for identifying these deleterious alleles in germplasm. Purging deleterious alleles through genomic selection has shown promise for improving crop performance and resilience. The distribution of deleterious alleles varies across wild relatives, landraces, and elite cultivars, with wild germplasm often carrying fewer fixed deleterious variants (Tang et al. 2022). Conservation of diverse germplasm is critical for maintaining a reservoir of beneficial alleles that can counteract the effects of deleterious mutations. The construction of super-pangenomes is critical to explore and exploit the diversity offered by wild relatives of plants to develop climate-suited varieties.

3.9.3 Development of Resistance Gene Catalogs

Resistance genes (R-genes) are specialized genes in plants that enable them to detect and respond to pathogens, such as bacteria, fungi, viruses, nematodes and insects, and these form a critical part of a plant's immune system. Resistance genes encode proteins that recognize specific molecules from pathogens (called effectors or avirulence factors), triggering defense responses that limit or prevent infection. The major classes of R-genes include NBS-LRR, receptor-like kinases (RLKs), receptor-like proteins (RLPs), and other types. These R-genes are crucial for reducing dependency on fungicides and pesticides, thereby ensuring food security by preventing devastating yield losses. They serve as a valuable resource for breeding programs and genetic engineering. Pangenomes analysis have resulted in establishing R-gene repertoires in plant species such as rice (Shang et al. 2022), tomato (Li et al. 2023), Arabidopsis (Lian et al. 2024), and wheat (Arora et al. 2019). R-genes are mostly found in clusters, and pangenome analysis is deployed to investigate the patterns of R-genes within a cluster and SVs associated with this important class of genes. The R-gene inventories in plant species serve as a resource for resistance breeding and the development of better resistance varieties that can withstand a broad range of biotic stresses. This is needed since the pathogens and their strains keep evolving into different populations and their resistances.

3.9.4 Identifying Genetic Variations Associated with Traits

Researchers have selected individuals/species with unique traits and superior phenotypes, and pangenome studies help identify the underlying genetic associations for traits and information needed for marker-assisted selection. Molecular breeding utilizes molecular markers, primarily SNPs and InDels, to identify causal links between specific genetic variants and agriculturally important traits, such as seed characteristics, yield, or disease resistance (Garg et al. 2025). SNPs and InDels are particularly suited for high-throughput genotyping, making them powerful tools for pinpointing relevant genomic regions in both experimental and natural populations. SVs' repositories for plant pangenomes have resulted in identifying the unexplored associations of SVs with biotic and abiotic stresses in these species (Zhou et al. 2022; He et al. 2023; Zhang et al. 2024a, b; Yang et al. 2025a). As more plant pangenomes are assembled, these may develop into comprehensive haplotype catalogs allowing breeders to retrieve the genome sequence of any given cultivar. While selecting crop varieties for study, key considerations include genotypes suitable for genetic transformation, parents of experimental populations, or those harboring beneficial genes or haplotypes of interest.

3.10 Trait Associations Identified with Plant Pangenomes and Super-Pangenomes: Case Studies

Pangenome resources serve as a useful platform for identifying genetic associations of traits other than traditional SNP-based association studies (Fig. 3.1). The pangenomes and super-pangenomes have been deployed in several plant species, and genomic variations associated with traits have been identified, which might be missed by the use of a single reference genome. For instance, the pearl millet pangenome identified that the endoplasmic reticulum (ER) stress response pathway plays a central role in pearl millet's heat tolerance mechanism (Yan et al. 2023). Pearl millet shows a quicker and more robust ER-related gene expression response to heat stress compared to maize and rice. Further, experimental validation showed that overexpression of one RWP-RK gene enhanced heat tolerance in transgenic plants and rapidly transactivated ER-related genes. The study identified specific SVs affecting gene expression associated with heat tolerance, particularly surrounding ER-related genes that shaped adaptation during domestication (Fig. 3.1). Cucumber pangenome analysis identified SVs associated with important agronomic traits such as warty fruits linked to the *CsTu* gene, flowering time associated with the *CsFT* locus, and root growth affecting *PELPK* genes (Li et al. 2022). The maize pangenome analysis revealed gene PAVs and their potential role during maize domestication, with teosinte-specific genes associated with stress responses, while maize-specific genes related to germination, nutrition, and flavor (Gui et al. 2022). Combining genetic analyses with multi-omics data, the analysis demonstrated the role of SVs in complex agronomic traits such as drought resistance. A recent wheat pangenome study reported chromosome-level genome assemblies of 17 wheat cultivars that represent the breeding history of China, complementing existing global wheat genomic resources (Jiao et al. 2025). The pangenome analysis identified that cultivars developed in the 1980s accumulated significant SVs linked to the

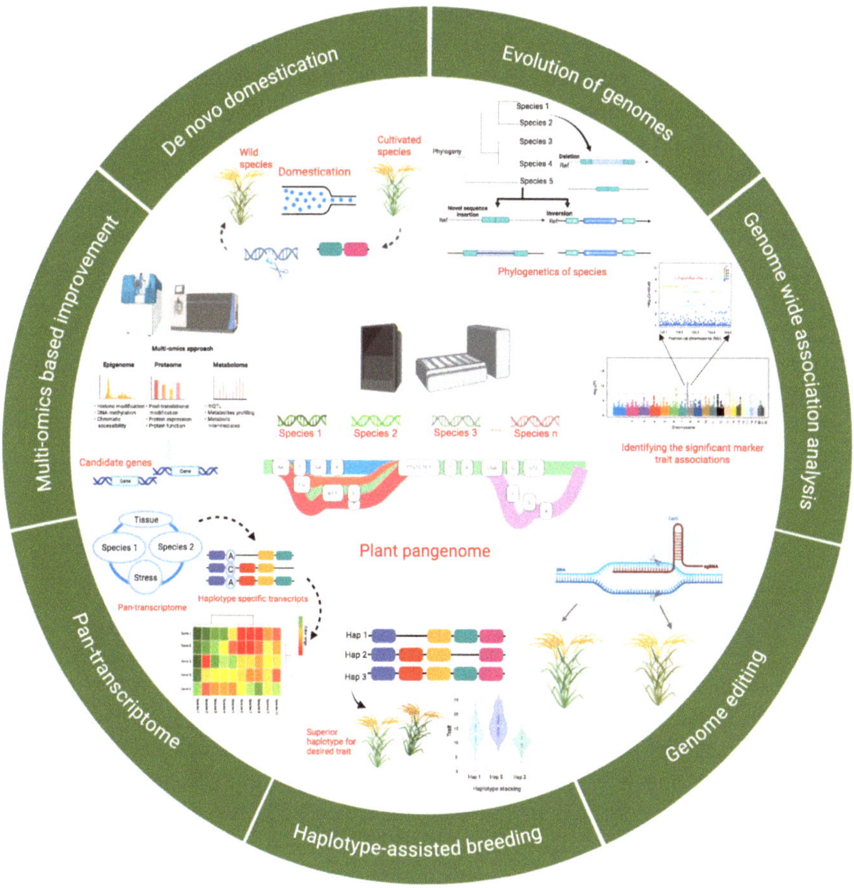

Fig. 3.1 Plant pangenomics and its applications. Plant pangenomes are used for diverse applications for crop improvement. It could be deployed to study the evolution of individuals and species, deciphering the simple to complex relationship between them. Pangenomes have been utilized to identify marker trait associations, shedding light on unexplored associations of SVs with traits. These markers have the potential to improve crop performance using genome editing approaches. The superior haplotypes identified using the pangenome analysis will help employ haplotype-assisted breeding in plant species. Pan-transcriptomes enhance the power of detecting significant genomic variations with trait deploying a multi-omics strategy by coupling other omics approaches such as epigenomics, proteomics and metabolomics. Further, exploring and exploiting the genetic diversity of wild species to improve the limited diversity of cultivated germplasm using approaches such as de novo domestication

incorporation of European and American varieties into Chinese breeding programs during that era. The study found that SVs in centromere-proximal regions are associated with reduced crossover events, which impacts breeding outcomes. Analysis revealed that common wheat evolved from spring to winter types through mutations and duplications of the *VRN-A1* gene as an adaptation to changing environments.

Super-pangenomes have further refined trait association analyses. In tomato, SVs affecting *TST2* (fruit sweetness), *LCYB* (flesh color), and flowering time genes were identified, including a rare gene from wild species that enhances fruit yield (Li et al. 2023). In rice, a graph-based super-pangenome spanning 251 accessions uncovered SVs linked to grain weight, submergence tolerance, and seed shattering, and characterized the pan-NLRome across accessions (Shang et al. 2022). The *Cicer* super-pangenome mapped SVs in genes regulating flowering, vernalization, and disease resistance (Khan et al. 2024). The grapevine pangenome identified alleles in sex determination and salt tolerance loci (Cochetel et al. 2023), while Populus species revealed a 180-bp PAV affecting *CUC2* gene expression and leaf serration (Shi et al. 2024). The Lactuca super-pangenome demonstrated that PAV–GWAS and CNV–GWAS revealed resistance loci overlooked by SNP–GWAS, particularly for downy mildew resistance (van Workum et al. 2024).

These case studies demonstrate the potential of super-pangenomes to unearth rare or wild-specific alleles with strong agronomic potential, opening the door to de novo domestication strategies and next-generation breeding (Fig. 3.1).

3.11 Plant Pan-Transcriptomics

While pangenomes reveal structural and sequence diversity, pan-transcriptomics extends this concept to gene expression, enabling insight into how genomic variation influences transcription and alternative splicing across individuals, tissues, or conditions. One of the first examples of pan-transcriptomics was in maize, where transcriptome data from 503 inbred lines identified 8681 representative transcript assemblies (Hirsch et al. 2014). Only 16.4% were universally expressed (core), while 82.7% were dispensable, contributing to developmental transitions and adaptation.

In alfalfa, long-read Iso-Seq and RNA-seq profiling across varieties and tissues identified stress-related long non-coding RNAs and genes regulating stress response (Medina et al. 2021). The potato pan-transcriptome revealed nearly 65,000 cultivar-specific transcripts, highlighting variation in gene expression among tetraploid cultivars (Petek et al. 2020). For tea (*Camellia sinensis*), a transcriptome built from 134 accessions identified metabolic pathway divergence between two main varieties (CSA and CSS), including regulatory genes linked to catechin types and flavonoid biosynthesis (Kong et al. 2022). A regulatory network was constructed connecting thousands of metabolites, genes, and expression QTLs.

Most recently, a rice pan-transcriptome uncovered alternative splicing as a key mechanism in cold stress response. Iso-Seq and RNA-seq of 11 cultivars identified *OsRS33* and *OsRS2Z38* as central splicing regulators, with allelic variation correlating with cold tolerance between *indica* and *japonica* lines (Zhong et al. 2024).

3.12 Integration of Multi-omics Strategy in Plants

The integration of multi-omics approaches within a pangenomic framework represents a transformative strategy for deciphering the molecular underpinnings of complex traits in plants. By incorporating multiple biological data layers such as genomics, transcriptomics, proteomics, metabolomics, and epigenomics, researchers can gain a comprehensive, systems-level understanding of plant biology that spans from DNA sequence variation to observable phenotypic outcomes.

This integrative approach enhances the detection and interpretation of SVs and PAVs, many of which are missed by traditional single-omics methodologies. Combining transcriptomic and proteomic data with genomic variation allows for improved functional annotation of variants, revealing which mutations influence gene expression, protein abundance, or biological pathways. Additionally, metabolomic profiling, when layered with genomic data, helps elucidate complex interactions such as epistasis and genotype-by-environment effects, thereby offering insight into how genetic variation translates into metabolic phenotypes and adaptive traits. Moreover, integrating multi-omics data enables the identification of population-specific molecular signatures associated with ecological adaptation or stress responses. This multi-layered framework significantly increases the statistical power of association studies by providing biological context and reducing the dimensionality of data. As a result, it addresses the limitations of isolated omics approaches and facilitates the translation of molecular findings into practical breeding strategies and functional validation pipelines.

In several major crops, particularly maize, rice, and brassica, transcriptomic data have been successfully integrated with pangenome analyses. These studies have provided gene expression evidence to support novel gene models and revealed SVs that influence gene regulation across tissues, developmental stages, and environmental conditions (Hufford et al. 2021; Shang et al. 2022; Li et al. 2024). Transcript-level variation has thus added another dimension to pangenome interpretation, especially in linking structural changes to phenotypic effects.

However, to date, comprehensive integration of more than two omics layers in a unified pangenomic context remains rare in plant science. No published studies have yet combined genomics, transcriptomics, proteomics, and metabolomics in a single framework to elucidate the molecular architecture of traits in plants. Some initial efforts across certain plant families, particularly in grasses, have begun to explore multi-omics strategies to better understand molecular functions and biological pathways, but such examples remain limited.

As sequencing and analytical technologies continue to evolve, the integration of multi-omics with pangenomics is poised to become central to systems-level plant biology. This fusion of data types offers not only deeper insights into complex traits but also more precise avenues for developing climate-resilient, nutrient-rich, and high-yielding crop varieties through next-generation breeding and genome editing.

3.13 Pangenome Databases

The rapid expansion of genomic data in the field of plant pangenomics presents a dual challenge: while it unlocks vast biological insights, it also creates significant barriers for data interpretation and application, especially for researchers lacking computational expertise (Chao et al. 2023). Extracting meaningful information from large-scale pangenome projects often requires advanced bioinformatics skills, underscoring the need for user-friendly, curated databases that simplify access to genomic diversity. There have been several efforts to develop pangenome databases to facilitate the use of large-scale genomics information underlying these data. In this direction, a Rice Gene Index (RGI), the first gene-based pangenome database for Asian rice, was developed (Yu et al. 2023). RGI integrated genomic information from 16 platinum standard reference genomes representing major Asian rice subpopulations. The database identified 119,783 non-redundant gene groups across these rice accessions, classified into 21,418 core genes, 40,141 dispensable genes, and 51,099 accession-specific genes. RGI introduced the ortholog gene index (OGI), a standardized numbering system for homologous gene groups across rice varieties. The platform provides multiple search methods, including keyword, sequence, and association-based, and visualization tools for exploring gene relationships at both local and global scales. RGI enables users to discover PAVs between rice varieties, such as the absence of the anaerobic germination tolerance gene OsTPP7 in IR64. BRAD V3.0 (Brassicaceae Database), an upgraded database for Brassicaceae plant species expanded to include 26 species and 36 genome assemblies (Chen et al. 2022). The new tools and modules included "MicroSynteny" for viewing synteny between genomic regions, "Polymorph" for retrieving and analysing genetic variation data, multiple sequence alignment and phylogenetic tree analysis, and improved syntenic gene retrieval across multiple genomes. The database integrates over a million syntenic genes, 1.83 million protein-coding genes with annotations and variation data from 1251 resequenced accessions. Similarly, there are several databases in plants that leverage pangenomic and multi-omics information, thereby enabling deeper molecular insights into genetic variations controlling traits.

3.14 Conclusions and Future Perspectives

The development of a plant pangenome is a resource-intensive endeavor that requires substantial computational infrastructure, skilled personnel, and financial investment. Acquiring diverse germplasm, particularly from global collections and wild species, necessitates extensive collaborations with gene banks and research institutions. As pangenomics expands beyond cultivated germplasm, field expeditions and collection efforts targeting wild relatives have become increasingly critical for capturing untapped genetic diversity. From a computational standpoint, high-performance computing (HPC) systems with robust processing capabilities and large-scale data storage are essential to manage and analyze pangenomic datasets. Cloud computing platforms are also being increasingly utilized to enable scalable, collaborative workflows. In parallel, dedicated server infrastructures are needed to host databases and support wide-scale data distribution. Sequencing costs remain a significant factor, especially when large populations are sequenced using long-read technologies to generate high-quality assemblies capable of resolving complex, repeat-rich genomic regions. Although many sequencing services are outsourced, the demand for high-throughput sample processing and robust library preparation pipelines remains high. The pangenome development process encompasses several bioinformatics-intensive stages, including genome assembly, whole-genome alignment, SV discovery, and graph-based genome construction. These tasks require the expertise of computational biologists and skilled bioinformaticians. Managing, integrating, and visualizing the massive datasets generated also calls for database administrators and software engineers. The financial burden associated with such initiatives is often met through competitive grant funding from national agricultural agencies, scientific foundations, and private industry partnerships. To facilitate and scale such efforts, international research consortia and public–private collaborations have been established, underpinned by open-data sharing agreements and cooperative frameworks.

Pangenomes and super-pangenomes have revolutionized plant genomics, offering a population-wide lens on SV, gene content diversity, and evolutionary dynamics. They have enabled the discovery of non-reference transcripts and genomic regions that are essential for understanding key traits such as stress tolerance, yield, and disease resistance traits often overlooked in single-reference analyses. The shift from reference-centric genomics to pangenome-based strategies marks a significant step toward comprehensive exploitation of natural diversity (Bayer et al. 2020). While considerable progress has been made for major crop species, a pressing challenge lies in extending these methodologies to understudied and orphan crops, where the genetic gains from pangenomic resources could have a transformative impact on local agriculture and food security (Sabety et al. 2024). Additionally, the development of tools and pipelines capable of handling large, polyploid, and repeat-rich genomes will be critical to scaling pangenome research across diverse plant families. There is also a need to explore the construction of super-pangenomes at higher taxonomic levels, such as across entire plant families or genera, which will require innovative methods to overcome the limitations of synteny and orthology-based pangenome models. Future efforts should prioritize the integration of transcriptomic data across developmental stages, tissues, and stress conditions to build comprehensive expression atlases. These resources will enhance gene annotations and deepen our understanding of gene regulation. Furthermore, the integration of multi-omics datasets, including proteomics, metabolomics, and epigenomics, will enable the generation of high-resolution genomic variation maps that link genetic diversity directly to phenotype.

In conclusion, plant pangenomics has significantly advanced our understanding of genomic diversity, domestication, and adaptation. It provides a roadmap for enriching existing germplasm with novel alleles from wild relatives, positioning plant breeders and geneticists to address the pressing global challenge of producing more resilient, productive crops in the face of environmental stress and a growing population. The next frontier lies in transforming these insights into practical applications for sustainable agriculture, food security, and climate-resilient crop development.

References

Alonge M et al (2020) Major impacts of widespread structural variation on gene expression and crop improvement in tomato. Cell 182:145–161.e23

Arora S et al (2019) Resistance gene cloning from a wild crop relative by sequence capture and association genetics. Nat Biotechnol 37:139–143

Bayer PE et al (2020) Plant pan-genomes are the new reference. Nat Plants 6:914–920

Bayer PE et al (2022) Wheat Panache: a pangenome graph database representing presence-absence variation across sixteen bread wheat genomes. Plant Genome 15:e20221

Beyer W et al (2019) Sequence tube maps: making graph genomes intuitive to commuters. Bioinformatics 35:5318–5320

Bohra A et al (2022) Reap the crop wild relatives for breeding future crops. Trends Biotechnol 40: 412–431

Bonnici V et al (2018) PanDelos: a dictionary-based method for pan-genome content discovery. BMC Bioinformatics 19:437

Buels R et al (2016) JBrowse: a dynamic web platform for genome visualization and analysis. Genome Biol 17:66

Chakraborty M et al (2019) Structural variants exhibit widespread allelic heterogeneity and shape variation in complex traits. Nat Commun 10:4872

Chao H et al (2023) Integrating omics databases for enhanced crop breeding. J Integr Bioinform 20. https://doi.org/10.1515/jib-2023-0012

Cheng H et al (2021) Haplotype-resolved de novo assembly using phased assembly graphs with hifiasm. Nat Methods 18:170–175

Cheng L et al (2025) Leveraging a phased pangenome for haplotype design of hybrid potato. Nature 640:408–417

Chen H et al (2022) BRAD V3. 0: an upgraded Brassicaceae database. Nucleic Acids Res 50: D1432–D1441

Chen J et al (2023) A complete telomere-to-telomere assembly of the maize genome. Nat Genet 55: 1221–1231

Chen S et al (2019) Paragraph: a graph-based structural variant genotyper for short-read sequence data. Genome Biol 20:291

Cochetel N et al (2023) A super-pangenome of the North American wild grape species. Genome Biol 24:290

Cook DE et al (2012) Copy number variation of multiple genes at Rhg1 mediates nematode resistance in soybean. Science 338:1206–1209

Della Coletta R et al (2021) How the pan-genome is changing crop genomics and improvement. Genome Biol 22:3

Deng Y et al (2022) A telomere-to-telomere gap-free reference genome of watermelon and its mutation library provide important resources for gene discovery and breeding. Mol Plant 15: 1268–1284

Denti L et al (2023) SVDSS: structural variation discovery in hard-to-call genomic regions using sample-specific strings from accurate long reads. Nat Methods 20:550–558

Diesh C et al (2023) JBrowse 2: a modular genome browser with views of synteny and structural variation. Genome Biol 24:74

Durant É et al (2021) Panache: a web browser-based viewer for linearized pangenomes. Bioinformatics 37:4556–4558
Dwivedi SL et al (2023) Evolutionary dynamics and adaptive benefits of deleterious mutations in crop gene pools. Trends Plant Sci 28:685–697
Ebler J et al (2022) Pangenome-based genome inference allows efficient and accurate genotyping across a wide spectrum of variant classes. Nat Genet 54:518–525
Eggertsson HP et al (2019) GraphTyper2 enables population-scale genotyping of structural variation using pangenome graphs. Nat Commun 10:5402
Fedoroff NV (2012) Transposable elements, epigenetics, and genome evolution. Science 338:758–767
Gao L et al (2019) The tomato pan-genome uncovers new genes and a rare allele regulating fruit flavor. Nat Genet 51:1044–1051
Garg V et al (2022) Chromosome-length genome assemblies of six legume species provide insights into genome organization, evolution, and agronomic traits for crop improvement. J Adv Res 42: 315–329
Garg V et al (2023) Near-gapless genome assemblies of Williams 82 and Lee cultivars for accelerating global soybean research. Plant Genome 16:e20382
Garg V et al (2024) Unlocking plant genetics with telomere-to-telomere genome assemblies. Nat Genet 56:1788–1799
Garg V et al (2025) An Australian chickpea pan-genome provides insights into genome organization and offers opportunities for enhancing drought adaptation for crop improvement. Plant Biotechnol J. https://doi.org/10.1111/pbi.70192
Garrison E et al (2024) Building pangenome graphs. Nat Methods 21:2008–2012
Garrison E, Guarracino A (2023) Unbiased pangenome graphs. Bioinformatics 39:btac743
Goel M et al (2019) SyRI: finding genomic rearrangements and local sequence differences from whole-genome assemblies. Genome Biol 20:277
Golicz AA et al (2016a) The pangenome of an agronomically important crop plant Brassica oleracea. Nat Commun 7:13390
Golicz AA et al (2020) Pangenomics comes of age: from bacteria to plant and animal applications. Trends Genet 36:132–145
Golicz AA et al (2016b) Towards plant pangenomics. Plant Biotechnol J 14:1099–1105
Gonnella G, Niehus N, Kurtz S (2019) GfaViz: flexible and interactive visualization of GFA sequence graphs. Bioinformatics 35:2853–2855
Gui S et al (2022) A pan-Zea genome map for enhancing maize improvement. Genome Biol 23:178
Guo N et al (2024) A graph-based pan-genome of Brassica oleracea provides new insights into its domestication and morphotype diversification. Plant Commun 5:100791
Gupta S et al (2024) The genome of *Haberlea rhodopensis* provides insights into the mechanisms for tolerance to multiple extreme environments. Cell Mol Life Sci 81:117
Heller D, Vingron M (2019) SVIM: structural variant identification using mapped long reads. Bioinformatics 35:2907–2915
Heller D, Vingron M (2021) SVIM-asm: structural variant detection from haploid and diploid genome assemblies. Bioinformatics 36:5519–5521
He Q et al (2023) A graph-based genome and pan-genome variation of the model plant Setaria. Nat Genet 55:1232–1242
Hickey G et al (2020) Genotyping structural variants in pangenome graphs using the vg toolkit. Genome Biol 21:35
Hickey G et al (2024) Pangenome graph construction from genome alignments with Minigraph-Cactus. Nat Biotechnol 42:663–673
Hirsch CN et al (2014) Insights into the maize pan-genome and pan-transcriptome. Plant Cell 26: 121–135
Hou X et al (2022) A near-complete assembly of an Arabidopsis thaliana genome. Mol Plant 15: 1247–1250

Hufford MB et al (2021) De novo assembly, annotation, and comparative analysis of 26 diverse maize genomes. Science 373:655–662
Hu Z et al (2017) EUPAN enables pan-genome studies of a large number of eukaryotic genomes. Bioinformatics 33:2408–2409
Jayakodi M et al (2024) Structural variation in the pangenome of wild and domesticated barley. Nature 636:654–662
Jiang T et al (2020) Long-read-based human genomic structural variation detection with cuteSV. Genome Biol 21:189
Jiao C et al (2025) Pan-genome bridges wheat structural variations with habitat and breeding. Nature 637:384–393
Karolchik D, Hinrichs AS, Kent WJ (2011) The UCSC genome browser. Curr Protoc Hum Genet 18:18.6.1–18.6.33
Khan AW et al (2020) Super-pangenome by integrating the wild side of a species for accelerated crop improvement. Trends Plant Sci 25:148–158
Khan AW et al (2024) Cicer super-pangenome provides insights into species evolution and agronomic trait loci for crop improvement in chickpea. Nat Genet 56:1225–1234
Kong W et al (2022) Pan-transcriptome assembly combined with multiple association analysis provides new insights into the regulatory network of specialized metabolites in the tea plant. Hortic Res 9:uhac100
Lian Q et al (2024) A pan-genome of 69 Arabidopsis thaliana accessions reveals a conserved genome structure throughout the global species range. Nat Genet 56:982–991
Li H (2018) Minimap2: pairwise alignment for nucleotide sequences. Bioinformatics 34:3094–3100
Li H et al (2022) Graph-based pan-genome reveals structural and sequence variations related to agronomic traits and domestication in cucumber. Nat Commun 13:682
Li H et al (2020) The design and construction of reference pangenome graphs with minigraph. Genome Biol 21:265
Li N et al (2023) Super-pangenome analyses highlight genomic diversity and structural variation across wild and cultivated tomato species. Nat Genet 55:852–860
Li R et al (2010) Building the sequence map of the human pan-genome. Nat Biotechnol 28:57–63
Liu Y et al (2020) Pan-genome of wild and cultivated soybeans. Cell 182:162–176.e13
Liu Z et al (2024) Grapevine pangenome facilitates trait genetics and genomic breeding. Nat Genet 56:2804–2814
Li X et al (2024) Large-scale gene expression alterations introduced by structural variation drive morphotype diversification in Brassica oleracea. Nat Genet 56:517–529
Li Y-H et al (2014) De novo assembly of soybean wild relatives for pan-genome analysis of diversity and agronomic traits. Nat Biotechnol 32:1045–1052
MacNish TR et al (2025) Brassica Panache: a multi-species graph pangenome representing presence absence variation across forty-one Brassica genomes. Plant Genome 18:e20535
Marçais G et al (2018) MUMmer4: a fast and versatile genome alignment system. PLoS Comput Biol 14:e1005944
Marx V (2023) Method of the year: long-read sequencing. Nat Methods 20:6–11
Medina CA et al (2021) Pan-transcriptome identifying master genes and regulation network in response to drought and salt stresses in Alfalfa (Medicago sativa L.). Sci Rep 11:17203
Mirus T et al (2024) GGTyper: genotyping complex structural variants using short-read sequencing data. Bioinformatics 40:ii11–ii19
Nattestad M, Schatz MC (2016) Assemblytics: a web analytics tool for the detection of variants from an assembly. Bioinformatics 32:3021–3023
Niu C et al (2022) Methylation of a MITE insertion in the MdRFNR1-1 promoter is positively associated with its allelic expression in apple in response to drought stress. Plant Cell 34:3983–4006
Petek M et al (2020) Cultivar-specific transcriptome and pan-transcriptome reconstruction of tetraploid potato. Sci Data 7:249

Qin P et al (2021) Pan-genome analysis of 33 genetically diverse rice accessions reveals hidden genomic variations. Cell 184:3542–3558.e16
Rautiainen M et al (2023) Telomere-to-telomere assembly of diploid chromosomes with Verkko. Nat Biotechnol 41:1474–1482
Rhie A et al (2021) Towards complete and error-free genome assemblies of all vertebrate species. Nature 592:737–746
Sabety J et al (2024) Unlocking diversity from wild relatives of perennial fruit crops in the pan-genomics era. Curr Opin Plant Biol 82:102652
Saxena RK et al (2014) Structural variations in plant genomes. Brief Funct Genomics 13:296–307
Schatz MC et al (2014) Whole genome de novo assemblies of three divergent strains of rice, Oryza sativa, document novel gene space of aus and indica. Genome Biol 15:506
Sedlazeck FJ et al (2018) Accurate detection of complex structural variations using single-molecule sequencing. Nat Methods 15:461–468
Shang L et al (2022) A super pan-genomic landscape of rice. Cell Res 32:878–896
Shang L et al (2023) A complete assembly of the rice Nipponbare reference genome. Mol Plant 16: 1232–1236
Shi T et al (2024) The super-pangenome of Populus unveils genomic facets for its adaptation and diversification in widespread forest trees. Mol Plant 17:725–746
Sibbesen JA et al (2018) Accurate genotyping across variant classes and lengths using variant graphs. Nat Genet 50:1054–1059
Smolka M et al (2024) Detection of mosaic and population-level structural variants with sniffles. Nat Biotechnol 42:1571–1580
Stein LD (2013) Using GBrowse 2.0 to visualize and share next-generation sequence data. Brief Bioinform 14:162–171
Tahir Ul Qamar M et al (2019) ppsPCP: a plant presence/absence variants scanner and pan-genome construction pipeline. Bioinformatics 35:4156–4158
Tang D et al (2022) Genome evolution and diversity of wild and cultivated potatoes. Nature 606: 535–541
Tao Y et al (2020) A graph-based pan-genome guides biological discovery. Mol Plant 13:1247–1249
Tettelin H et al (2005) Genome analysis of multiple pathogenic isolates of Streptococcus agalactiae: implications for the microbial "pan-genome". Proc Natl Acad Sci USA 102:13950–13955
Varshney RK et al (2021) A chickpea genetic variation map based on the sequencing of 3,366 genomes. Nature 599:622–627
Wang J et al (2023) A pangenome analysis pipeline provides insights into functional gene identification in rice. Genome Biol 24:19
Wang L et al (2017) The interplay of demography and selection during maize domestication and expansion. Genome Biol 18:215
Wang W et al (2018) Genomic variation in 3,010 diverse accessions of Asian cultivated rice. Nature 557:43–49
Wick RR et al (2015) Bandage: interactive visualization of de novo genome assemblies. Bioinformatics 31:3350–3352
van Workum D-JM et al (2024) Lactuca super-pangenome reduces bias towards reference genes in lettuce research. BMC Plant Biol 24:1019
Wu S et al (2023) A Citrullus genus super-pangenome reveals extensive variations in wild and cultivated watermelons and sheds light on watermelon evolution and domestication. Plant Biotechnol J 21:1926–1928
Xie Y et al (2023) Transposable element-initiated enhancer-like elements generate the subgenome-biased spike specificity of polyploid wheat. Nat Commun 14:7465
Yang L et al (2025a) GWAS meta-analysis using a graph-based pan-genome enhanced gene mining efficiency for agronomic traits in rice. Nat Commun 16:3171
Yang Q et al (2025b) SVLearn: a dual-reference machine learning approach enables accurate cross-species genotyping of structural variants. Nat Commun 16:2406

Yang T et al (2022) Improved pea reference genome and pan-genome highlight genomic features and evolutionary characteristics. Nat Genet 54:1553–1563
Yan H et al (2023) Pangenomic analysis identifies structural variation associated with heat tolerance in pearl millet. Nat Genet 55:507–518
Yano R et al (2025) The genomic landscape of gene-level structural variations in Japanese and global soybean Glycine max cultivars. Nat Genet 57:973–985
Yuan Y et al (2021) Current status of structural variation studies in plants. Plant Biotechnol J 19: 2153–2163
Yu Z et al (2023) Rice Gene Index: a comprehensive pan-genome database for comparative and functional genomics of Asian rice. Mol Plant 16:798–801
Zhang C et al (2023) The T2T genome assembly of soybean cultivar ZH13 and its epigenetic landscapes. Mol Plant 16:1715–1718
Zhang Y et al (2024a) Structural variation reshapes population gene expression and trait variation in 2,105 Brassica napus accessions. Nat Genet 56:2538–2550
Zhang Y et al (2024b) Telomere-to-telomere Citrullus super-pangenome provides direction for watermelon breeding. Nat Genet 56:1750–1761
Zhong Y et al (2024) Pan-transcriptomic analysis reveals alternative splicing control of cold tolerance in rice. Plant Cell 36:2117–2139
Zhou Y et al (2019) The population genetics of structural variants in grapevine domestication. Nat Plants 5:965–979
Zhou Y et al (2022) Graph pangenome captures missing heritability and empowers tomato breeding. Nature 606:527–534
Zhuang W et al (2019) The genome of cultivated peanut provides insight into legume karyotypes, polyploid evolution and crop domestication. Nat Genet 51:865–876
Zhu X et al (2025) Graph-based pangenome provides insights into structural variations and genetic basis of metabolic traits in potato. Mol Plant 18:590–602

Chapter 4
Exploring Genetic Diversity Using Genomic Approaches

Peterson W. Wambugu, Quinn R. Campbell, Marie-Noelle Ndjiondjop, and Robert J. Henry

Abstract Advances in sequencing technology have generated large amounts of high-quality genomic, amplicon, and transcriptomic data that are enabling analysis of genetic variation in plants. These genomic technologies are supporting conservation and enabling analysis and use of crop diversity in plant breeding. Identification of superior alleles and their transfer using modern genomic tools, such as genome editing, enable rapid development of new varieties. Genomics also provides opportunities for enhancing food security by domesticating new crops and exploring better environments for crop production in the face of climate change. The genetic diversity of wild plant populations is expected to change over time due to climatic and anthropogenic factors. Analysis of spatial and temporal patterns of plant diversity is providing insights into the genetic mechanisms of climate adaptation.

Keywords Genetic diversity · Plant genomics · Food security · Conservation genomics · Domestication

P. W. Wambugu (✉)
Kenya Agricultural and Livestock Research Organization, Genetic Resources Research Institute, Nairobi, Kenya
e-mail: peterson.wambugu@kalro.org

Q. R. Campbell
Department of Tropical Plant and Soil Science, University of Hawai'i at Mānoa, Honolulu, HI, USA

M.-N. Ndjiondjop
M'bé Research Station, Africa Rice Center (AfricaRice), Bouaké, Côte d'Ivoire

R. J. Henry
Queensland Alliance for Agriculture and Food Innovation, University of Queensland, Brisbane, QLD, Australia

ARC Centre of Excellence for Plant Success in Nature and Agriculture, University of Queensland, Brisbane, QLD, Australia
e-mail: robert.henry@uq.edu.au

V. Garg et al. (eds.), *DNA of Sustainability*, Sustainability Sciences in Asia and Africa,
https://doi.org/10.1007/978-981-95-5109-5_4

4.1 Introduction

Significant increases in food production will be needed to meet the food demands of the rapidly growing global human population. Plant genetic resources, which are the foundation of agriculture as well as food and nutritional security, provide the means for achieving these increases. The increased use of the genetic diversity found in agricultural species helps to provide a range of responses to biotic and abiotic stresses, thereby forming the basis for adapting agriculture to climate change (Henry and Nevo 2014). If these resources are to play an important role in agricultural production, it is imperative that they are properly conserved, accessible, well-characterized, and sustainably utilized. To efficiently deploy plant variation in plant improvement programs, it is important to understand the available genetic diversity and the underlying genetic mechanisms controlling traits of interest.

The current genomic revolution has been characterized by faster and cheaper next-generation sequencing (NGS), providing tools that assist in cost-effective study of genetic variation and the delivery of new varieties at unprecedented speed. These advances in DNA sequencing are rapidly replacing some of the traditional genetic approaches used in assessing and applying plant diversity (Allendorf et al. 2010). Genomics enables the assay of thousands of markers across the genome, thus providing greater resolution of neutral and functional diversity compared to traditional genetic approaches (Shafer et al. 2015). It provides powerful tools to study genetic diversity in terms of genes/alleles and their combinations as a basis for their deployment in plant improvement and conservation management decision-making. Compared with previous genetic studies that mainly used neutral genetic markers, there is now a greater focus and capacity for the identification and deployment of functionally important diversity (Wambugu et al. 2018b). However, the potential of genomics in assessing genetic diversity has not been fully exploited (Stange et al. 2021). In addition, much of the available genetic diversity, particularly in crop wild relatives, remains underutilized. This diversity also remains poorly collected, creating many conservation gaps that can be better understood using genomic tools. The genomic tools available for exploring crop diversity and deploying it for crop improvement are discussed in this chapter.

4.2 Importance of Reference and Pan-Genome Sequences

Reference genome sequences have been developed for most major crop species, apart from a few with large and polyploid genomes (Doležel et al. 2018; Kyriakidou et al. 2018). However, the highly complex (12-ploid) sugarcane genome was recently reported (Healey et al. 2024). The increased availability of publicly accessible reference genomic sequences across the tree of life is revolutionizing many areas of biological research. The availability of diverse reference sequences is essential for building the pan-genome. Reference sequences and pan-genomes are arguably the most important resources for supporting diversity assessment, conservation, and utilization. The availability of pan-genome sequences is addressing the inherent limitation of using a single reference when conducting a wide range of genetic analyses. Pan-genomes allow a comprehensive assessment of the diversity inherent in a taxon by including the variation missing in a reference sequence. Reference sequences may, for example, lack the fast-evolving disease-resistance genes that could have been lost during the evolutionary process (Sakai and Itoh 2010). Improved algorithms have been developed, making it less computationally challenging to identify non-reference variants (Krannich et al. 2021). Genome analysis has identified useful genes in non-reference sequences in rice that control important traits (Woldegiorgis et al. 2022; Shivute et al. 2024). Advances in long-read sequencing technology are enabling de novo assembly of complex and repetitive genomes (Mccoy et al. 2014). These techniques now allow the assembly of chromosome-level genomes from sequence data alone (Wijesundara et al. 2024). Haplotype-resolved plant genomes can now be assembled, allowing routine analysis of the genome at the haplotype level (Abdullah et al. 2024). Comparative analysis of multiple reference sequences of a particular species is being used to identify and correct misannotations, thereby improving their quality (Monnahan et al. 2020). The increased availability of whole-genome sequences and well-annotated reference sequences is enabling the discovery of useful genes and alleles and allowing prediction of their potential biological roles.

4.3 Genomic Approaches for Germplasm Characterization

Genomic approaches used in genetic characterization broadly involve either whole-genome sequencing, anonymous sequencing, or targeted sequencing of a subset of the genome. Targeted sequencing focuses on specific genomic regions, with the commonly used approaches being organelle sequencing, transcriptome sequencing, exome sequencing, gene-enriched genome sequencing, and amplicon sequencing (Henry et al. 2012). Anonymous sequencing involves the use of reduced-representation libraries such as Restriction Site Associated DNA (RAD) (Baird et al. 2008) and Genotyping by Sequencing (GBS) (Elshire et al. 2011) to randomly assay sections of the genome. Anonymous sequencing approaches require no prior sequence information and are finding great application in conservation, particularly for species with limited genomic resources. The various approaches have their own merits and demerits in terms of cost, accuracy, and applicability. The choice of an appropriate genomic approach should be guided by the study objective, available genomic approaches, funding, and knowledge of the genetic mechanisms underlying the trait being analyzed (Fuentes-Pardo and Ruzzante 2017). For example, compared with targeted sequencing or whole-genome sequencing, anonymous sequencing has less chance of identifying functionally important diversity. In addition, reduced-representation sequencing (RRS) approaches would be more efficient and cost-effective for analyzing neutral diversity compared with whole-genome sequencing.

4.3.1 Reduced Representation Sequencing

In the case of non-model organisms, particularly those whose reference genome sequences are lacking, the use of genome reduction approaches is common. However, with the current advances in sequencing and bioinformatics approaches, it is now possible to map the RRS-generated reads against reference sequences, thus improving the quality of the assembly (Rochette et al. 2019). Although RRS targets only a subset of the genome, typically focusing on regions of interest, it generates enough genome data to allow effective and efficient estimation of genetic diversity and other genetic applications. This makes RRS a valuable tool in evolutionary genetics and population studies, particularly in species with low diversity. Some of the common RRS approaches include DArTseq (Ref), ddRADseq (Peterson et al. 2012), RADseq (Baird et al. 2008), and Genotyping by sequencing (GBS) (Elshire et al. 2011). DArTseq has been employed to analyze genetic diversity and population structure, develop core and mini-core collections, and develop species diagnostic markers of *Oryza* germplasm collection at AfricaRice genebank (Ndjiondjop et al. 2017, 2018). GBS has been used in marker-assisted selection, thus hastening the development of new varieties (He et al. 2014). One of the potential limitations of these RRS approaches is the reduced capacity to identify adaptive variation due to the small proportion of the genome that is analyzed. In the case of RADSeq, the sparse distribution of markers particularly contributes to the reduced chances of detecting adaptive loci (Lowry et al. 2017).

4.3.2 Transcriptome Analysis

Genetic diversity and environment are the key factors that determine gene expression patterns. RNA sequencing (RNA-seq) provides a robust approach to identify candidate genes and characterize functional variation. Gene expression analysis has provided insights into genetic diversity in responses to drought in loblolly pine (*Pinus taeda* L.) (Li et al. 2021), salt stress in rice (Razzaque et al. 2019), and *Phytophthora nicotianae* resistance in tobacco (Meng et al. 2021). Transcriptome analysis has increased our understanding of functional diversity in plant microbiota and their role in shaping their hosts' responses (Chialva et al. 2019). Transcriptome sequencing of cypress species (*Cupressus gigantea*) populations has shed light on the importance of introgression as a source of climate-adaptive variation (Ma et al. 2019). Transcriptomic data provide an effective and robust alternative approach for genomic analysis of species with inadequate genomic resources (Lin et al. 2024).

4.3.3 Amplicon Sequencing

Amplicon sequencing is a targeted, PCR-based, NGS method that enables the analysis of genetic variation in selected genomic regions of interest. Careful library preparation and choice of primers are recommended, as this can be a major source of errors (Schirmer et al. 2015). Over the years, advances in sequencing technology and bioinformatics have significantly improved the capacity and accuracy of amplicon sequencing (Karst et al. 2021; Özkurt et al. 2022). Sequencing 18 starch-related genes identified important functional SNP diversity associated with various starch properties in rice (Kharabian-Masouleh et al. 2012). Amplicon sequencing is finding greater application in the analysis of microbes than plants. The choice on whether to undertake amplicon sequencing or whole-genome sequencing will depend on factors such as whether the candidate gene or genomic region of interest is known and the resources available.

4.3.4 *Whole-Genome Sequencing*

Whole-genome sequencing is useful in assessing genetic diversity, defining genetic relationships, and analyzing population structure. Whole-genome re-sequencing is commonly used to assess the genetic variability between individuals. With increased genomic advances such as improved long-read sequencing technologies, there is increased availability of quality whole-genome data that are enabling various forms of genetic analysis (Murigneux et al. 2020). Where no reference sequences exist, de novo assembly of different species and re-sequencing of their populations is contributing greatly to enhanced knowledge of their diversity. Generally, the most common approach of assessing genetic diversity from genomic sequence data is aligning the randomly generated sequencing reads against a reference sequence. Polymorphisms such as SNPs, indels, copy number variations, etc., are then called using different bioinformatics software that employ various algorithms. However, de novo sequencing will provide greater accuracy in the identification and analysis of structural differences within a species (Henry 2022). Once variations have been identified, typically through low-coverage whole-genome re-sequencing, detailed analyses of specific target regions of interest can then be conducted (Natarajan et al. 2016; Yang et al. 2017). This could involve examining the functional significance of variations, for example, by analyzing their association with traits and their frequency in larger populations. SNP genotyping panels developed from the whole-genome data allow a detailed assessment of diversity in selected loci, particularly in crops with limited genomic resources (Semalaiyappan et al. 2022). However, with decreasing sequencing costs allowing cost-efficient deep-coverage whole-genome sequencing of a large number of samples, SNP panels seem to be less popular nowadays. Although the cost of DNA sequencing has decreased dramatically over the last decade, the cost of computing infrastructure, bioinformatics software, and the technical skills for whole-genome data analysis is still prohibitive for many research and conservation programs, particularly in developing countries.

4.4 Genetic Mapping Approaches

Genome sequencing has opened almost unlimited possibilities in genetic mapping, which have gone a long way in addressing previously intractable genetic mapping challenges. Traditional QTL mapping approaches were slow and had less resolution. Although there has been increased capacity to identify DNA polymorphisms through genomic technologies, major gaps exist in linking them to phenotypic traits so that they can be of greater value in crop improvement (Wambugu et al. 2013). Integrating phenotypic, genomic, and transcriptome data is contributing to increased knowledge of mechanisms underlying important traits (Fig. 4.1) (Kumar et al. 2023). Linking these data sets with environmental data is also providing useful insights into trait variation. However, many adaptive traits are polygenic in nature, where they are controlled by many genes of small effects, and it may be challenging to identify the functional alleles (Rockman 2012). NGS is enabling the rapid identification of causal loci to single-nucleotide resolution.

4.4.1 Genome-Wide Association Studies

These studies are important in identifying genetic variants showing statistical association with the trait of interest. They act as a powerful tool for dissecting the genetic architecture of complex traits. Genome-wide association mapping has demonstrated that it has sufficient resolution to identify both minor and major effect loci (Mccouch et al. 2016). GWAS takes advantage of historical recombination events spanning

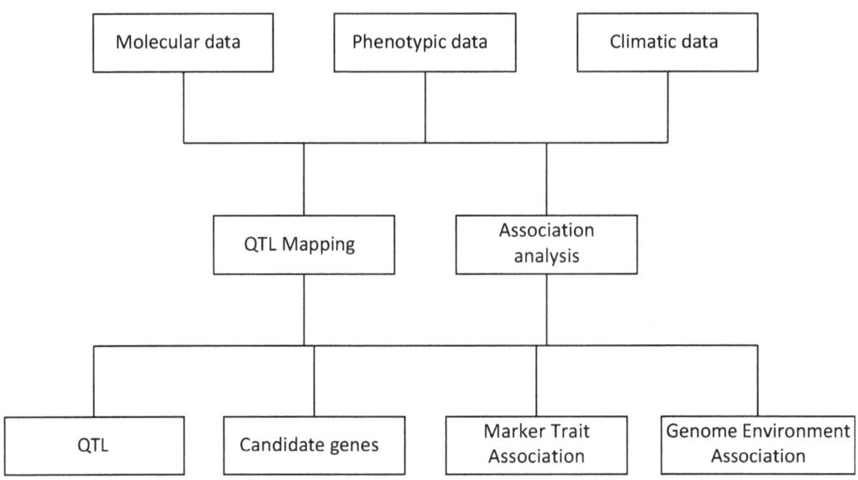

Fig. 4.1 Schematic representation of genetic mapping approaches

many generations. Although there has been increased capacity to identify DNA polymorphisms through genomic technologies, major gaps exist in linking them to phenotypic traits so that they can be of greater value in crop improvement (Wambugu et al. 2013). Genome-wide association studies are providing insights into the value of these polymorphisms by providing information on genotype–phenotype relationships. Analysis of genotype–phenotype relationships is enabling the identification of marker–trait associations (MTA). Strong MTA have been identified for several traits in rice, namely grain length and panicle characteristics (Mccouch et al. 2016; Crowell et al. 2016). This information is providing a strong foundation for conducting marker-assisted selection. Sequencing of these candidate genes, coupled with well-developed analytical pipelines, will enable the identification and unlocking of beneficial trait variation.

To increase the chance of identifying marker–trait relationships, a well-assembled, genetically representative diversity panel is important. This will enable the identification of important alleles appearing at low frequency in the population, including population-specific alleles. Pleiotropic, trait-associated genetic markers (Lei et al. 2020; Saeidnia et al. 2022) are particularly valuable, as they can aid in selecting genotypes with beneficial trait combinations, thus potentially accelerating genetic gain. Association studies involving samples obtained from wide eco-geographical regions assist in identifying important trait variation and its genetic basis. However, such samples usually have strong population structure, which could confound identification of trait-associated genetic markers. There, however, exist models that address the confounding effects of population structure, thus preventing spurious associations (Segura et al. 2012). Combining Focused Identification of Germplasm Strategy (FIGS) and GWAS is an important strategy to not only identify new genetic variation but also to determine the genetic basis of the trait of interest (Mccouch et al. 2012). Capacity to generate high-quality phenotypic and agronomic data, however, remains a challenge in identifying strong marker–trait relationships.

4.4.2 Bulk Segregant Analysis

Over the last decade, the cost of high-throughput sequencing has decreased by several orders of magnitude. However, despite this dramatic decrease in sequencing costs, whole-genome sequencing of many individuals is still not affordable for many resource-constrained labs. Consequently, to enhance cost-effectiveness, DNA pooling approaches have become popular in genetic mapping and population genetic studies (Schlötterer et al. 2014). Bulk segregant analysis is one of the most popular DNA pooling approaches. This trait-based genetic mapping method involves analyzing bulks of individuals that exhibit extreme contrasting phenotypes for the trait of interest. Sequencing of the bulks and analysis of the polymorphisms assist in identifying genetic markers and candidate genes associated with the target trait. BSA has been demonstrated to have the capacity to accelerate the unravelling of the genetic basis of complex traits (Song et al. 2017). The reduced costs of library preparation and the low number of samples to sequence make it highly cost-effective. To increase resolution and accuracy, additional bulks from random individuals can be created to compare with the two bulks constituted from contrasting phenotypes (Zou et al. 2016). NGS-based BSA can also be conducted even in species with no reference genome sequences (Nordström et al. 2013), making it suitable for species with inadequate genomic resources. However, this may present some complexities and is challenging computationally (Li and Xu 2022).

Previously, BSA suffered from the lack of a sufficient number of markers (Ehrenreich et al. 2009), but with the advent and continued popularity of whole-genome sequencing, the number of markers available is no longer a limiting factor. NGS-based BSA operates on the principle of allele frequency, with the allele associated with the trait being enriched in one of the bulks, while the alleles with no association with the trait show random segregation in both bulks (Wambugu et al. 2018a). Several variant forms of this genetic mapping approach have been developed and applied to various model and non-model species. These include SHOREmap, used for identification of loci controlling growth habit and color of leaves in *Arabidopsis thaliana* (Schneeberger et al. 2009); Next Generation Mapping (NGM), used to identify genes involved in cell-wall biology in *Arabidopsis thaliana* (Austin et al. 2011); MutMap, used to map genes for pale-green leaves and dwarfism in rice (Abe et al. 2012), MutMap-Gap for mapping blast-resistance gene in rice (Takagi et al. 2013b), MutMap+, for mapping causal mutations for lethality in early stages of plant development (Fekih et al. 2013) and Isogenic Mapping-by-Sequencing for mapping suppressor mutants (Hartwig et al. 2012). NGS-based BSA has also been referred to as QTL-seq (Takagi et al. 2013a).

NGS-based BSA has been used in both populations with natural phenotypes as well as those subjected to mutagenesis (Schneeberger et al. 2009; Austin et al. 2011; Abe et al. 2012). Though NGS-aided BSA has only been conducted on bi-parental populations, it has been suggested that it can also be applied on unstructured populations that may have undergone some artificial or natural selection (Takagi et al. 2013a). It might, for instance, be useful in studying the genetic basis of climate change adaptation. Such a study might involve pooling individuals subjected to different climatic conditions over a period of time or those along an environmental gradient. In addition, NGS-aided BSA has been coupled with RNA-seq and various genome complexity reduction sequencing techniques such as RAD-seq. The first application of NGS-aided BSA was in yeast, where it was used to identify and validate genes responsible for ethanol tolerance (Swinnen et al. 2012), vacuole inheritance (Birkeland et al. 2010), pseudohyphal growth (Song et al. 2014), glycerol and ethanol yield (Hubmann et al. 2013), and xylose metabolism (Wenger et al. 2010). It has been successfully used in mapping QTLs in tomato (Illa-Berenguer et al. 2015), sorghum (Han et al. 2015), chickpea (Das et al. 2015), cucumber (Lu et al. 2014), rice (Yang et al. 2013), and pigeon pea (Singh et al. 2016).

4.5 Application of Genomics in Diversity Assessment

4.5.1 Plant Genetic Resources Conservation

Huge amounts of plant diversity have been conserved in the world's plant genetic repositories. Sequencing of these genetic resources is supporting utilization, research, and conservation management decisions. In addition to its technical and scientific application, genome-scale data is also being used to advance conservation policy (Funk et al. 2019). Genome-scale analysis of populations is enabling greater understanding of the diversity inherent in taxa. This population genomic analysis is informing conservation priorities by identifying populations of conservation concern and value in terms of useful allelic diversity, adaptive capacity, threat status, and conservation gaps. Genomics has been used to identify biodiversity hot spots and rare alleles (Supple and Shapiro 2018), thereby providing useful information to support prioritization of target areas for germplasm conservation. NGS-based DNA barcoding approaches are enabling accurate species identification, which is critical in undertaking proper genetic and taxonomic diversity assessment. This information is, for example, vital in guiding conservation gap analysis. Genomic technologies are supporting the identification of genetic redundancy in genebanks and thereby guiding rationalization of germplasm collections (Singh et al. 2018). With continued advances in genomic technologies, it is possible that future germplasm acquisition in genebanks will prioritize materials based on their diversity levels, with functional diversity receiving higher priority. Identification of alleles that are specific to certain populations (Mccouch et al. 2016) will guide the development of sampling plans during germplasm collecting. Genomic data is supporting genebank managers in conducting targeted selection of germplasm from genebank collections for use in breeding and research (Wambugu et al. 2018a). This is making it possible for genebank managers to link genebank accessions to traits, thus enabling targeted use (Anglin et al. 2018). Similarly, increasing accession-level information by integrating genomic, ecological, and phenotypic data is also a powerful way of enabling more targeted selection of germplasm from the huge diversity conserved in genebanks.

Ex situ genebank accessions capture a sample of the genetic diversity of a particular population at a particular time, but practices of regeneration can lead to genetic drift or inbreeding, leading to losses or shifts in genetic diversity over time (Gómez et al. 2005; Cadima Fuentes et al. 2017). SNP markers are being used for quality control in germplasm conservation and are allowing the assessment of the effectiveness of various genebank operations in maintaining genetic diversity, integrity, and identity (Gouda et al. 2021). Similar markers have also been developed for undertaking quality control in sorghum breeding programs (Gimode et al. 2024). Genome analysis is providing great understanding on temporal and spatial patterns of genetic diversity distribution (Fitzgerald et al. 2011; Shapter et al. 2012). This information is important in guiding the development of sampling pattern for ex situ conservation, selecting priority in situ conservation sites, and establishment of novel protected areas. Large-scale sequencing and genotyping (Wang et al. 2018) of genetic materials conserved in genebanks has significantly increased our understanding of the diversity conserved in genebanks, thus enhancing their utilization. It has also provided an impetus for the discovery of novel diversity associated with useful traits and enabled a wide range of genetic analyses. Germplasm users have shown great interest in accessions with publicly available genomic sequence and characterization data (Halewood et al. 2020). Many approaches that aim to prioritize samples for genomic analysis have been proposed. These include development of core collections, mini-core collections, trait-specific subsets, and use of the FIGS. (Bari et al. 2012; Khazaei et al. 2013). The materials can then be subjected to intensive genotyping or resequencing, thus enabling allele mining.

4.5.2 Climate Change Adaptation

Crop diversity underpins the resilience and adaptive capacity of our agricultural production systems. With continued changes in climatic conditions, it is imperative that new sources of adaptive variation are found. Climate-adaptive variation has mainly been analyzed in the nuclear genome, but chloroplast genes with adaptive capacity have also been identified (Gao et al. 2019). The genetic resources that withstand the effects of climate change have great potential for contributing valuable traits for crop improvement. Genomic analysis of samples collected along an environmental gradient will provide insights into loci responsible for climate-change adaptation (Henry 2014). Analysis of such genomes provides information on the level of differentiation and assists in identifying genomic regions with unusually high genetic divergence. Analysis of the diversity found in these loci will aid in identifying the functionally important alleles. This information is important in making decisions on what genetic resources to prioritize for conservation. One of the options for adapting to climate change is to explore the production of crops in new environments (Henry 2020). Genetic diversity is expected to provide the adaptive capacity for such a shift.

Analysis of genome–environment associations (GEA) is enabling the identification of useful sources of variation for climate adaptation (Cortés et al. 2022). Genetic markers associated with various bioclimatic variables have been identified in *Glycine soja*, soybean's wild relative (Anderson et al. 2016). The capacity provided by environment-associated SNP markers to predict phenotypes has great potential for accelerating crop breeding programs, thus reducing the time required to develop new varieties (Lasky et al. 2015). Population genomic analysis of *Populus koreana* provided insights into the vulnerability of existing populations, thereby informing conservation and climate-mitigation decisions (Sang et al. 2022). In the absence of information on functional diversity, genome-wide estimates of SNP variation can be used as a proxy indicator of potential adaptive capacity (He et al. 2016). There are, however, conflicting reports on the relationship between functional and neutral diversity and their role in ecological adaptation and fitness (Holderegger et al.

2006; Whitlock 2014; Yıldırım et al. 2018). Although most of this SNP variation may constitute neutral diversity, the higher the number of heterozygous variants, the greater the chance that some will have a role in adaptation.

4.5.3 Domestication

The availability of large-scale genomic data and high-quality reference sequences for both cultivated and wild species is providing great insights into the domestication process and diversification history. Knowledge of the genetic mechanisms underlying the domestication process presents an opportunity for domesticating new crops. Genomic approaches that allow modification of domestication genes from other species are making this feasible (Henry 2021). Genebanks hold a wide diversity of various types of germplasm, including some species that may now be extinct. Studying such germplasm may contribute to understanding the evolutionary processes during domestication. Comparative analysis of the genomes of progenitors and their domesticates is facilitating the identification of selective-sweep regions (Wang et al. 2014). These highly differentiated regions possessing signatures of selection usually harbor domestication loci. In some other cases, however, such selective sweeps have not been associated with any domestication genes and could thus be false positives (Harris et al. 2018). While domestication was primarily associated with selection for human-preferred traits, it also negatively impacted some useful traits such as disease resistance, adaptive capacity, and nutritional quality (Singh and Van Der Knaap 2022). Domestication, for example, led to the loss of important disease-resistance genes from African and Asian rice (Sakai and Itoh 2010; Zhang et al. 2014). Whole-genome resequencing has identified several NBS-LRR genes in wild relatives that are responsible for defense against pathogens (Shivute et al. 2024). The loss of these genes could be attributed to pleiotropic effects or close genetic linkage between different loci. Advances in genetics and genomics are enabling understanding of the genetic basis of this potentially useful variation, thereby allowing its targeted reintroduction into cultivated species (Wambugu et al. 2021; Singh and Van Der Knaap 2022). Restoring the intrinsic resistance found in crop wild relatives has, for example, been suggested as one sustainable way of enhancing resistance against pests in domesticated crops (Karlsson Green et al. 2020). Development of cross-specific markers, such as SNPs, will assist in monitoring the introgression of specific alleles into crop wild relatives, thus avoiding linkage drag.

4.5.4 Ecosystem Restoration

Climate change has led to widespread degradation of habitats, thereby threatening the survival of humankind. Deploying wide inter- and intra-specific diversity is critical in ensuring successful and sustainable ecosystem restoration. Such restoration programs should leverage the rich genetic diversity held in various ex situ conservation facilities (Wambugu et al. 2023). Population genomic analysis of germplasm obtained from different provenances is enabling the identification of suitably adapted materials for restoring degraded ecosystems (Breed et al. 2019). Landscape genomics is also being used to inform important seed-sourcing decisions. Analysis of GEA has assisted in identification of SNPs associated with various climatic variables in various plant species namely sorghum (Lasky et al. 2015; Menamo et al. 2021), *Eucalyptus microcarpa* (Jordan et al. 2017), *Arthropodium fimbriatum* (Jordan et al. 2023), *Panicum hallii* (Gould et al. 2018) and *Phaseolus vulgaris* L. (Cortés and Blair 2018). These GEA studies are important in predicting plant responses to various climatic conditions (Lasky et al. 2015). This information is also important in making predictions about the optimal environmental conditions for other genotypes within the same species (Manel et al. 2018). Information on GEA is especially valuable for sourcing germplasm for ecological restoration because it helps identify genotypes that have potential ecological adaptation to future climate conditions (Rossetto et al. 2019). Analysis of allele-frequency changes in adaptive SNPs in *Eucalyptus microcarpa* provided insights into the adaptive capacity of populations from various climatic gradients (Jordan et al. 2017). Analysis of regions flanking environment-associated markers will assist in identifying candidate genes for adaptation. Genome data has potential for assisting genebank managers in selecting germplasm with the right adaptive capacity from the rich ecotypic diversity held in genebanks.

4.5.5 Nutritional Quality–Eating and Cooking Quality

Genomic technologies are providing increased understanding of the genetic basis of important traits and may therefore potentially accelerate genetic gain and thus hasten the breeding process. Sequencing of bulks (Wambugu et al. 2018a) with high and low amylose has been used to dissect the genetic basis of amylose content in African rice. It also allowed identification of novel variation in starch genes in African rice, paving the way for the use of this variation to improve the starch properties of Asian rice.

4.5.6 Pest and Disease Resistance

Climate change has led to the increased emergence of pests and diseases, including new pathogenic strains that are threatening global food security and leading to loss of biodiversity (Singh et al. 2023). One effective method for managing pests and diseases sustainably is by identifying sources of durable resistance and incorporating this genetic diversity into new varieties. Recent structural and functional analysis of the *Citrus australasica* genome has revealed valuable disease-resistant genes, facilitating accelerated genetic improvement of this species (Nakandala et al. 2024).

4.6 Analysis of Temporal and Spatial Patterns of Genetic Diversity

Genetic resources, whether located within genebanks, on farms, or in the wild, are dynamic populations with the potential to change over time. Optimal use of these resources depends on the characterization and understanding of all available genetic variation. Understanding how the genetic diversity within these populations shifts over time is essential to their effective deployment (Vigouroux et al. 2011; Bellon et al. 2017; Mclean-Rodríguez et al. 2021). While the exact patterns and processes of temporal changes in genetic diversity are not well understood, global genebank collections are relatively well documented, extensive, and accessible, providing a powerful resource to study these temporal changes (Wambugu et al. 2024). Genomics is creating opportunities for developing robust molecular-based tools and approaches for spatial and temporal monitoring of genetic diversity in populations (Wambugu and Henry 2022). Comparative genomic analysis of relatively static historical genebank accessions with their modern source populations that have undergone decades of changes can illuminate how genetic diversity shifts with time and in the face of diverse biotic and abiotic pressures. This will provide insights into genomic loci impacted by climate change.

Studies of temporal changes in genetic diversity have provided a complex picture of the processes and patterns underlying this change, providing crucial information for conservation and use of valuable genetic diversity. In situ populations face pressures like climate change, natural selection, farmer selection, genetic drift, and shifts in land use, which may be driving shifts in genetic diversity and phenotypic traits (Nevo et al. 2012; Bellon et al. 2017; Mclean-Rodríguez et al. 2021). These changes may lead to a loss of genetic diversity (Hammer et al. 1996; Nevo et al. 2012), a reduction in population differentiation (Thormann et al. 2017), or extinction (Goettsch et al. 2021). Analysis of molecular data and genetic demographic parameters will provide information on the factors underlying the changes in genetic variation.

Crop genetic erosion is a major focus of many studies of temporal change in genetic diversity, motivated by widespread declines in landrace cultivation and declines in habitat for crop wild relatives. Genomic data is enabling identification of species and populations that are about undergo extinction due to reduction in genetic diversity, thereby assisting in making necessary conservation decisions (Teixeira and Nazareno 2021). Molecular analysis of wild emmer wheat populations in Israel showed a significant loss of alleles in seven populations collected in 2008 compared to those collected in 1980, while three populations showed an increase in alleles (Nevo et al. 2012). Shifts in maize landrace diversity in Mexico have been thoroughly studied, yielding evidence of complex shifts in diversity over time. While the local genetic diversity of landraces has decreased within farms and regions of Mexico (Dyer et al. 2014), there was not a significant change in the number of races grown or their frequency between 1950, 1975, and 2005 across Mexico (Perales and Golicher 2014). A study of landrace diversity in Morelos, Mexico, showed a slight decrease in the number of SNPs and pairwise genetic distance, but no decrease in heterozygosity between 1967 and 2017 (Mclean-Rodríguez et al. 2021), while also identifying several loci under selection. While there is substantial evidence of global crop genetic erosion, there is also clear evidence for the increased adaptive potential of both on-farm and wild diversity in response to global changes.

Understanding the potential for in situ populations to adapt to global changes may be beneficial for breeding programs and germplasm exchange programs aimed at adapting to climate change and other emerging threats. Populations experiencing evolutionary pressure may produce novel genetic diversity or undergo increases in rare adaptive alleles that provide tolerance or resistance, which increase the odds of capture or discovery (Bellon et al. 2017). Genome analysis is enabling the identification of such novel genetic diversity. Thus, studies of temporal shifts under abiotic or biotic stress may aid in the discovery of useful genetic diversity generated or increasing under emerging threats. Many studies to date have provided evidence of adaptive responses in crop landraces and wild relatives in response to climate change. A study on Sahelian pearl millet landraces collected from the same locations in 1973 and 2003 displayed shifts toward a shorter lifecycle and smaller plant and spike size over the study period, as well as an increase in frequency of an early-flowering allele (Vigouroux et al. 2011). Studies of crop wild relatives and landraces in maize, barley, sorghum, sunflower, and pearl millet have also provided evidence of shifts in genetic diversity and phenotypic traits associated with climate change (Rojas-Barrera et al. 2019; Olodo et al. 2020; Spear et al. 2023). Population genomic analysis of wild and cultivated species may reveal the increase or emergence of rare adaptive alleles, which could aid in efforts to adapt crops to climate change. Conservation of such diversity is of great priority.

4.7 Deploying Diversity Crop Improvement

Selecting genetic resources obtained from environments that match the one being bred for increases the chances of getting materials with the necessary adaptation. To take advantage of advanced genomic technologies such as gene editing, it is important to have an understanding of the functional significance of various alleles associated with a target trait (Henry 2021). This will enable the selection of the best allele for breeding genotypes suitable for the target environment. To enhance genetic gain, it is important to expand the source of diversity used for crop improvement. Wild genetic resources hold a lot of unutilized diversity that has great potential for crop improvement (Wambugu et al. 2013). This diversity has great potential in enhancing climate resilience, food security and environmental sustainability. Resequencing of populations of crop wild relatives is assisting in the identification of functionally important novel diversity. However, limited understanding of the potential genetic value of this plant diversity remains the greatest hindrance to its use in breeding and research.

Linkage drag has also been a major challenge in the use of wild genetic resources for crop improvement. In addition, useful trait diversity is, in some cases, masked by genes with deleterious and contrasting effects (Mccouch et al. 2020). Genomic analysis is revealing recombination breakpoints, thereby enabling the breaking of tight genetic linkage (Anirban et al. 2023). Pre-breeding helps in developing recombinant populations that assist in transferring useful diversity from wild species into cultivated crop genetic backgrounds (Mccouch et al. 2020). Predictive algorithms that have the capacity to identify deleterious mutations, particularly in putative trait-associated genomic regions identified through GWAS and QTL analysis, may help address this challenge (Kovalev et al. 2018). Genome sequencing also offers the capacity to identify maladaptive genes (Leroy et al. 2018). The capacity offered by technologies to monitor introgression ensures that only the desired alleles are transferred from the wild species. Genome editing enables rapid and more targeted transfer of genetic diversity without the time-consuming backcrossing and selfing to fix the desired alleles and eliminate unwanted ones (Swarup et al. 2021). Genomics is providing tools to assess the phylogenetic relationships between species, thus providing insights into the most readily available diversity for crop improvement (Wambugu et al. 2015). This information is also important in developing effective strategies for the use of this diversity in developing improved varieties.

References

Abdullah M, Masouleh A, Okemo P, Furtado A, Henry RJ (2024) An improved haplotype resolved genome reveals more rice genes. Trop Plants 3:e009

Abe A, Kosugi S, Yoshida K, Natsume S, Takagi H, Kanzaki H, Matsumura H, Yoshida K, Mitsuoka C, Tamiru M, Innan H, Cano L, Kamoun S, Terauchi R (2012) Genome sequencing reveals agronomically important loci in rice using MutMap. Nat Biotechnol 30:174–178

Allendorf FW, Hohenlohe PA, Luikart G (2010) Genomics and the future of conservation genetics. Nat Rev Genet 11:697

Anderson JE, Kono TJ, Stupar RM, Kantar MB, Morrell PL (2016) Environmental association analyses identify candidates for abiotic stress tolerance in Glycine soja, the wild progenitor of cultivated soybeans. G3 (Bethesda) 6:835–843

Anglin NL, Amri A, Kehel Z, Ellis D (2018) A case of need: linking traits to genebank accessions. Biopreserv Biobank 16:337–349

Anirban A, Hayward A, Hong HT, Masouleh AK, Henry RJ, O'hare TJ (2023) Breaking the tight genetic linkage between the a1 and sh2 genes led to the development of anthocyanin-rich purple-pericarp super-sweetcorn. Sci Rep 13:1050

Austin RS, Vidaurre D, Stamatiou G, Breit R, Provart NJ, Bonetta D, Zhang J, Fung P, Gong Y, Wang PW, Mccourt P, Guttman DS (2011) Next-generation mapping of Arabidopsis genes. Plant J 67:715–725

Baird NA, Etter PD, Atwood TS, Currey MC, Shiver AL, Lewis ZA, Selker EU, Cresko WA, Johnson EA (2008) Rapid SNP discovery and genetic mapping using sequenced RAD markers. PLoS One 3:e3376

Bari A, Street K, Mackay M, Endresen DTF, De Pauw E, Amri A (2012) Focused identification of germplasm strategy (FIGS) detects wheat stem rust resistance linked to environmental variables. Genet Resour Crop Evol 59:1465–1481

Bellon MR, Dulloo E, Sardos J, Thormann I, Burdon JJ (2017) In situ conservation—harnessing natural and human-derived evolutionary forces to ensure future crop adaptation. Evol Appl 10: 965–977

Birkeland SR, Jin N, Ozdemir AC, Lyons RH Jr, Weisman LS, Wilson TE (2010) Discovery of mutations in Saccharomyces cerevisiae by pooled linkage analysis and whole-genome sequencing. Genetics 186:1127–1137

Breed MF, Harrison PA, Blyth C, Byrne M, Gaget V, Gellie NJC, Groom SVC, Hodgson R, Mills JG, Prowse TA, Steane DA, Mohr JJ (2019) The potential of genomics for restoring ecosystems and biodiversity. Nat Rev Genet 20:615–628

Cadima Fuentes X, Van Treuren R, Hoekstra R, Van Den Berg RG, Sosef MSM (2017) Genetic diversity of Bolivian wild potato germplasm: changes during ex situ conservation management and comparisons with resampled in situ populations. Genet Resour Crop Evol 64:331–344

Chialva M, Ghignone S, Novero M, Hozzein WN, Lanfranco L, Bonfante P (2019) Tomato RNA-seq data mining reveals the taxonomic and functional diversity of root-associated microbiota. Microorganisms 8:38

Cortés AJ, Blair MW (2018) Genotyping by sequencing and genome-environment associations in wild common bean predict widespread divergent adaptation to drought. Front Plant Sci 9:128

Cortés AJ, López-Hernández F, Blair MW (2022) Genome–environment associations, an innovative tool for studying heritable evolutionary adaptation in orphan crops and wild relatives. Front Genet 13

Crowell S, Korniliev P, Falcão A, Ismail A, Gregorio G, Mezey J, McCouch S (2016) Genome-wide association and high-resolution phenotyping link Oryza sativa panicle traits to numerous trait-specific QTL clusters. Nat Commun 7(1):10527. https://doi.org/10.1038/ncomms10527

Das S, Upadhyaya HD, Bajaj D, Kujur A, Badoni S, Laxmi, Kumar V, Tripathi S, Gowda CLL, Sharma S, Singh S, Tyagi AK, Parida SK (2015) Deploying QTL-seq for rapid delineation of a potential candidate gene underlying major trait-associated QTL in chickpea. DNA Res 22:193–203

Doležel J, Čížková J, Šimková H, Bartoš J (2018) One major challenge of sequencing large plant genomes is to know how big they really are. Int J Mol Sci 19:3554

Dyer GA, López-Feldman A, Yúnez-Naude A, Taylor JE (2014) Genetic erosion in maize's center of origin. Proc Natl Acad Sci 111:14094–14099

Ehrenreich IM, Gerke JP, Kruglyak L (2009) Genetic dissection of complex traits in yeast: insights from studies of gene expression and other phenotypes in the BYxRM cross. Cold Spring Harb Symp Quant Biol 74:145–153

Elshire RJ, Glaubitz JC, Sun Q, Poland JA, Kawamoto K, Buckler ES, Mitchell SE (2011) A robust, simple genotyping-by-sequencing (GBS) approach for high diversity species. PLoS One 6: e19379

Fekih R, Takagi H, Tamiru M, Abe A, Natsume S, Yaegashi H, Sharma S, Sharma S, Kanzaki H, Matsumura H, Saitoh H, Mitsuoka C, Utsushi H, Uemura A, Kanzaki E, Kosugi S, Yoshida K, Cano L, Kamoun S, Terauchi R (2013) MutMap+: genetic mapping and mutant identification without crossing in rice. PLoS One 8:e68529

Fitzgerald TL, Shapter FM, Mcdonald S, Waters DL, Chivers IH, Drenth A, Nevo E, Henry RJ (2011) Genome diversity in wild grasses under environmental stress. Proc Natl Acad Sci USA 108:21140–21145

Fuentes-Pardo AP, Ruzzante DE (2017) Whole-genome sequencing approaches for conservation biology: advantages, limitations and practical recommendations. Mol Ecol 26:5369–5406

Funk WC, Forester BR, Converse SJ, Darst C, Morey S (2019) Improving conservation policy with genomics: a guide to integrating adaptive potential into U.S. Endangered Species Act decisions for conservation practitioners and geneticists. Conserv Genet 20:115–134

Gao LZ, Liu YL, Zhang D et al (2019) Evolution of Oryza chloroplast genomes promoted adaptation to diverse ecological habitats. Commun Biol 2(1):278. https://doi.org/10.1038/s42003-019-0531-2

Gimode DM, Ochieng G, Deshpande S, Manyasa EO, Kondombo CP, Mikwa EO, Avosa MO, Kunguni JS, Ngugi K, Sheunda P, Jumbo MB, Odeny DA (2024) Validation of sorghum quality control (QC) markers across African breeding lines. Plant Genome 17:e20438

Goettsch B, Urquiza-Haas T, Koleff P, Acevedo F, Aguilar-Meléndez A, Alavez V, Alejandre-Iturbide G, Cuevas F, Pérez C, Carr J, Castellanos-Morales G, López J, Contreras Toledo A, Correa M, De La Cruz L, Debouck D, Delgado-Salinas A, Gomez-Ruiz E, Ledesma M, Jenkins R (2021) Extinction risk of Mesoamerican crop wild relatives. Plants, People, Planet 3:1–21

Gómez OJ, Blair MW, Frankow-Lindberg BE, Gullberg U (2005) Comparative study of common bean (Phaseolus vulgaris L.) landraces conserved ex situ in genebanks and in situ by farmers. Genet Resour Crop Evol 52:371–380

Gouda AC, Warburton ML, Djedatin GL, Kpeki SB, Wambugu PW, Gnikoua K, Ndjiondjop MN (2021) Development and validation of diagnostic SNP markers for quality control genotyping in a collection of four rice (Oryza) species. Sci Rep 11:18617

Gould BA, Palacio-Mejia JD, Jenkins J, Mamidi S, Barry K, Schmutz J, Juenger TE, Lowry DB (2018) Population genomics and climate adaptation of a C4 perennial grass, Panicum hallii (Poaceae). BMC Genomics 19:792

Halewood M, Jamora N, Noriega IL, Anglin NL, Wenzl P, Payne T, Ndjiondjop M-N, Guarino L, Kumar PL, Yazbek M, Muchugi A, Azevedo V, Tchamba M, Jones CS, Venuprasad R, Roux N, Rojas E, Lusty C (2020) Germplasm acquisition and distribution by CGIAR genebanks. Plants 9:1296

Hammer K, Knüpffer H, Xhuveli L, Perrino P (1996) Estimating genetic erosion in landraces — two case studies. Genet Resour Crop Evol 43:329–336

Han Y, Lv P, Hou S, Li S, Ji G, Ma X, Du R, Liu G (2015) Combining next generation sequencing with bulked segregant analysis to fine map a stem moisture locus in sorghum (Sorghum bicolor L. Moench). PLoS One 10:e0127065

Harris AM, Garud NR, Degiorgio M (2018) Detection and classification of hard and soft sweeps from unphased genotypes by multilocus genotype identity. Genetics 210:1429–1452

Hartwig B, James GV, Konrad K, Schneeberger K, Turck F (2012) Fast isogenic mapping-by-sequencing of ethyl methanesulfonate-induced mutant bulks. Plant Physiol 160:591–600

He JF, Zhao XQ, Laroche A, Lu ZX, Liu HK, Li ZQ (2014) Genotyping-by-sequencing (GBS), an ultimate marker-assisted selection (MAS) tool to accelerate plant breeding. Front Plant Sci 5: 484

He X, Johansson ML, Heath DD (2016) Role of genomics and transcriptomics in selection of reintroduction source populations. Conserv Biol 30:1010–1018

Healey A, Garsmeur O, Lovell JT, Shengquiang S, Sreedasyam A, Jenkins J, Plott CB, Piperidis N, Pompidor N, Llaca V, Metcalfe C, Doležel J, Cápal P, Carlson JW, Hoarau JY, Hervouet C, Zini C, Dievart A, Lipzen A, Williams M, Boston LB, Webber J, Keymanesh K, Tejomurthula S, Rajasekar S, Suchecki R, Furtado A, May G, Parakkal P, Simmons BA, Barry K, Henry RJ, Grimwood J, Aitken K, Schmutz J, D'Hont A (2024) The complex polyploid genome architecture of sugarcane. Nature. https://doi.org/10.1038/s41586-024-07231-4

Henry RJ (2014) Genomics strategies for germplasm characterization and the development of climate resilient crops. Front Plant Sci 5:68

Henry RJ (2020) Innovations in plant genetics adapting agriculture to climate change. Curr Opin Plant Biol 56:168–173

Henry R (2021) Genomics of grain quality in cereals. In: Crop breeding and applied biotechnology, Brazilian Association of Plant Breeding, Brazil, vol 21.

Henry RJ (2022) Wild rice research: advancing plant science and food security. Mol Plant 15:563–565

Henry RJ, Nevo E (2014) Exploring natural selection to guide breeding for agriculture. Plant Biotechnol J 12:655–662

Henry RJ, Edwards M, Waters DLE, Gopala Krishnan S, Bundock P, Sexton TR, Masouleh AK, Nock CJ, Pattemore J (2012) Application of large-scale sequencing to marker discovery in plants. J Biosci 37:829–841

Holderegger R, Kamm U, Gugerli F (2006) Adaptive vs. neutral genetic diversity: implications for landscape genetics. Landsc Ecol 21:797–807

Hubmann G, Mathé L, Foulquié-Moreno MR, Duitama J, Nevoigt E, Thevelein JM (2013) Identification of multiple interacting alleles conferring low glycerol and high ethanol yield in Saccharomyces cerevisiae ethanolic fermentation. Biotechnol Biofuels 6:87–87

Illa-Berenguer E, Van Houten J, Huang Z, Van Der Knaap E (2015) Rapid and reliable identification of tomato fruit weight and locule number loci by QTL-seq. Theor Appl Genet 128(7): 1329–1342

Jordan R, Hoffmann A, Dillon S, Prober S (2017) Evidence of genomic adaptation to climate in Eucalyptus microcarpa: implications for adaptive potential to projected climate change. Mol Ecol 26:6002–6020

Jordan R, Price M, Harrison PA, Prober SM, Vaillancourt RE, Steane D (2023) Landscape genomics reveals signals of climate adaptation and a cryptic lineage in Arthropodium fimbriatum. Conserv Genet 24:473–487

Karlsson Green K, Stenberg JA, Lankinen Å (2020) Making sense of Integrated Pest Management (IPM) in the light of evolution. Evol Appl 13:1791–1805

Karst SM, Ziels RM, Kirkegaard RH, Sørensen EA, Mcdonald D, Zhu Q, Knight R, Albertsen M (2021) High-accuracy long-read amplicon sequences using unique molecular identifiers with Nanopore or PacBio sequencing. Nat Methods 18:165–169

Kharabian-Masouleh A, Waters DLE, Reinke RF, Ward R, Henry RJ (2012) SNP in starch biosynthesis genes associated with nutritional and functional properties of rice. Sci Rep 2:557

Khazaei H, Street K, Bari A, Mackay M, Stoddard FL (2013) The FIGS (focused identification of germplasm strategy) approach identifies traits related to drought adaptation in Vicia faba genetic resources. PLoS One 8:e63107

Kovalev MS, Igolkina AA, Samsonova MG, Nuzhdin SV (2018) A pipeline for classifying deleterious coding mutations in agricultural plants. Front Plant Sci 9:1734

Krannich T, White WTJ, Niehus S, Holley G, Halldórsson BV, Kehr B (2021) Population-scale detection of non-reference sequence variants using colored de Bruijn graphs. Bioinformatics 38: 604–611

Kumar A, Thomas J, Gill N, Dwiningsih Y, Ruiz C, Famoso A, Pereira A (2023) Molecular mapping and characterization of QTLs for grain quality traits in a RIL population of US rice under high nighttime temperature stress. Sci Rep 13:4880

Kyriakidou M, Tai HH, Anglin NL, Ellis D, Strömvik MV (2018) Current strategies of polyploid plant genome sequence assembly. Front Plant Sci 9:1660

Lasky JR, Upadhyaya HD, Ramu P, Deshpande S, Hash CT, Bonnette J, Juenger TE, Hyma K, Acharya C, Mitchell SE, Buckler ES, Brenton Z, Kresovich S, Morris GP (2015) Genome-environment associations in sorghum landraces predict adaptive traits. Sci Adv 1:e1400218

Lei L, Wang L, Wang S, Wu J (2020) Marker-trait association analysis of seed traits in accessions of common bean (Phaseolus vulgaris L.) in China. Front Genet 11:698. https://doi.org/10.3389/fgene.2020.00698

Leroy G, Carroll EL, Bruford MW et al (2018) Next-generation metrics for monitoring genetic erosion within populations of conservation concern. Evol Appl 11:1066–1083. https://doi.org/10.1111/eva.12564

Li Z, Xu Y (2022) Bulk segregation analysis in the NGS era: a review of its teenage years. Plant J 109:1355–1374

Li J, West JB, Hart A, Wegrzyn JL, Smith MA, Domec J-C, Loopstra CA, Casola C (2021) Extensive variation in drought-induced gene expression changes between loblolly pine genotypes. Front Genet 12:661440

Lin L, Cai L, Huang H, Ming S, Sun W (2024) Transcriptome data reveals the conservation genetics of Cypripedium forrestii, a plant species with extremely small populations endemic to Yunnan, China. Front Plant Sci 15:1303625

Lowry DB, Hoban S, Kelley JL, Lotterhos KE, Reed LK, Antolin MF, Storfer A (2017) Breaking RAD: an evaluation of the utility of restriction site-associated DNA sequencing for genome scans of adaptation. Mol Ecol Resour 17:142–152

Lu H, Lin T, Klein J, Wang S, Qi J, Zhou Q, Sun J, Zhang Z, Weng Y, Huang S (2014) QTL-seq identifies an early flowering QTL located near Flowering Locus T in cucumber. Theor Appl Genet 127:1491–1499

Ma Y, Wang J, Hu Q, Li J, Sun Y, Zhang L, Abbott RJ, Liu J, Mao K (2019) Ancient introgression drives adaptation to cooler and drier mountain habitats in a cypress species complex. Commun Biol 2:213

Manel S, Andrello M, Henry K, Verdelet D, Darracq A, Guerin P-E, Desprez B, Devaux P (2018) Predicting genotype environmental range from genome–environment associations. Mol Ecol 27:2823–2833

Mccouch SR, Mcnally KL, Wang W, Sackville Hamilton R (2012) Genomics of gene banks: a case study in rice. Am J Bot 99:407–423

Mccouch SR, Wright MH, Tung C-W, Maron LG, Mcnally KL, Fitzgerald M, Singh N, Declerck G, Agosto-Perez F, Korniliev P, Greenberg AJ, Naredo MEB, Mercado SMQ, Harrington SE, Shi Y, Branchini DA, Kuser-Falcão PR, Leung H, Ebana K, Yano M, Eizenga G, Mcclung A, Mezey J (2016) Open access resources for genome-wide association mapping in rice. Nat Commun 7:10532

Mccouch S, Navabi ZK, Abberton M, Anglin NL, Barbieri RL, Baum M, Bett K, Booker H, Brown GL, Bryan GJ, Cattivelli L, Charest D, Eversole K, Freitas M, Ghamkhar K, Grattapaglia D, Henry R, Valadares Inglis MC, Islam T, Kehel Z, Kersey PJ, King GJ, Kresovich S, Marden E, Mayes S, Ndjiondjop MN, Nguyen HT, Paiva SR, Papa R, Phillips PWB, Rasheed A, Richards C, Rouard M, Amstalden Sampaio MJ, Scholz U, Shaw PD, Sherman B, Staton SE, Stein N, Svensson J, Tester M, Montenegro Valls JF, Varshney R, Visscher S, Von Wettberg E, Waugh R, Wenzl P, Rieseberg LH (2020) Mobilizing crop biodiversity. Mol Plant 13:1341–1344

Mccoy RC, Taylor RW, Blauwkamp TA, Kelley JL, Kertesz M, Pushkarev D, Petrov DA, Fiston-Lavier A-S (2014) Illumina TruSeq synthetic long-reads empower de novo assembly and resolve complex, highly-repetitive transposable elements. PLoS One 9:e106689

Mclean-Rodríguez FD, Costich DE, Camacho-Villa TC, Pè ME, Dell'acqua M (2021) Genetic diversity and selection signatures in maize landraces compared across 50 years of in situ and ex situ conservation. Heredity 126:913–928

Menamo T, Kassahun B, Borrell A, Jordan D, Tao Y, Hunt C, Mace E (2021) Genetic diversity of Ethiopian sorghum reveals signatures of climatic adaptation. Theor Appl Genet 134:1–12

Meng H, Sun M, Jiang Z, Liu Y, Sun Y, Liu D, Jiang C, Ren M, Yuan G, Yu W, Feng Q, Yang A, Cheng L, Wang Y (2021) Comparative transcriptome analysis reveals resistant and susceptible genes in tobacco cultivars in response to infection by Phytophthora nicotianae. Sci Rep 11:809

Monnahan PJ, Michno JM, O'connor C, Brohammer AB, Springer NM, Mcgaugh SE, Hirsch CN (2020) Using multiple reference genomes to identify and resolve annotation inconsistencies. BMC Genomics 21:281

Murigneux V, Rai SK, Furtado A, Bruxner TJC, Tian W, Harliwong I, Wei H, Yang B, Ye Q, Anderson E, Mao Q, Drmanac R, Wang O, Peters BA, Xu M, Wu P, Topp B, Coin LJM, Henry RJ (2020) Comparison of long-read methods for sequencing and assembly of a plant genome. Gigascience 9:giaa146

Nakandala U, Furtado A, Masouleh AK, Smith MW, Williams DC, Henry RJ (2024) The genome of Citrus australasica reveals disease resistance and other species specific genes. BMC Plant Biol 24:260

Natarajan S, Kim HT, Thamilarasan SK, Veerappan K, Park JI, Nou IS (2016) Whole genome re-sequencing and characterization of powdery mildew disease-associated allelic variation in melon. PLoS One 11:e0157524

Ndjiondjop M-N, Semagn K, Gouda AC, Kpeki SB, Dro Tia D, Sow M, Goungoulou A, Sie M, Perrier X, Ghesquiere A, Warburton ML (2017) Genetic variation and population structure of Oryza glaberrima and development of a mini-core collection using DArTseq. Front Plant Sci 8: 1748

Ndjiondjop MN, Semagn K, Zhang J, Gouda AC, Kpeki SB, Goungoulou A, Wambugu P, Dramé KN, Bimpong IK, Zhao D (2018) Development of species diagnostic SNP markers for quality control genotyping in four rice (Oryza L.) species. Mol Breed 38:131–131

Nevo E, Fu Y-B, Pavlicek T, Khalifa S, Tavasi M, Beiles A (2012) Evolution of wild cereals during 28 years of global warming in Israel. Proc Natl Acad Sci 109:3412–3415

Nordström KJV, Albani MC, James GV, Gutjahr C, Hartwig B, Turck F, Paszkowski U, Coupland G, Schneeberger K (2013) Mutation identification by direct comparison of whole-genome sequencing data from mutant and wild-type individuals using k-mers. Nat Biotechnol 31:325–330

Olodo KF, Barnaud A, Kane NA, Mariac C, Faye A, Couderc M, Zekraouï L, Dequincey A, Diouf D, Vigouroux Y, Berthouly-Salazar C (2020) Abandonment of pearl millet cropping and homogenization of its diversity over a 40 year period in Senegal. PLoS One 15:e0239123

Özkurt E, Fritscher J, Soranzo N, Ng DYK, Davey RP, Bahram M, Hildebrand F (2022) LotuS2: an ultrafast and highly accurate tool for amplicon sequencing analysis. Microbiome 10:176

Perales H, Golicher D (2014) Mapping the diversity of maize races in Mexico. PLoS One 9: e114657

Peterson BK, Weber JN, Kay EH, Fisher HS, Hoekstra HE (2012) Double digest RADseq: an inexpensive method for de novo SNP discovery and genotyping in model and non-model species. PLoS One 7:e37135

Razzaque S, Elias SM, Haque T, Biswas S, Jewel GMNA, Rahman S, Weng X, Ismail AM, Walia H, Juenger TE, Seraj ZI (2019) Gene expression analysis associated with salt stress in a reciprocally crossed rice population. Sci Rep 9:8249

Rochette NC, Rivera-Colón AG, Catchen JM (2019) Stacks 2: analytical methods for paired-end sequencing improve RADseq-based population genomics. Mol Ecol 28:4737–4754

Rockman MV (2012) The QTN PROGRAM and the alleles that matter for evolution: all that's gold does not glitter. Evolution 66:1–17

Rojas-Barrera IC, Wegier A, Sánchez González JDJ, Owens GL, Rieseberg LH, Piñero D (2019) Contemporary evolution of maize landraces and their wild relatives influenced by gene flow with modern maize varieties. Proc Natl Acad Sci 116:21302–21311

Rossetto M, Bragg J, Kilian A, Mcpherson H, Van Der Merwe M, Wilson PD (2019) Restore and renew: a genomics-era framework for species provenance delimitation. Restor Ecol 27:538–548

Saeidnia F, Majidi MM, Mirlohi A, Ahmadi B (2022) Association analysis revealed loci linked to post-drought recovery and traits related to persistence of smooth bromegrass (Bromus inermis). Plos One 17(12):e0278687. https://doi.org/10.1371/journal.pone.0278687

Sakai H, Itoh T (2010) Massive gene losses in Asian cultivated rice unveiled by comparative genome analysis. BMC Genomics 11:121

Sang Y, Long Z, Dan X, Feng J, Shi T, Jia C, Zhang X, Lai Q, Yang G, Zhang H, Xu X, Liu H, Jiang Y, Ingvarsson PK, Liu J, Mao K, Wang J (2022) Genomic insights into local adaptation and future climate-induced vulnerability of a keystone forest tree in East Asia. Nat Commun 13: 6541

Schirmer M, Ijaz UZ, D'amore R, Hall N, Sloan WT, Quince C (2015) Insight into biases and sequencing errors for amplicon sequencing with the Illumina MiSeq platform. Nucleic Acids Res 43:e37–e37

Schlötterer C, Tobler R, Kofler R, Nolte V (2014) Sequencing pools of individuals - mining genome-wide polymorphism data without big funding. Nat Rev Genet 15:749–763

Schneeberger K, Ossowski S, Lanz C, Jørgensen J-E, Petersen AH, Nielsen KL, Weigel D, Juul T, Andersen SU (2009) SHOREmap: simultaneous mapping and mutation identification by deep sequencing. Nat Methods 6:550–551

Segura V, Vilhjálmsson BJ, Platt A, Korte A, Seren Ü, Long Q, Nordborg M (2012) An efficient multi-locus mixed-model approach for genome-wide association studies in structured populations. Nat Genet 44:825–830

Semalaiyappan J, Selvanayagam S, Rathore A, Gupta SK, Chakraborty A, Gujjula KR, Haktan S, Viswanath A, Malipatil R, Shah P, Govindaraj M, Ignacio JC, Reddy S, Singh AK, Thirunavukkarasu N (2022) Development of a new AgriSeq 4K mid-density SNP genotyping panel and its utility in pearl millet breeding. Front Plant Sci 13:1068883

Shafer ABA, Wolf JBW, Alves PC, Bergström L, Bruford MW, Brännström I, Colling G, Dalén L, De Meester L, Ekblom R, Fawcett KD, Fior S, Hajibabaei M, Hill JA, Hoezel AR, Höglund J, Jensen EL, Krause J, Kristensen TN, Krützen M, Mckay JK, Norman AJ, Ogden R, Österling EM, Ouborg NJ, Piccolo J, Popović D, Primmer CR, Reed FA, Roumet M, Salmona J, Schenekar T, Schwartz MK, Segelbacher G, Senn H, Thaulow J, Valtonen M, Veale A, Vergeer P, Vijay N, Vilà C, Weissensteiner M, Wennerström L, Wheat CW, Zieliński P (2015) Genomics and the challenging translation into conservation practice. Trends Ecol Evol 30:78–87

Shapter FM, Fitzgerald TL, Waters DLE, Mcdonald S, Chivers IH, Nevo E, Henry R (2012) Analysis of adaptive ribosomal gene diversity in wild plant populations from contrasting climatic environments. Plant Signal Behav 7:602–604

Shivute FN, Zhong Y, Wu J, Bao Y, Wang W, Liu X, Lu Z, Yu H (2024) Genome-wide and pan-genomic analysis reveals rich variants of NBS-LRR genes in a newly developed wild rice line from Oryza alta Swallen. Front Plant Sci 15:1345708

Singh J, Van Der Knaap E (2022) Unintended consequences of plant domestication. Plant Cell Physiol 63:1573–1583

Singh VK, Khan AW, Saxena RK, Kumar V, Kale SM, Sinha P, Chitikineni A, Pazhamala LT, Garg V, Sharma M, Sameer Kumar CV, Parupalli S, Vechalapu S, Patil S, Muniswamy S, Ghanta A, Yamini KN, Dharmaraj PS, Varshney RK (2016) Next-generation sequencing for identification of candidate genes for Fusarium wilt and sterility mosaic disease in pigeonpea (Cajanus cajan). Plant Biotechnol J 14(5):1183–1194

Singh N, Wu S, Raupp WJ, Sehgal S, Arora S, Tiwari V, Vikram P, Singh S, Chhuneja P, Gill BS, Poland J (2018) Efficient curation of genebanks using next generation sequencing reveals substantial duplication of germplasm accessions. bioRxiv, 410779

Singh BK, Delgado-Baquerizo M, Egidi E, Guirado E, Leach JE, Liu H, Trivedi P (2023) Climate change impacts on plant pathogens, food security and paths forward. Nat Rev Microbiol 21: 640–656

Song QX, Johnson C, Wilson TE, Kumar A (2014) Pooled segregant sequencing reveals genetic determinants of yeast pseudohyphal growth. PLoS Genet 10:e1004570

Song J, Li Z, Liu Z, Guo Y, Qiu L-J (2017) Next-generation sequencing from bulked-segregant analysis accelerates the simultaneous identification of two qualitative genes in soybean. Front Plant Sci 8:919

Spear MM, Levi SJ, Etterson JR, Gross BL (2023) Resurrecting urban sunflowers: phenotypic and molecular changes between antecedent and modern populations separated by 36 years. Mol Ecol 32:5241–5259

Stange M, Barrett RDH, Hendry AP (2021) The importance of genomic variation for biodiversity, ecosystems and people. Nat Rev Genet 22:89–105

Supple MA, Shapiro B (2018) Conservation of biodiversity in the genomics era. Genome Biol 19: 131

Swarup S, Cargill EJ, Crosby K, Flagel L, Kniskern J, Glenn KC (2021) Genetic diversity is indispensable for plant breeding to improve crops. Crop Sci 61:839–852

Swinnen S, Schaerlaekens K, Pais T, Claesen J, Hubmann G, Yang Y, Demeke M, Foulquíe-Moreno MR, Goovaerts A, Souvereyns K, Clement L, Dumortier F, Thevelein JM (2012) Identification of novel causative genes determining the complex trait of high ethanol tolerance in yeast using pooled-segregant whole-genome sequence analysis. Genome Res 22:975–984

Takagi H, Abe A, Yoshida K, Kosugi S, Natsume S, Mitsuoka C, Uemura A, Utsushi H, Tamiru M, Takuno S, Innan H, Cano LM, Kamoun S, Terauchi R (2013a) QTL-seq: rapid mapping of quantitative trait loci in rice by whole genome resequencing of DNA from two bulked populations. Plant J 74:174–183

Takagi H, Uemura A, Yaegashi H, Tamiru M, Abe A, Mitsuoka C, Utsushi H, Natsume S, Kanzaki H, Matsumura H, Saitoh H, Yoshida K, Cano LM, Kamoun S, Terauchi R (2013b) MutMap-gap: whole-genome resequencing of mutant F2 progeny bulk combined with de novo assembly of gap regions identifies the rice blast resistance gene Pii. New Phytol 200:276–283

Teixeira TM, Nazareno AG (2021) One step away from extinction: a population genomic analysis of a narrow endemic, tropical plant species. Front Plant Sci 12:730258

Thormann I, Reeves P, Thumm S, Reilley A, Engels JMM, Biradar CM, Lohwasser U, Börner A, Pillen K, Richards CM (2017) Genotypic and phenotypic changes in wild barley (Hordeum vulgare subsp. spontaneum) during a period of climate change in Jordan. Genet Resour Crop Evol 64:1295–1312

Vigouroux Y, Ali A, Ndjeunga J, Luong V, Thuillet A-C, Saïdou A-A, Bezançon G, Mariac C, De Mita S, Pham J-L, Gérard B, Kapran I, Sagnard F, Deu M, Chantereau J (2011) Selection for earlier flowering crop associated with climatic variations in the Sahel. PLoS One 6:e19563

Wambugu PW, Henry R (2022) Supporting in situ conservation of the genetic diversity of crop wild relatives using genomic technologies. Mol Ecol 31:2207–2222

Wambugu P, Furtado A, Waters D, Nyamongo D, Henry R (2013) Conservation and utilization of African Oryza genetic resources. Rice 6:29–29

Wambugu PW, Brozynska M, Furtado A, Waters DL, Henry RJ (2015) Relationships of wild and domesticated rices (Oryza AA genome species) based upon whole chloroplast genome sequences. Sci Rep 5:13957

Wambugu P, Ndjiondjop M-N, Furtado A, Henry R (2018a) Sequencing of bulks of segregants allows dissection of genetic control of amylose content in rice. Plant Biotechnol J 16:100–110

Wambugu PW, Ndjiondjop M-N, Henry RJ (2018b) Role of genomics in promoting the utilization of plant genetic resources in genebanks. Brief Funct Genomics 17:198–206

Wambugu PW, Ndjiondjop M-N, Henry R (2021) Genetics and genomics of African rice (Oryza glaberrima Steud) domestication. Rice 14:6

Wambugu PW, Nyamongo DO, Kirwa EC (2023) Role of seed banks in supporting ecosystem and biodiversity conservation and restoration. Diversity 15:896

Wambugu P, Ndjiondjop M-N, Rangan P, Henry R (2024) Application of genomics in supporting efficient conservation and utilization of plant genetic resources. In: Sustainable utilization and conservation of plant genetic diversity. Springer, Singapore, pp 339–361

Wang MH, Yu Y, Haberer G, Marri PR, Fan CZ, Goicoechea JL, Zuccolo A, Song X, Kudrna D, Ammiraju JSS, Cossu RM, Maldonado C, Chen J, Lee S, Sisneros N, De Baynast K, Golser W, Wissotski M, Kim W, Sanchez P, Ndjiondjop MN, Sanni K, Long MY, Carney J, Panaud O, Wicker T, Machado CA, Chen MS, Mayer KFX, Rounsley S, Wing RA (2014) The genome sequence of African rice (Oryza glaberrima) and evidence for independent domestication. Nat Genet 46:982–988

Wang W, Mauleon R, Hu Z, Chebotarov D, Tai S, Wu Z, Li M, Zheng T, Fuentes RR, Zhang F, Mansueto L, Copetti D, Sanciangco M, Palis KC, Xu J, Sun C, Fu B, Zhang H, Gao Y, Zhao X, Shen F, Cui X, Yu H, Li Z, Chen M, Detras J, Zhou Y, Zhang X, Zhao Y, Kudrna D, Wang C, Li R, Jia B, Lu J, He X, Dong Z, Xu J, Li Y, Wang M, Shi J, Li J, Zhang D, Lee S, Hu W, Poliakov A, Dubchak I, Ulat VJ, Borja FN, Mendoza JR, Ali J, Li J, Gao Q, Niu Y, Yue Z, Naredo MEB, Talag J, Wang X, Li J, Fang X, Yin Y, Glaszmann J-C, Zhang J, Li J, Hamilton RS, Wing RA, Ruan J, Zhang G, Wei C, Alexandrov N, Mcnally KL, Li Z, Leung H (2018) Genomic variation in 3,010 diverse accessions of Asian cultivated rice. Nature 557:43–49

Wenger JW, Schwartz K, Sherlock G (2010) Bulk segregant analysis by high-throughput sequencing reveals a novel xylose utilization gene from Saccharomyces cerevisiae. PLoS Genet 6: e1000942

Whitlock R (2014) Relationships between adaptive and neutral genetic diversity and ecological structure and functioning: a meta-analysis. J Ecol 102:857–872

Wijesundara UK, Masouleh AK, Furtado A, Dillon NL, Henry RJ (2024) Chromosome level genome of mango exclusively from long read sequence data. Plant Genome. https://doi.org/10.1002/tpg2.20441

Woldegiorgis ST, Wu T, Gao L, Huang Y, Zheng Y, Qiu F, Xu S, Tao H, Harrison A, Liu W, He H (2022) Identification of heat-tolerant genes in non-reference sequences in Rice by integrating pan-genome, transcriptomics, and QTLs. Genes (Basel) 13

Yang Z, Huang D, Tang W, Zheng Y, Liang K, Cutler AJ, Wu W (2013) Mapping of quantitative trait loci underlying cold tolerance in rice seedlings via high-throughput sequencing of pooled extremes. PLoS One 8:e68433

Yang X, Xia X, Zhang Z, Nong B, Zeng Y, Xiong F, Wu Y, Gao J, Deng G, Li D (2017) QTL mapping by whole genome re-sequencing and analysis of candidate genes for nitrogen use efficiency in rice. Front Plant Sci 8:1634

Yıldırım Y, Tinnert J, Forsman A (2018) Contrasting patterns of neutral and functional genetic diversity in stable and disturbed environments. Ecol Evol 8:12073–12089

Zhang Q-J, Zhu T, Xia E-H, Shi C, Liu Y-L, Zhang Y, Liu Y, Jiang W-K, Zhao Y-J, Mao S-Y, Zhang L-P, Huang H, Jiao J-Y, Xu P-Z, Yao Q-Y, Zeng F-C, Yang L-L, Gao J, Tao D-Y, Wang Y-J, Bennetzen JL, Gao L-Z (2014) Rapid diversification of five Oryza AA genomes associated with rice adaptation. Proc Natl Acad Sci 111:E4954–E4962

Zou C, Wang P, Xu Y (2016) Bulked sample analysis in genetics, genomics and crop improvement. Plant Biotechnol J 14:1941–1955

Part II
Enabling Technologies and Tools

Chapter 5
Challenges and Opportunities for Genomic Data Resources

Enrique Lopez-Gomez, Amanda Bullones, Javier Santos-del-Rio, Valeria Serrano-García, and Noe Fernandez-Pozo

Abstract Advances in sequencing technologies have enabled the generation of vast amounts of genomics data and resources that drive plant research and breeding. The number of sequenced plant species has grown remarkably due to the price drop in sequencing, along with the gain in completeness of genomic references. This was possible thanks to the improvement in sequencing quality of long-reads technology, reaching even telomere-to-telomere genome sequences in non-model species. However, this progress also has significant challenges, such as the management and analysis of these data, the limited capacity for data storage, the unification of nomenclatures, the need for qualified personnel, the proliferation of multiple genome versions, and access to information dispersed in numerous databases. Despite these challenges, these data are paving the way to new frontiers, such as the development of pan-genomic references and the use of artificial intelligence to analyze genetic variability in thousands of varieties. These advances are greatly accelerating research in plant biology and plant breeding, with a significant impact on agriculture and plant science.

Keywords Bioinformatics · Plants · Crops · Genomics · Resources · Sequencing · Pan-genome · Databases

---

E. Lopez-Gomez · V. Serrano-García · N. Fernandez-Pozo (✉)
Instituto de Hortofruticultura Subtropical y Mediterránea "La Mayora", Consejo Superior de Investigaciones Científicas-Universidad de Málaga (IHSM, CSIC-UMA), Málaga, Spain
e-mail: noe.fernandez.pozo@csic.es

A. Bullones · J. Santos-del-Rio
Instituto de Hortofruticultura Subtropical y Mediterránea "La Mayora", Consejo Superior de Investigaciones Científicas-Universidad de Málaga (IHSM, CSIC-UMA), Málaga, Spain

Departamento de Biología Molecular y Bioquímica, Universidad de Málaga (UMA), Málaga, Spain

V. Garg et al. (eds.), *DNA of Sustainability*, Sustainability Sciences in Asia and Africa,
https://doi.org/10.1007/978-981-95-5109-5_5

5.1 Advancements of Sequencing Technologies Are Responsible for a Plethora of Genomic Resources for Crops

The genome of *Arabidopsis thaliana*, with a size of 135 Mb, was the first plant genome sequenced (The Arabidopsis Genome Initiative 2000). At that time, with the available sequencing technologies, it was an enormous global effort involving research groups from the United States, Japan, and several European countries, each taking responsibility for sequencing specific chromosomes. Sanger sequencing was used to sequence the genome from clones of Bacterial Artificial Chromosomes (BACs), cosmids, Transformation-Competent Artificial Chromosomes (TACs), Yeast Artificial Chromosomes (YACs), and P1 phage, and assembled together with the support of physical and genetic maps. Finally, a total of 47,778 BAC clones were used in order to obtain a total size of 125 Mb (equivalent to 115,409,949 bp) (The Arabidopsis Genome Initiative 2000). The project needed 10 years to be completed and required a total estimated cost of around $70 million (Theologis 2001). In comparison with other plants, *A. thaliana* is a small plant with a short life cycle (6 weeks), which is easy to maintain in a laboratory. Additionally, it has a very small genome, organized into five chromosomes, for which it is an excellent model for functional genomics, becoming the main plant genome reference from which most of the gene functional annotations have been validated by experimental procedures.

The sequencing of other plant genomes in the following years used similar technologies, as seen in the cases of rice (*Oryza sativa*) in 2002 with 420–466 Mb (Yu et al. 2002; Goff et al. 2002), poplar (*Populus trichocarpa*) with 410 Mb in 2006 (Tuskan et al. 2006), grapevine (*Vitis vinifera*) with 487 Mb in 2007 (The French–Italian Public Consortium for Grapevine Genome Characterization 2007), papaya (*Carica papaya*) with 372 Mb in 2008 (Ming et al. 2008), maize (*Zea mays*) with 2.3 Gb in 2009 (Schnable et al. 2009), or soybean (*Glycine max*) with 1.1 Gb in 2010 (Schmutz et al. 2010) (Fig. 5.1). However, the financial effort was unaffordable for most of the plant genomes. Fortunately, the explosion of Next Generation Sequencing (NGS) technologies drastically reduced the cost of sequencing, which has been highly used since 2010. The advent of NGS technologies like Illumina, in combination with other methods, had a profound impact on plant genomics, enabling the sequencing of numerous plant genomes that were previously out of reach (Claros et al. 2012). Some interesting examples could be bread wheat (*Triticum aestivum*) with a hexaploid genome of 17 Gb, a very large genome, which was sequenced in 2014 (The International Wheat Genome Sequencing Consortium (IWGSC) et al. 2014), or garlic (*Allium sativum*) with 16.24 Gb, sequenced in 2020 (Sun et al. 2020) (Fig. 5.1). Another example of the fast improvement and reduction of cost in sequencing technologies is the project 10KP (10,000 Plants) Genome Sequencing Project (Cheng et al. 2018), which aims to sequence 10,000 genomes of plants, representing the diversity of every major clade of embryophytes, with 6000 species of land plants planned to be sequenced.

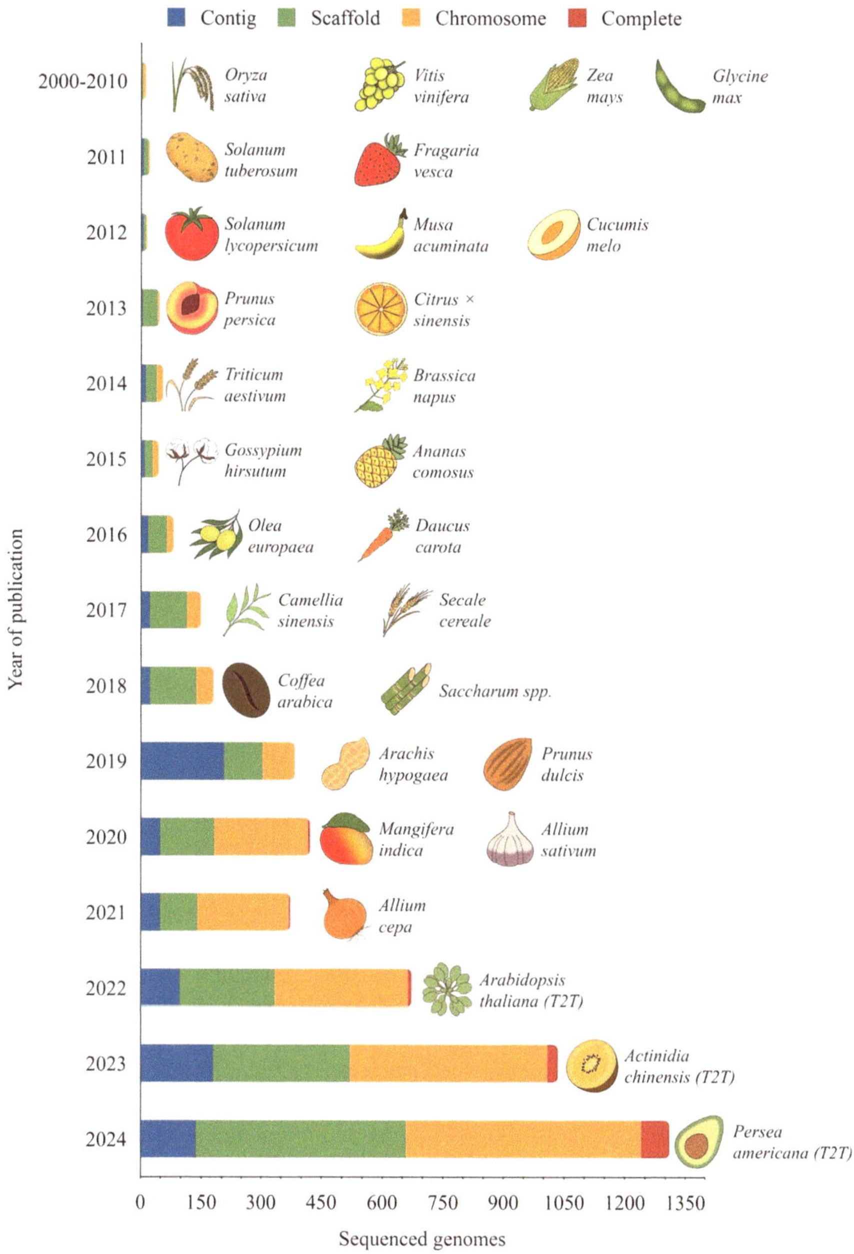

Fig. 5.1 Number of plant genomes (Embryophyta) sequenced every year from 2000 to 2024 based on the NCBI Genome DB. Some examples of important crops sequenced every year are included

Based on the land plant genomes available at the National Center for Biotechnology Information (NCBI), we can observe an increasing number of genomes sequenced per year, with a trend to produce higher quality genomes, most of which are organized in chromosomes, and even completing the full sequence as the technologies advance every year (Fig. 5.1). 2017 marks the milestone of sequencing more than 100 genomes per year (149 genomes), including the publication of genomes such as the tea plant (*Camellia sinensis*), with 3.02 Gb (Xia et al. 2017). 2020 is the first year in which the number of chromosome-level genomes is greater than the number of scaffold-level genomes, with 235 and 133, respectively. Moreover, this was the year when the first complete genomes were published: two for rice and two for potato (Zhang et al. 2016; Zhou et al. 2020; Tiwari et al. 2021), increasing to a total of 111 complete genomes available in 2024. Additionally, from 2022, more than 500 genomes were sequenced and assembled every year, leading to a total of 1310 plant assembled genomes in 2024 (data updated as of October 1st, 2024).

A review of plant genomes sequenced from 2000 to 2023 (Xie et al. 2024) shows that from 2000 to 2010, only 25 genomes of 23 plant species (Viridiplantae) were sequenced. Then, NGS technologies and the improvement of computational algorithms allowed the sequencing of 1119 genomes from 759 different plant species from 2011 to 2020, providing high-quality genomic resources in plants. Interestingly, from 2021 to 2023, with 2373 plant genomes belonging to 1031 species (793 newly sequenced species), the number of sequenced genomes has increased significantly, sequencing more genomes in these 3 years than in the previous 20 years (Xie et al. 2024). Nine hundred twenty-one of these genomes were part of pan-genome projects. Additionally, the quality of these plant genomes was also improved in comparison with the ones from previous years, with 73.2% of assemblies at the chromosome level from 2020 to 2023 in comparison to 47.3% in the previous 20 years (Xie et al. 2024). The reason for the increase in genomic quality and quantity was the development of the third-generation of sequencing technologies. The improvement of long-read sequencing quality, such as PacBio's HiFi technology (Hon et al. 2020), has enabled the generation of sequences of 10–25 kb with a low error rate (0.01%). This has greatly facilitated sequence assembly, often resulting in the complete assembly of entire chromosomes using this technology alone. Additionally, the emergence of novel technologies such as Hi-C (Lieberman-Aiden et al. 2009), which is based on the three-dimensional conformation of the chromatin, allows ordering and orienting contigs and scaffolds without the need for genetic maps. The combination of these technologies has revolutionized plant genomics, enabling the sequencing of non-model plants that, just a few years ago, lacked genomic references. Now, relatively small research communities can afford to produce fully annotated genomes for non-model plants, including haplotype-resolved assemblies, or even complete gapless reference genomes spanning from telomere to telomere (T2T) (Nath et al. 2022; Wijesundara et al. 2024). It's important to note that many plant species, including some major crops, are sometimes trees that require several years to produce fruits and, consequently, the offspring necessary to generate genetic maps. Therefore, these new technologies represent a huge advancement in genome sequencing and assembly, significantly reducing time and cost to generate genomes, not only from the sequencing point of view but also in terms of calculation time for computers and workload for bioinformaticians.

For example, the recent T2T genome of *A. thaliana* (Wang et al. 2022) constitutes a major improvement over the original version from 2000. The application of technologies such as Oxford Nanopore and PacBio HiFi long-read sequencing in the T2T version could be estimated to cost less than $50,000 for a genome size of 135 Mb. This is a huge reduction in the expenses of this project: it is more than 1000 times cheaper than the estimation of $70 million spent on the original assembly. This cost reduction is proof of a higher efficiency, which saves time and resources while producing more complete and accurate genome assemblies. Unlike the original sequencing, which involved extensive international collaboration and traditional Sanger sequencing, the T2T version provides nearly complete chromosome sequences, including previously unresolved regions like centromeres. Among other examples of plant genomes completely sequenced are the genomes of rice (Li et al. 2021a; Song et al. 2021), kiwifruit (Yue et al. 2023), maize (Chen et al. 2023), soybean (Zhang et al. 2023), avocado (Yang et al. 2024), wild strawberry (Zhou et al. 2023), grapevine (Shi et al. 2023a), and mango (Wijesundara et al. 2024).

On the other hand, advancements in sequencing technologies have brought not only more and better genomic references, but also valuable diversity data of thousands of genotypes from some of the most important crops (Table 5.1). This information is of great interest to functional genomics and breeding in order to understand the differences observed in varieties from the same species. Furthermore, it allows the use of pan-genomic references that represent the genetic variance of a species based on a larger number of accessions, instead of a single genotype (Bauchet et al. 2019). Pan-genomic references usually include genetic variation from varieties and wild relatives with interesting agricultural traits, increasing the gene pool to make informed decisions in plant breeding. Due to its relevance, one of the main objectives of plant pan-genome research has been to study the genetic variations of major crop species. The first plant pan-genome based on high-quality genomes was released in 2014 for wild soybean (Li et al. 2014). In the following years, the pan-genomes of many agronomically important crops, such as wild cabbage, tomato, soybean, and chickpea, have been released (Table 5.1). Additionally, the pan-genomes of several major cereals were released, such as the bread wheat, whose pan-genome integrated the information of 18 representative wheat cultivars, rice, with a pan-genome reference based on the information of 3010 resequenced accessions, and maize, constructed using 26 inbred lines (Table 5.1).

Many crop genome assemblies are cost-prohibitive due to their length and complexity, but mapping resequencing reads to a reference genome eases that problem and enables the construction of pan-genomes lowering the sequencing depth and cost, such as, for example, the case of tomato, with 725 accessions (Gao et al. 2019), and chickpea, with 3366 samples (Varshney et al. 2021), where hundreds or thousands of accessions were resequenced and connected to a

Table 5.1 Examples of crop pan-genomes based on a high number of accessions in the last decade

Year	Species	Sample size	Method	References
2014	*Glycine soja*	7	De novo assembly	Li et al. (2014)
2016	*Brassica oleracea*	9	De novo assembly	Golicz et al. (2016)
2017	*Triticum aestivum*	18	Resequencing	Montenegro et al. (2017)
2018	*Oryza sativa*	3010	Resequencing	Wang et al. (2018)
2019	*Solanum lycopersicum*	725	Resequencing	Gao et al. (2019)
2020	*Glycine max*	26 + 2872	De novo assembly + resequencing	Liu et al. (2020)
2021	*Zea mays*	26	De novo assembly	Hufford et al. (2021)
2021	*Oryza* spp.	31 + 2	De novo assembly + resequencing	Qin et al. (2021)
2021	*Cicer arietinum*	3366	Resequencing	Varshney et al. (2021)
2021	*Fragaria* spp.	5 + 128	De novo assembly + resequencing	Qiao et al. (2021)
2021	*Gossypium* spp.	10 + 1951	De novo assembly + resequencing	Li et al. (2021b)
2022	*Solanum lycopersicum*	32 + 806	De novo assembly + resequencing	Zhou et al. (2022)
2022	*Zea* spp.	721	Resequencing	Gui et al. (2022)
2023	*Solanum* section *Lycopersicon*	11 + 2	De novo assembly + resequencing	Li et al. (2023)
2024	*Cicer* spp.	8	De novo assembly	Khan et al. (2024)

Note: De novo assemblies are built from scratch using sequence reads without a reference genome. Pan-genomes based on resequencing use a high-quality reference genome for aligning new sequence data. Both methods can be complemented with other steps to build the pan-genome reference

high-quality genomic reference using methods like iterative assembly (Hu et al. 2024) (Table 5.1). However, there are other methods to construct a pan-genome. Due to the rapid decrease in sequencing costs, a de novo assembly approach for all accessions is possible (Hufford et al. 2021). This method has the advantage of reducing the potential genome size or structure biases, and it minimizes the possibility of misalignment, although it is computationally intensive (Khan et al. 2020). Recently, graph-based pan-genomes have also gained popularity as a way to iteratively upgrade the pan-genomes with newly identified structural variations (SVs) (Shi et al. 2023b), which are intuitively stored in nodes and paths (Wang et al. 2023a). Also, the aforementioned T2T genomes can be of great help for graph-based pan-genome construction, as they can be considered consistent backbone sequences for the graphs, improving the cross-comparisons (Shi et al. 2023b). However, the computational costs of analyzing a complex genome graph are still very high (Sherman and Salzberg 2020).

As the availability of single-species pan-genomes continues rising, the production of multi-species pan-genomes representing higher taxonomic groups becomes an important next step in understanding genomic diversity (Danilevicz et al. 2020), these are called super pan-genomes. In crop super pan-genomes, the wild relatives tend to contribute more to the sequence diversity because wild alleles can be lost in cultivars due to genetic bottlenecks during crop domestication (Shi and Lai 2015). Some of these representative super pan-genomes include the *Oryza* spp., the *Fragaria* spp., and the recently released *Cicer* spp. super pan-genome (Table 5.1).

5.2 Genomics Portals and Bioinformatics Tools to Access the Data

The huge accumulation of genomics sequences is a great resource that has driven plant genomics for two decades. However, managing and providing access to all this data also presents challenges for the research community. The development of databases and genomic portals has been key to addressing these challenges. These resources provide centralized access to genomic data, enabling researchers to analyze, visualize, and retrieve the genetic information available across a wide range of species, becoming essential tools in modern biology. These databases have their roots in the early days of genome sequencing, when the volume of data required the development of specialized systems to store and study genetic sequences. As sequencing technologies have advanced, databases have grown in complexity and size to include not only genetic sequences, but also functional annotations and tools to analyze these data.

In Table 5.2, there is an overview of key generalist genomics and bioinformatics resources. The NCBI maintains more than 40 databases for the scientific community (Smith 2008). NCBI's GenBank is one of the largest collections of publicly available genetic sequences, granting access to annotated sequences from species of all kingdoms. While sequences in GenBank come directly from the authors, RefSeq provides reference sequences that are selected, reviewed, and updated to accurately represent genes, genomes, and proteins of specific organisms (O'Leary et al. 2016). There are also databases to gather and submit raw sequencing data, containing millions of datasets from genomics and transcriptomics sequencing projects. These databases include the Sequence Read Archive (SRA) at NCBI, the European Nucleotide Archive (ENA) at European Bioinformatics Institute (EBI), or the DDBJ Sequence Read Archive (DRA) at DNA Data Bank of Japan (DDBJ). NCBI, EBI, and DDBJ are part of the International Nucleotide Sequence Database Collaboration (INSDC) (https://www.insdc.org/), exchanging data on a daily basis to ensure worldwide coverage (Benson et al. 2012). Additionally, the China National Center for Bioinformation (CNCB) is making efforts to facilitate the exchange and accessibility of global data as part of its National Genomics Data Center (NGDC) (CNCB-NGDC Members and Partners et al. 2024). Another resource at the NCBI, GEO

Table 5.2 Examples of major generalist sites for genomics resources

Name	Description	Website
INSDC	Collaboration between the NCBI, EBI, and DDBJ	https://insdc.org/
CNCB	China National Center for Bioinformatics, which includes the NGDC resources	https://www.cncb.ac.cn/
GenBank	Public genetic and annotated sequences	https://ncbi.nlm.nih.gov/genbank/
RefSeq	Reference sequences that are selected, reviewed, and updated	https://ncbi.nlm.nih.gov/refseq/
SRA/ENA/DRA/GSA	Raw data sequences repository	https://ncbi.nlm.nih.gov/sra/ https://ebi.ac.uk/ena/ https://www.ddbj.nig.ac.jp/dra/ https://ngdc.cncb.ac.cn/gsa/
GEO	Database of curated gene expression DataSets	https://ncbi.nlm.nih.gov/gds/
UniProt	Resource of protein sequence and functional information	https://uniprot.org/
CoGe	Platform for comparative genomics	https://genomevolution.org/
Gene Ontology	Resource of gene ontologies	https://geneontology.org/
MetaCyc	Metabolic pathway database	https://metacyc.org/
KEGG	Metabolic pathway database	https://www.kegg.jp/

(Gene Expression Omnibus), provides gene expression and microarray hybridization data from a multitude of experimental conditions in numerous species (Edgar 2002).

Although there are many databases for nucleotide sequences, there are also specific databases for other biological data. The Universal Protein Resource (UniProt) is a general reference platform for protein sequences and their annotations (Table 5.2). UniProt is divided into Swiss-Prot, where proteins are manually curated based on literature data, and TrEMBL, which contains automatically annotated proteins (The UniProt Consortium 2015). It is updated every 8 weeks, and it currently contains more than 245 million entries. Most of them are from TrEMBL, but more than half a million are manually curated entries included in Swiss-Prot.

Another interesting resource is CoGe (Lyons and Freeling 2008), an online resource designed for comparative genomics. It stores several versions of multiple genomes from many organisms and provides bioinformatics tools to assist genomic research, allowing users to add their own data. Also, very useful is the Gene Ontology (GO), available at the GO Knowledgebase, which provides a controlled and hierarchical vocabulary to define gene products, incorporating many collections from plant, animal, and microbial genomes (Ashburner et al. 2000). This ontology is widely used for gene functional annotations and gene set enrichment analysis. On the other hand, resources such as KEGG (Kyoto Encyclopedia of Genes and Genomes) or MetaCyc are databases for metabolic pathways and the genes and metabolites involved in them (Kanehisa 2000; Caspi et al. 2014) (Table 5.2).

Table 5.3 Examples of plant-specific genomics portals

Name	Description	Website
Phytozome	Database of genomes and annotated plants	https://phytozome-next.jgi.doe.gov/
Ensembl Plants	Genome data of plants with tools for the manipulation, analysis, and visualization of that data	http://plants.ensembl.org/
PLAZA	Plant comparative genomics resource	https://bioinformatics.psb.ugent.be/plaza/
Plant Metabolic Network	Resource with multi-species and species/taxon-specific metabolic databases	https://plantcyc.org/
Gramene	Database of comparative plant genomics for crops and model organisms	https://www.gramene.org/
Sol Genomics Network	Main genomics hub for tomato and other Solanaceae species	https://solgenomics.net/
BreedBase	System to manage breeding programs	https://breedbase.org/
Genome Database for Rosaceae	Resource for Rosaceae research discovery, and crop improvement	https://rosaceae.org/
Citrus Genome Database	Resource for citrus genomics, genetics, breeding, and disease research	https://citrusgenomedb.org/
Pulse Crop Database	Genomic, genetic, and breeding resources for pulse crop improvement	https://pulsedb.org/
CuGenDBv2	Resource for the cucurbit research and breeding community	http://cucurbitgenomics.org/v2/
GreenPhylDB	Resource for comparative and functional genomics in plants	https://www.greenphyl.org/
Legumepedia	Genome assemblies and annotations for legume species	https://cegresources.icrisat.org/legumepedia/

As plant genome sequencing projects expanded, the need for specific plant databases that integrate and compare data across plant species increased, too. An overview of some of the main genomic plant databases is shown in Table 5.3. The aim of these plant databases is to provide access to genomics data and annotations through bioinformatics tools. One of the most comprehensive of these databases is Phytozome, a centralized resource for plant comparative genomics, including multiple bioinformatics tools and resources for each species (Goodstein et al. 2012). Phytozome allows researchers to explore gene families and download and compare data across different plant species, from whole genomes to functional annotations. The most recent genome release (June 2024) includes 401 plant species. Another important resource is Ensembl Plants, with 153 plant species (release 59), including genome sequences, protein annotations, transcriptional data, genetic variation, and comparative results from different taxa (Kersey et al. 2016). PLAZA is an access point for plant comparative genomics (Proost et al. 2010) that integrates data from multiple plant genomes and offers resources like gene family finder or synteny analysis. Similarly, GreenPhylDB focuses on comparative and functional genomics in plants, with 19 pan-genomes and 27 reference genomes of 46 plant species (Guignon et al. 2021). For metabolic pathways in plants, the Plant Metabolomic Network (PMN) includes PlantCyc, a multi-species platform based on MetaCyc, with 155 plant species/taxon-specific databases (Hawkins et al. 2021). In comparison, Plant Reactome also serves as a database of pathways from plants (Fabregat et al. 2016), but provides a broader range of biological pathways, including molecular information of DNA replication, metabolism, signal transduction, gene expression, diseases, transmembrane transport of small molecules, and vesicle-mediated transport. However, it includes fewer plant species compared to PMN.

Some genomics portals are focused on specific plant families or closely related groups of plants that are of special interest to the research communities and breeders working on them. For example, Gramene, which is part of Ensembl, was created to serve resources of the grass family (Poaceae), providing comparative genome tools for crops such as rice, maize, wheat, and barley (Table 5.3). In Gramene, there is a pan-genome section including links to pan-genome resources for rice, maize, sorghum, and grapevine (Ware et al. 2002). On the other hand, the Citrus Pan-genome to Breeding Database (CPBD) (Liu et al. 2022) is another database that collects the latest genomes and transcriptomes for citrus. As a more generalist website, PlantPan (https://ngdc.cncb.ac.cn/plantpan/) offers nine pan-genomes for different crop species such as potato, strawberry, or tomato.

Another interesting example is the Sol Genomics Network (SGN), the main hub for tomato genomics, which provides diverse tools and resources for species of the Solanaceae family, such as tomato, potato, pepper, eggplant, tobacco, and *Nicotiana benthamiana* (Fernandez-Pozo et al. 2015a) (Table 5.3). It is worth highlighting one of its tools, such as the SGN VIGS Tool, which is useful to design constructs for Virus-Induced Gene Silencing, a technique for gene silencing primarily used in plants, with more than 60 plant species available in the tool (Fernandez-Pozo et al. 2015b). The code of the SGN has also been used to create other genomics portals, such as Citrus Greening Solutions (https://citrusgreening.org/), and has evolved to Breedbase (Morales et al. 2022), which is a breeding management software used in many plant species. It allows users to design field layouts, collect phenotypic data, manage genotyping samples, store high-density genotypic data, and provide tools for Genomic Selection, GWAS, and marker-assisted breeding among many other things (Morales et al. 2022). An illustrative example of the implementation of BreedBase is CassavaBase (Fernandez-Pozo et al. 2015a).

Another important resource is the Genome Database for Rosaceae (GDR), including data and tools for some plant species of great importance, such as strawberries and other red fruits, apples, pears, and many drupes from the Prunus genus (peach, apricot, cherry, almond, etc.) (Jung et al. 2019) (Table 5.3). The GDR is based on Tripal (Spoor et al. 2019), the most popular open-source software to develop genomic portals. It uses the Chado database schema to store the genomic data, the content management features of Drupal, and GMOD tools. Other important plant databases based on Tripal are the Citrus Genome Database (CGD), for citrus species, Pulse Crop Database (PCD), for pulses such as lentils, peas, chickpeas, and beans, and CuGenDBv2, for cucurbits such as cucumber, melon, watermelon, and

Table 5.4 Examples of genomics portals dedicated to a single plant species

Name	Description	Website
TAIR	Resources of the *A. thaliana* genome	https://arabidopsis.org/
MaizeGDB	Focused on *Z. mays*	https://maizegdb.org/
SpudDB	The potato genome sequence	https://spuddb.uga.edu/
Rice Genome Annotation Project	Provides the genome sequence from the Nipponbare subspecies of rice and annotation of the chromosomes	https://rice.uga.edu/
CassavaBase	Database to improve cassava breeding and share data and resources	https://cassavabase.org/
OliveTreeDB	Genomic portal for *Olea europaea* cv. Picual	https://genomaolivar.dipujaen.es/
CottonGen	Genomic, genetic, and breeding resources for cotton research discovery and crop improvement	https://cottongen.org/
Coffee Genome Hub	Database with genomic resources for coffee with tools to facilitate translational and applied research	https://coffee-genome-hub.southgreen.fr/
Banana Genome Hub	Centralizes databases of genetic and genomic data for the *Musa* spp.	https://banana-genome-hub.southgreen.fr/
Sugarcane Genome Hub	Centralized platform that offers access to sugarcane genome sequences and genomic resources	https://sugarcane-genome.cirad.fr/
MangoBase	Platform with multiple bioinformatics tools and resources dedicated to mango genomics	https://mangobase.org/
AvoBase	Bioinformatics tools to access avocado genomics data	https://avocado.uma.es/

pumpkins. Also, for more resources on pulses, the Legumepedia provides an extensive repository of the latest genome assemblies and annotations of different legume species (Garg et al. 2022) (Table 5.3).

The publication of the complete genome sequence of *A. thaliana*, the first plant reference genome, marked the beginning of the plant genome era (Sun et al. 2022). Since this publication, the need for a specific database has been crucial. To achieve this objective, the project called The Arabidopsis Information Resource (TAIR) was born (Huala 2001). Following TAIR, many species-specific plant databases have been developed to manage the details of the resources, tools, news, and events interesting to the research communities working in them (Table 5.4). The MSU Rice Genome Annotation Project (Kawahara et al. 2013) provides sequence and annotation data for the rice genome, released in 2002 (Goff et al. 2002). The Maize Genome Sequencing Project published the complete genome sequence of *Z. mays* (Schnable et al. 2009) in MaizeGDB (Woodhouse et al. 2021), and it was included in Gramene (Ware et al. 2002). The Potato Genome Sequencing Consortium (PGSC) released the first draft of the potato genome (The Potato Genome Sequencing Consortium 2011), available on the SpudDB website (Hirsch et al. 2014).

Table 5.4 presents other examples of species-specific genomic portals, such as OliveTreeDB (Jiménez-Ruiz et al. 2020) for olive, AvoBase for avocado, CottonGen for cotton, or the Coffee Genome Hub for coffee.

One of the major challenges in relation to genomics portals and databases is obtaining funding for servers, curators, and developers to maintain and update these resources. Two of the most striking examples are the cases of the well-known repositories SRA, the main repository worldwide to store and share DNA/RNA raw reads, and TAIR, the main site for the model plant Arabidopsis. On February 16, 2011, the NCBI announced the closure of the SRA due to budget constraints (GB Editorial Team 2011). That year, the SRA contained more than 100 Terabases of sequencing data from NGS, 84% of them from Illumina (Kodama et al. 2012). The SRA stores nowadays more than 31 petabytes of data (https://ncbi.nlm.nih.gov/sra/docs/sragrowth/), which is predicted by the SRA Data Working Group to soon exceed the NIH budget for storage and maintenance, suggesting the search for new solutions to compress and store data that can be sustainable in the future (https://dpcpsi.nih.gov/council/sradwg). On the other hand, TAIR suffered similar problems: due to budget reduction, they started a subscription system in 2014 (Reiser et al. 2016), while Araport was launched the same year as an open-access alternative to TAIR (Pasha et al. 2020). However, 4 years later, again due to financial limitations, Araport funding was discontinued, and the project was taken over by TAIR, the National Center for Genome Resources (NCGR), and the Bio-Analytic Resource for Plant Biology (BAR).

In 1998, Ellis and Kalumbi analyzed 321 public biological databases listed in the biological database catalog DBcat through a survey (Ellis and Kalumbi 1998). A 48% response rate was achieved, and the authors determined that 68% of 153 databases had less than 5 years of funding. Fifteen years later, Attwood et al. (2015) checked that around 60% of the 321 biological databases were closed within a period of 18 years, and 14% of them were archived. Ironically, DBcat itself closed in 2006. Nowadays, some genomic databases are no longer being updated. For example, PlantGDB (https://goblinp.luddy.indiana.edu/) is a reference genomic database for plants, which includes more than 100 plant species but has not been updated since July 2015. The development and maintenance of genomic databases is a significant challenge, and it requires specialized infrastructure and bioinformatics skills. To deal with challenges like this, systems such as EasyGDB were designed to simplify the process of developing and maintaining genome databases (Fernandez-Pozo and Bombarely 2022). It provides a pre-built genomics portal with bioinformatic tools and an expression atlas ready to import the data, and customize style and navigation with simple configuration files. The main advantage of EasyGDB is its file system, which is able to store all types of data, greatly simplifying data maintenance and updates, not requiring experienced bioinformaticians to manage a relational database. Then, just by placing the data files in their corresponding folders, the data will be available in the interface. Some examples of sites using this system are MangoBase (Gómez-Ollé et al. 2023) and OliveAtlas (Bullones et al. 2023).

Another challenge produced by sequencing technologies is to analyze the huge amount of data produced. Currently, the bottleneck to analyze these data is the lack of enough bioinformaticians to manage them. A decade ago, sequencing a plant genome was a big effort that often required the collaboration of several research groups and took several years to complete. However, with recent advancements in sequencing technologies and data generation, the pace of producing plant genomics resources has accelerated dramatically. In contrast, now it is possible that, while working on a genomic reference, multiple improved reference genomes and annotations can be released. For example, the first draft genome for the avocado cultivar Hass was published in August 2019 (Rendón-Anaya et al. 2019), another draft followed in December 2019 (Talavera et al. 2019), a haplotype resolved assembly genome organized in chromosomes was published in 2022 (Nath et al. 2022), and a T2T complete genome was published in 2024 (Yang et al. 2024), all for the same cultivar of a non-model species. On the other hand, one of the challenges solved by the improvement of the long-read sequencing technology is the reduction in computational time for genome assembly. Before their arrival, plant genome assemblies, in some cases with genome sizes over 1 Gb, could require a supercomputer with large amounts of shared RAM memory (up to 1 Tb) and many weeks of calculation, requiring multiple steps and the combination of several sequencing technologies. However, nowadays, with high quality long-reads it is possible to assemble small and medium genomes in a laptop in a few days, and it is possible to obtain haplotype resolved assemblies for huge genomes such as the hexaploid genome of California redwood, with around 30 Gb, in 3 days in a supercomputer (using 80 CPUs and a maximum of 700 GB of RAM) (Cheng et al. 2021).

Despite the opportunities described above, there are some challenges to be faced. Some of them are the classic problems we are already familiar with in data science: data dispersion over many repositories, lack of documentation, and data scarcity (Ceri and Pinoli 2020). Good data management and maintenance are essential for the scientific community to reuse these resources after publication, and to this end, the use of the FAIR principles is crucial (Bauchet et al. 2019). FAIR is an acronym for Findability, Accessibility, Interoperability, and Reusability, which aims to ensure transparency, reproducibility, and reusability (Wilkinson et al. 2016). However, sometimes it is possible to find multiple gene nomenclatures for the same genome in several databases, which is very confusing for researchers and might produce mistakes if they are misinterpreted (Fernandez-Pozo et al. 2022; Ambrosino et al. 2018). For example, 33,929 genes of *Citrus clementina* can be found in Phytozome and the Citrus Genome Database, following a nomenclature such as Ciclev10000001m. On the other hand, 34,557 genes are found in Ensembl Plants with IDs like ESR32424 and 32,586 in NCBI with different identifiers for its different databases. Occasionally, it may happen that some genomic references only exist in NCBI, such is the case of the genome sequence of the mango cultivar Alphonso (Wang et al. 2020). In this case, in the absence of a proper gene nomenclature, the only identifiers available are the ones for the NCBI database, which are uninformative in regards to the gene position in the genome and can be very confusing due to the different identifiers for gene locus (123218660), transcripts (XM_044640188.1), which can have multiple splice variants, and proteins (XP_044496123.1). A good example of an informative gene nomenclature is the one used for *A. thaliana*, similarly used in many other plant species. This gene nomenclature (AT1G01010.1) starts with some initials for the species name (AT), followed by the chromosome number (1), the gene number starting with a "G" (G1010), and the transcript version (.1). Additionally, this nomenclature contains more information because the gene numbers are sorted from 5′ to 3′ of the chromosome (Berardini et al. 2015), and increasing by tens to allow the insertion of more genes between identifiers in future versions. Moreover, this nomenclature maintains the same identifiers for genes, transcripts, and proteins, facilitating their use in many formats and tools.

Management, analysis, and integration of the data are major challenges in bioinformatics and plant genomics (Ambrosino et al. 2018). Genomic data resource generation also presents challenges for data curation, whose costs and complexity are often underestimated (Ceri and Pinoli 2020). In comparison to the fast sequencing rate, human-curated results represent the slowest phase of omics research (Ambrosino et al. 2018). For example, a quick search for Embryophyta proteins in UniProt returns 15.9 million entries, of which only 0.25% (40,054) are manually curated. Most of these results are from *A. thaliana* (16,389), rice (5063), maize (850), and tomato (506). These challenges become even clearer in the case of non-model plant species, where the numbers turn from thousands to a few tens or hundreds.

5.3 Gene Expression Atlases for Functional Genomics

Expression data are useful to know when and where the genes are expressed, and they are essential for functional genomics. This knowledge is really helpful to assist breeding in order to make informed decisions selecting genes that can improve stress resistance, productivity, fruit quality, and harvest time, among many other traits. Expression atlases make the information accessible to the scientific community, providing interactive bioinformatics tools to access different omics quantitative data, such as RNA-seq, expression microarrays, proteomics, or metabolomics.

The best-known gene expression atlas for plants is probably the Arabidopsis eFP Browser available at the Bio-Analytic Resource for Plant Biology (BAR). There, it is possible to find tools and expression browsers for over 30 other plant species like maize, tomato, soybean, potato, rice, wheat, and sugarcane (Waese and Provart 2017) (Table 5.5). It includes three different types of tools to view expression patterns as electronic Fluorescent Pictographs (eFP) or heatmaps for exploring gene expression and co-expression patterns, promoter analyses, functional classification enrichment analysis, and protein-protein interactions (Waese and Provart 2017). Another very useful resource is the Tomato Expression Atlas (Fernandez-Pozo et al. 2017) (Table 5.5). It provides RNA-seq data to deeply study the tomato fruit development and ripening, with ten data points from anthesis to the red ripe stage for all the fruit parts (pericarp, seeds, locular material, septum, placenta, and columella), three of the developmental stages providing data at three latitudinal

Table 5.5 Examples of expression atlases

Name	Specie(s) included	Website
The Bio-Analytic Resource for Plant Biology	>30 plant species	https://bar.utoronto.ca/
EBI Expression Atlas	27 plant species	https://ebi.ac.uk/gxa/
Tomato Expression Atlas	*Solanum lycopersicum*	https://tea.solgenomics.net/
Wheat Expression Browser	*Triticum aestivum*	https://wheat-expression.com/
OliveAtlas	*Olea europaea*	https://oliveatlas.uma.es/
Soybean Expression Atlas	*Glycine max*	https://soyatlas.venanciogroup.uenf.br/
3D Flower Meristem	*A. thaliana* flower meristem	http://threed-flower-meristem.herokuapp.com/
Single Cell Expression Atlas	4 plant species	https://ebi.ac.uk/gxa/sc/
Root Cell Atlas	*A. thaliana* (root)	https://rootcellatlas.org/
ScPlantDB	16 plant species	https://biobigdata.nju.edu.cn/scplantdb/

sections. Additionally, it includes high-resolution laser capture microdissection data for the pericarp cell/tissue-types (outer epidermis, collenchyma, parenchyma, vascular tissue, and inner epidermis) (Shinozaki et al. 2018). The Tomato Expression Atlas provides multiple visualization tools, such as cartoons, heatmaps, barplots, and a 3D "Expression Cube" that allows users to explore and compare the gene expression in all tissues at all development/ripening stages for multiple genes, finding the co-expressed genes with the most similar expression profiles to the query gene. More examples of gene expression atlases for plants can be found in Table 5.5.

A great advance related to gene expression was the arrival of single-cell sequencing, which offered a notably higher resolution. Single-cell technology has existed for nearly 15 years now (Tang et al. 2009) and has been an extraordinarily useful resource to date. However, the first single-cell analyses in plants did not arrive until recently and were performed in *A. thaliana* roots (Denyer et al. 2019; Shulse et al. 2019) to provide a spatiotemporal perspective of root cell types to help find candidates for developmental regulators in the cell differentiation process and to study the effects of sucrose on cell-type identity, for example. Not only roots, but also Arabidopsis leaves have been analyzed to detect expression gradients in various cell types and to generate a single-cell spatial transcriptome profile capable of distinguishing cell types and subtypes (Xia et al. 2022).

Some interesting single-cell applications in plants have already emerged, and they have brought valuable information about different species. For example, crops of interest are being studied at the single-cell level: in maize, single-cell genomics was used in six organs to identify transcription factors defining diverse cell identities and coordinating chromatin organization (Marand et al. 2021), finding important cell-type-specific regulatory elements responsible for phenotype-associated genetic variants, which have been selected in modern maize breeding. Another experiment in rice studied the transition from inflorescence to floret during early reproductive development, providing single-cell-level insights into axillary meristem differentiation (Zong et al. 2022).

Similar to what has occurred to genomic information in recent times, plant-related single-cell experiments and data are beginning to accumulate at a very high speed, and there is a need for databases to store all this information. Arabidopsis had one of the first plant single-cell expression atlas with the Root Cell Atlas (Table 5.5). Another example is the Plant Cell Atlas (https://plantcellatlas.org) (Rhee et al. 2019), which was developed with the goal of creating a community resource able to store descriptions of the state of different plant cell types at high resolutions and continues its effort to use and improve existing architectures and frameworks, keeping its community-centered structure as its main philosophy (Fahlgren et al. 2023) and working as a sort of hub to easily connect with different single-cell tools, databases and publications. The EBI also introduced a single-cell expression atlas (Papatheodorou et al. 2019) with data from Arabidopsis, rice, tomato, and maize. Another interesting example focused on plants is scPlantDB (He et al. 2024) (Table 5.5). This interactive website stores information about single-cell transcriptomic profiles from 67 high-quality datasets across 16 plant species.

5.4 A Bright Future for Plant Genomics and Breeding

There have been huge advances in the generation and processing of genomic data: T2T sequencing to obtain complete genomes, (super) pan-genomes that aim to capture the genetic diversity of a species or a genus, the detection of gene expression at single-cell resolution, as well as different kinds of bioinformatics tools and databases to analyze, store and organize the information accessible and updated. Nonetheless, these are not the last steps of the achievements of plant genomics research and breeding; the future is very promising. For example, Oxford Nanopore Technologies (ONT) showcased a method to read intact protein strands with a nanopore sensor array (Motone et al. 2024), aiming to enable high-throughput protein sequencing (Wanunu 2022). Also, using CRISPR–Cas9 technology, genome editing can rapidly improve existing cultivars and allow fast domestication of wild plants, editing a small number of key genes involved in productivity, resistance, and other interesting traits (Huang et al. 2022). Furthermore, thanks to the enormous amount of data and knowledge accumulated over the last two decades and the advances in computer science, it is now possible to use Artificial Intelligence (AI) in plant genomics to obtain new valuable biological information, and in plant breeding to optimize crop management (Farooq et al. 2024).

Artificial intelligence brings a crucial scope in order to improve productivity and sustainability in farming practices. AI-driven techniques, such as remote sensors for soil moisture content or automated irrigation with the help of GPS, were already described as highly effective (Talaviya et al. 2020). AI is also able to integrate critical spatiotemporal data in models of climate effects on emerging disease risk, or generate maps of geographic priorities for surveillance and mitigation strategies (Garrett et al. 2022). Nowadays, for example, phytopathology studies can go as far as identifying 100 possible variables that are later fed to 21 different classification learners and cross-validated to generate a highly reliable and accurate model to predict the presence of *Xylella fastidiosa* in plants (Martinetti and Soubeyrand 2019).

Another example of the uses of AI could be the acceleration of plant breeding programs in combination with Genomic Selection (GS) (Berardini et al. 2015). Many conventional GS approaches employ linear parametric regression, and their computational efficiency makes them the best choice. However, they have some limitations: only additive effects are taken into account, and they assume phenotypes are normally distributed and continuous (González-Camacho et al. 2018). As an alternative, Machine Learning techniques may improve the performance of GS if computational requirements are met (Negus et al. 2024), especially taking into consideration that they capture nonlinear patterns more efficiently than statistical classical methods when the dataset is large enough (Montesinos-López et al. 2021).

Some really interesting examples of AI in plant breeding are in phenomics, developing an image-based wheat spike counter to predict grain yield (Li et al. 2022) or an accurate identifier of plant leaf diseases (Geetharamani and Arun Pandian 2019). Both experiments were successful because of the high complexity and dimensions of the datasets. Actually, AI can obtain better results by integrating data from different sources (Farooq et al. 2024), as it was proven when predicting soybean yield production (Maimaitijiang et al. 2020; Yoosefzadeh-Najafabadi et al. 2021), including multi-omic data to detect live genes and dead pseudogenes in maize (Schrag et al. 2018) or to predict fixation of mutations retained by natural selection (Sartor et al. 2019). Moreover, GWAS and QTL mapping studies will greatly benefit from the era of AI, with the example of QTG-Finder2, which is useful to explain GWAS results by applying machine learning methods (Lin et al. 2020). It incorporates information on orthologous genes from model plants to enrich the feature set for comprehensive gene discovery and the prioritization of genes identified through GWAS in non-model species.

One of the latest efforts to try to take advantage of all these recent findings has been the use of a deep neural network genomic prediction method (Wang et al. 2023b) to integrate multi-omics data in plants. It has some interesting features, such as the ability to deal with small datasets competitively and its low computation time, as well as performing better than at least five classical models, whether they are linear, or based on machine learning or deep learning.

In recent years, genomics research has experienced remarkable progress. Crops are more important than ever due to climate change threats and the exponential growth of the world population. Beyond genomics, a wide range of omics technologies have been developed and explored in plants, providing invaluable insights that are already reshaping agriculture. We expect this to be just the tip of the iceberg, as genomics and breeding data increase at a higher rate each year and new techniques and tools continue to be developed, the future looks even brighter than the groundbreaking advances we have seen so far. Some of the key agents of this promising perspective include the power of AI to process vast amounts of data and genome editing to rapidly adapt crops to meet new challenges. Additionally, the unprecedented level of resolution we can now achieve allows us to address the complexity of cellular types very accurately. Altogether, these innovations will boost omics research beyond the already ambitious expectations of previous years, bringing new possibilities to the future of agriculture.

References

Ambrosino L, Colantuono C, Monticolo F et al (2018) Bioinformatics resources for plant genomics: opportunities and bottlenecks in the -omics era. Curr Issues Mol Biol 27:71–88. https://doi.org/10.21775/cimb.027.071

Ashburner M, Ball CA, Blake JA et al (2000) Gene Ontology: tool for the unification of biology. Nat Genet 25:25–29. https://doi.org/10.1038/75556

Attwood TK, Agit B, Ellis LBM (2015) Longevity of biological databases. EMBnetjournal 21. https://doi.org/10.14806/ej.21.0.803

Bauchet GJ, Bett KE, Cameron CT et al (2019) The future of legume genetic data resources: challenges, opportunities, and priorities. Legum Sci 1:e16. https://doi.org/10.1002/leg3.16

Benson DA, Cavanaugh M, Clark K et al (2012) GenBank. Nucleic Acids Res 41:D36–D42. https://doi.org/10.1093/nar/gks1195

Berardini TZ, Reiser L, Li D et al (2015) The Arabidopsis information resource: making and mining the "gold standard" annotated reference plant genome. Genesis 53:474–485. https://doi.org/10.1002/dvg.22877

Bullones A, Castro AJ, Lima-Cabello E et al (2023) OliveAtlas: a gene expression atlas tool for *Olea europaea*. Plants 12:1274. https://doi.org/10.3390/plants12061274

Caspi R, Altman T, Billington R et al (2014) The MetaCyc database of metabolic pathways and enzymes and the BioCyc collection of Pathway/Genome Databases. Nucleic Acids Res 42: D459–D471. https://doi.org/10.1093/nar/gkt1103

Ceri S, Pinoli P (2020) Data science for genomic data management: challenges, resources, experiences. SN Comput Sci 1:5. https://doi.org/10.1007/s42979-019-0005-0

Chen J, Wang Z, Tan K et al (2023) A complete telomere-to-telomere assembly of the maize genome. Nat Genet 55:1221–1231. https://doi.org/10.1038/s41588-023-01419-6

Cheng S, Melkonian M, Smith SA et al (2018) 10KP: a phylodiverse genome sequencing plan. Gigascience 7. https://doi.org/10.1093/gigascience/giy013

Cheng H, Concepcion GT, Feng X et al (2021) Haplotype-resolved de novo assembly using phased assembly graphs with hifiasm. Nat Methods 18:170–175. https://doi.org/10.1038/s41592-020-01056-5

Claros MG, Bautista R, Guerrero-Fernández D et al (2012) Why assembling plant genome sequences is so challenging. Biology 1:439–459. https://doi.org/10.3390/biology1020439

CNCB-NGDC Members and Partners, Bai X, Bao Y et al (2024) Database resources of the National Genomics Data Center, China National Center for Bioinformation in 2024. Nucleic Acids Res 52:D18–D32. https://doi.org/10.1093/nar/gkad1078

Danilevicz MF, Tay Fernandez CG, Marsh JI et al (2020) Plant pangenomics: approaches, applications and advancements. Curr Opin Plant Biol 54:18–25. https://doi.org/10.1016/j.pbi.2019.12.005

Denyer T, Ma X, Klesen S et al (2019) Spatiotemporal developmental trajectories in the *Arabidopsis* root revealed using high-throughput single-cell RNA sequencing. Dev Cell 48: 840–852.e5. https://doi.org/10.1016/j.devcel.2019.02.022

Edgar R (2002) Gene Expression Omnibus: NCBI gene expression and hybridization array data repository. Nucleic Acids Res 30:207–210. https://doi.org/10.1093/nar/30.1.207

Ellis LBM, Kalumbi D (1998) The demise of public data on the web? Nat Biotechnol 16:1323–1324. https://doi.org/10.1038/4296

Fabregat A, Sidiropoulos K, Garapati P et al (2016) The Reactome pathway Knowledgebase. Nucleic Acids Res 44:D481–D487. https://doi.org/10.1093/nar/gkv1351

Fahlgren N, Kapoor M, Yordanova G et al (2023) Toward a data infrastructure for the Plant Cell Atlas. Plant Physiol 191:35–46. https://doi.org/10.1093/plphys/kiac468

Farooq MA, Gao S, Hassan MA et al (2024) Artificial intelligence in plant breeding. Trends Genet 40:891–908. https://doi.org/10.1016/j.tig.2024.07.001

Fernandez-Pozo N, Bombarely A (2022) EasyGDB: a low-maintenance and highly customizable system to develop genomics portals. Bioinformatics 38:4048–4050. https://doi.org/10.1093/bioinformatics/btac412

Fernandez-Pozo N, Menda N, Edwards JD et al (2015a) The SOL Genomics Network (SGN)—from genotype to phenotype to breeding. Nucleic Acids Res 43:D1036–D1041. https://doi.org/10.1093/nar/gku1195

Fernandez-Pozo N, Rosli HG, Martin GB et al (2015b) The SGN VIGS Tool: user-friendly software to design virus-induced gene silencing (VIGS) constructs for functional genomics. Mol Plant 8: 486–488. https://doi.org/10.1016/j.molp.2014.11.024

Fernandez-Pozo N, Zheng Y, Snyder SI et al (2017) The tomato expression atlas. Bioinformatics 33:2397–2398. https://doi.org/10.1093/bioinformatics/btx190

Fernandez-Pozo N, Haas FB, Gould SB et al (2022) An overview of bioinformatics, genomics, and transcriptomics resources for bryophytes. J Exp Bot 73:4291–4305. https://doi.org/10.1093/jxb/erac052

Gao L, Gonda I, Sun H et al (2019) The tomato pan-genome uncovers new genes and a rare allele regulating fruit flavor. Nat Genet 51:1044–1051. https://doi.org/10.1038/s41588-019-0410-2

Garg V, Dudchenko O, Wang J et al (2022) Chromosome-length genome assemblies of six legume species provide insights into genome organization, evolution, and agronomic traits for crop improvement. J Adv Res 42:315–329. https://doi.org/10.1016/j.jare.2021.10.009

Garrett KA, Bebber DP, Etherton BA et al (2022) Climate change effects on pathogen emergence: artificial intelligence to translate big data for mitigation. Annu Rev Phytopathol 60:357–378. https://doi.org/10.1146/annurev-phyto-021021-042636

GB Editorial Team (2011) Closure of the NCBI SRA and implications for the long-term future of genomics data storage. Genome Biol 12:402. https://doi.org/10.1186/gb-2011-12-3-402

Geetharamani G, Arun Pandian J (2019) Identification of plant leaf diseases using a nine-layer deep convolutional neural network. Comput Electr Eng 76:323–338. https://doi.org/10.1016/j.compeleceng.2019.04.011

Goff SA, Ricke D, Lan T-H et al (2002) A draft sequence of the rice genome (*Oryza sativa* L. ssp. *japonica*). Science 296:92–100. https://doi.org/10.1126/science.1068275

Golicz AA, Bayer PE, Barker GC et al (2016) The pangenome of an agronomically important crop plant *Brassica oleracea*. Nat Commun 7:13390. https://doi.org/10.1038/ncomms13390

Gómez-Ollé A, Bullones A, Hormaza JI et al (2023) MangoBase: a genomics portal and gene expression atlas for *Mangifera indica*. Plants 12:1273. https://doi.org/10.3390/plants12061273

González-Camacho JM, Ornella L, Pérez-Rodríguez P et al (2018) Applications of machine learning methods to genomic selection in breeding wheat for rust resistance. Plant Genome 11:170104. https://doi.org/10.3835/plantgenome2017.11.0104

Goodstein DM, Shu S, Howson R et al (2012) Phytozome: a comparative platform for green plant genomics. Nucleic Acids Res 40:D1178–D1186. https://doi.org/10.1093/nar/gkr944

Gui S, Wei W, Jiang C et al (2022) A pan-*Zea* genome map for enhancing maize improvement. Genome Biol 23:178. https://doi.org/10.1186/s13059-022-02742-7

Guignon V, Toure A, Droc G et al (2021) GreenPhylDB v5: a comparative pangenomic database for plant genomes. Nucleic Acids Res 49:D1464–D1471. https://doi.org/10.1093/nar/gkaa1068

Hawkins C, Ginzburg D, Zhao K et al (2021) Plant Metabolic Network 15: a resource of genome-wide metabolism databases for 126 plants and algae. J Integr Plant Biol 63:1888–1905. https://doi.org/10.1111/jipb.13163

He Z, Luo Y, Zhou X et al (2024) scPlantDB: a comprehensive database for exploring cell types and markers of plant cell atlases. Nucleic Acids Res 52:D1629–D1638. https://doi.org/10.1093/nar/gkad706

Hirsch CD, Hamilton JP, Childs KL et al (2014) Spud DB: a resource for mining sequences, genotypes, and phenotypes to accelerate potato breeding. Plant Genome 7: plantgenome2013.12.0042. https://doi.org/10.3835/plantgenome2013.12.0042

Hon T, Mars K, Young G et al (2020) Highly accurate long-read HiFi sequencing data for five complex genomes. Sci Data 7:399. https://doi.org/10.1038/s41597-020-00743-4

Hu H, Li R, Zhao J et al (2024) Technological development and advances for constructing and analyzing plant pangenomes. Genome Biol Evol 16:evae081. https://doi.org/10.1093/gbe/evae081

Huala E (2001) The Arabidopsis Information Resource (TAIR): a comprehensive database and web-based information retrieval, analysis, and visualization system for a model plant. Nucleic Acids Res 29:102–105. https://doi.org/10.1093/nar/29.1.102

Huang X, Huang S, Han B et al (2022) The integrated genomics of crop domestication and breeding. Cell 185:2828–2839. https://doi.org/10.1016/j.cell.2022.04.036

Hufford MB, Seetharam AS, Woodhouse MR et al (2021) De novo assembly, annotation, and comparative analysis of 26 diverse maize genomes. Science 373:655–662. https://doi.org/10.1126/science.abg5289

Jiménez-Ruiz J, Ramírez-Tejero JA, Fernández-Pozo N et al (2020) Transposon activation is a major driver in the genome evolution of cultivated olive trees (*Olea europaea* L.). Plant Genome 13:e20010. https://doi.org/10.1002/tpg2.20010

Jung S, Lee T, Cheng C-H et al (2019) 15 years of GDR: new data and functionality in the Genome Database for Rosaceae. Nucleic Acids Res 47:D1137–D1145. https://doi.org/10.1093/nar/gky1000

Kanehisa M (2000) KEGG: Kyoto encyclopedia of genes and genomes. Nucleic Acids Res 28:27–30. https://doi.org/10.1093/nar/28.1.27

Kawahara Y, De La Bastide M, Hamilton JP et al (2013) Improvement of the *Oryza sativa* Nipponbare reference genome using next generation sequence and optical map data. Rice 6:4. https://doi.org/10.1186/1939-8433-6-4

Kersey PJ, Allen JE, Armean I et al (2016) Ensembl Genomes 2016: more genomes, more complexity. Nucleic Acids Res 44:D574–D580. https://doi.org/10.1093/nar/gkv1209

Khan AW, Garg V, Roorkiwal M et al (2020) Super-pangenome by integrating the wild side of a species for accelerated crop improvement. Trends Plant Sci 25:148–158. https://doi.org/10.1016/j.tplants.2019.10.012

Khan AW, Garg V, Sun S et al (2024) *Cicer* super-pangenome provides insights into species evolution and agronomic trait loci for crop improvement in chickpea. Nat Genet 56:1225–1234. https://doi.org/10.1038/s41588-024-01760-4

Kodama Y, Shumway M, Leinonen R et al (2012) The sequence read archive: explosive growth of sequencing data. Nucleic Acids Res 40:D54–D56. https://doi.org/10.1093/nar/gkr854

Li Y, Zhou G, Ma J et al (2014) *De novo* assembly of soybean wild relatives for pan-genome analysis of diversity and agronomic traits. Nat Biotechnol 32:1045–1052. https://doi.org/10.1038/nbt.2979

Li K, Jiang W, Hui Y et al (2021a) Gapless *indica* rice genome reveals synergistic contributions of active transposable elements and segmental duplications to rice genome evolution. Mol Plant 14:1745–1756. https://doi.org/10.1016/j.molp.2021.06.017

Li J, Yuan D, Wang P et al (2021b) Cotton pan-genome retrieves the lost sequences and genes during domestication and selection. Genome Biol 22:119. https://doi.org/10.1186/s13059-021-02351-w

Li L, Hassan MA, Yang S et al (2022) Development of image-based wheat spike counter through a Faster R-CNN algorithm and application for genetic studies. Crop J 10:1303–1311. https://doi.org/10.1016/j.cj.2022.07.007

Li N, He Q, Wang J et al (2023) Super-pangenome analyses highlight genomic diversity and structural variation across wild and cultivated tomato species. Nat Genet 55:852–860. https://doi.org/10.1038/s41588-023-01340-y

Lieberman-Aiden E, Van Berkum NL, Williams L et al (2009) Comprehensive mapping of long-range interactions reveals folding principles of the human genome. Science 326:289–293. https://doi.org/10.1126/science.1181369

Lin F, Lazarus EZ, Rhee SY (2020) QTG-Finder2: a generalized machine-learning algorithm for prioritizing QTL causal genes in plants. G3 (Bethesda) 10:2411–2421. https://doi.org/10.1534/g3.120.401122

Liu Y, Du H, Li P et al (2020) Pan-genome of wild and cultivated soybeans. Cell 182:162–176.e13. https://doi.org/10.1016/j.cell.2020.05.023

Liu H, Wang X, Liu S et al (2022) Citrus Pan-Genome to Breeding Database (CPBD): a comprehensive genome database for citrus breeding. Mol Plant 15:1503–1505. https://doi.org/10.1016/j.molp.2022.08.006

Lyons E, Freeling M (2008) How to usefully compare homologous plant genes and chromosomes as DNA sequences. Plant J 53:661–673. https://doi.org/10.1111/j.1365-313X.2007.03326.x

Maimaitijiang M, Sagan V, Sidike P et al (2020) Soybean yield prediction from UAV using multimodal data fusion and deep learning. Remote Sens Environ 237:111599. https://doi.org/10.1016/j.rse.2019.111599

Marand AP, Chen Z, Gallavotti A et al (2021) A cis-regulatory atlas in maize at single-cell resolution. Cell 184:3041–3055.e21. https://doi.org/10.1016/j.cell.2021.04.014

Martinetti D, Soubeyrand S (2019) Identifying lookouts for epidemio-surveillance: application to the emergence of *Xylella fastidiosa* in France. Phytopathology 109:265–276. https://doi.org/10.1094/PHYTO-07-18-0237-FI

Ming R, Hou S, Feng Y et al (2008) The draft genome of the transgenic tropical fruit tree papaya (*Carica papaya* Linnaeus). Nature 452:991–996. https://doi.org/10.1038/nature06856

Montenegro JD, Golicz AA, Bayer PE et al (2017) The pangenome of hexaploid bread wheat. Plant J 90:1007–1013. https://doi.org/10.1111/tpj.13515

Montesinos-López OA, Montesinos-López A, Pérez-Rodríguez P et al (2021) A review of deep learning applications for genomic selection. BMC Genomics 22:19. https://doi.org/10.1186/s12864-020-07319-x

Morales N, Ogbonna AC, Ellerbrock BJ et al (2022) Breedbase: a digital ecosystem for modern plant breeding. G3 (Bethesda) 12:jkac078. https://doi.org/10.1093/g3journal/jkac078

Motone K, Kontogiorgos-Heintz D, Wee J et al (2024) Multi-pass, single-molecule nanopore reading of long protein strands. Nature 633:662–669. https://doi.org/10.1038/s41586-024-07935-7

Nath O, Fletcher SJ, Hayward A et al (2022) A haplotype resolved chromosomal level avocado genome allows analysis of novel avocado genes. Hortic Res 9:uhac157. https://doi.org/10.1093/hr/uhac157

Negus KL, Li X, Welch SM et al (2024) The role of artificial intelligence in crop improvement. In: Advances in agronomy. Elsevier, pp 1–66. https://doi.org/10.1016/bs.agron.2023.11.001

O'Leary NA, Wright MW, Brister JR et al (2016) Reference sequence (RefSeq) database at NCBI: current status, taxonomic expansion, and functional annotation. Nucleic Acids Res 44:D733–D745. https://doi.org/10.1093/nar/gkv1189

Papatheodorou I, Moreno P, Manning J et al (2019) Expression Atlas update: from tissues to single cells. Nucleic Acids Res:gkz947. https://doi.org/10.1093/nar/gkz947

Pasha A, Subramaniam S, Cleary A et al (2020) Araport lives: an updated framework for Arabidopsis bioinformatics. Plant Cell 32:2683–2686. https://doi.org/10.1105/tpc.20.00358

Proost S, Van Bel M, Sterck L et al (2010) PLAZA: a comparative genomics resource to study gene and genome evolution in plants. Plant Cell 21:3718–3731. https://doi.org/10.1105/tpc.109.071506
Qiao Q, Edger PP, Xue L et al (2021) Evolutionary history and pan-genome dynamics of strawberry (*Fragaria* spp.). Proc Natl Acad Sci USA 118:e2105431118. https://doi.org/10.1073/pnas.2105431118
Qin P, Lu H, Du H et al (2021) Pan-genome analysis of 33 genetically diverse rice accessions reveals hidden genomic variations. Cell 184:3542–3558.e16. https://doi.org/10.1016/j.cell.2021.04.046
Reiser L, Berardini TZ, Li D et al (2016) Sustainable funding for biocuration: The Arabidopsis Information Resource (TAIR) as a case study of a subscription-based funding model. Database 2016:baw018. https://doi.org/10.1093/database/baw018
Rendón-Anaya M, Ibarra-Laclette E, Méndez-Bravo A et al (2019) The avocado genome informs deep angiosperm phylogeny, highlights introgressive hybridization, and reveals pathogen-influenced gene space adaptation. Proc Natl Acad Sci USA 116:17081–17089. https://doi.org/10.1073/pnas.1822129116
Rhee SY, Birnbaum KD, Ehrhardt DW (2019) Towards building a plant cell atlas. Trends Plant Sci 24:303–310. https://doi.org/10.1016/j.tplants.2019.01.006
Sartor RC, Noshay J, Springer NM et al (2019) Identification of the expressome by machine learning on omics data. Proc Natl Acad Sci USA 116:18119–18125. https://doi.org/10.1073/pnas.1813645116
Schmutz J, Cannon SB, Schlueter J et al (2010) Genome sequence of the palaeopolyploid soybean. Nature 463:178–183. https://doi.org/10.1038/nature08670
Schnable PS, Ware D, Fulton RS et al (2009) The B73 maize genome: complexity, diversity, and dynamics. Science 326:1112–1115. https://doi.org/10.1126/science.1178534
Schrag TA, Westhues M, Schipprack W et al (2018) Beyond genomic prediction: combining different types of *omics* data can improve prediction of hybrid performance in maize. Genetics 208:1373–1385. https://doi.org/10.1534/genetics.117.300374
Sherman RM, Salzberg SL (2020) Pan-genomics in the human genome era. Nat Rev Genet 21:243–254. https://doi.org/10.1038/s41576-020-0210-7
Shi J, Lai J (2015) Patterns of genomic changes with crop domestication and breeding. Curr Opin Plant Biol 24:47–53. https://doi.org/10.1016/j.pbi.2015.01.008
Shi X, Cao S, Wang X et al (2023a) The complete reference genome for grapevine (*Vitis vinifera* L.) genetics and breeding. Hortic Res 10:uhad061. https://doi.org/10.1093/hr/uhad061
Shi J, Tian Z, Lai J et al (2023b) Plant pan-genomics and its applications. Mol Plant 16:168–186. https://doi.org/10.1016/j.molp.2022.12.009
Shinozaki Y, Nicolas P, Fernandez-Pozo N et al (2018) High-resolution spatiotemporal transcriptome mapping of tomato fruit development and ripening. Nat Commun 9:364. https://doi.org/10.1038/s41467-017-02782-9
Shulse CN, Cole BJ, Ciobanu D et al (2019) High-throughput single-cell transcriptome profiling of plant cell types. Cell Rep 27:2241–2247.e4. https://doi.org/10.1016/j.celrep.2019.04.054
Smith KA (2008) Laws, leaders, and legends of the modern National Library of Medicine. J Med Libr Assoc 96:121–133. https://doi.org/10.3163/1536-5050.96.2.121
Song J-M, Xie W-Z, Wang S et al (2021) Two gap-free reference genomes and a global view of the centromere architecture in rice. Mol Plant 14:1757–1767. https://doi.org/10.1016/j.molp.2021.06.018
Spoor S, Cheng C-H, Sanderson L-A et al (2019) Tripal v3: an ontology-based toolkit for construction of FAIR biological community databases. Database 2019:baz077. https://doi.org/10.1093/database/baz077
Sun X, Zhu S, Li N et al (2020) A chromosome-level genome assembly of garlic (*Allium sativum*) provides insights into genome evolution and allicin biosynthesis. Mol Plant 13:1328–1339. https://doi.org/10.1016/j.molp.2020.07.019

Sun Y, Shang L, Zhu Q-H et al (2022) Twenty years of plant genome sequencing: achievements and challenges. Trends Plant Sci 27:391–401. https://doi.org/10.1016/j.tplants.2021.10.006

Talavera A, Soorni A, Bombarely A et al (2019) Genome-Wide SNP discovery and genomic characterization in avocado (*Persea americana* Mill.). Sci Rep 9:20137. https://doi.org/10.1038/s41598-019-56526-4

Talaviya T, Shah D, Patel N et al (2020) Implementation of artificial intelligence in agriculture for optimisation of irrigation and application of pesticides and herbicides. Artif Intell Agric 4:58–73. https://doi.org/10.1016/j.aiia.2020.04.002

Tang F, Barbacioru C, Wang Y et al (2009) mRNA-Seq whole-transcriptome analysis of a single cell. Nat Methods 6:377–382. https://doi.org/10.1038/nmeth.1315

The Arabidopsis Genome Initiative (2000) Analysis of the genome sequence of the flowering plant *Arabidopsis thaliana*. Nature 408:796–815. https://doi.org/10.1038/35048692

The French–Italian Public Consortium for Grapevine Genome Characterization (2007) The grapevine genome sequence suggests ancestral hexaploidization in major angiosperm phyla. Nature 449:463–467. https://doi.org/10.1038/nature06148

The International Wheat Genome Sequencing Consortium (IWGSC), Mayer KFX, Rogers J et al (2014) A chromosome-based draft sequence of the hexaploid bread wheat (*Triticum aestivum*) genome. Science 345:1251788. https://doi.org/10.1126/science.1251788

The Potato Genome Sequencing Consortium (2011) Genome sequence and analysis of the tuber crop potato. Nature 475:189–195. https://doi.org/10.1038/nature10158

The UniProt Consortium (2015) UniProt: a hub for protein information. Nucleic Acids Res 43: D204–D212. https://doi.org/10.1093/nar/gku989

Theologis A (2001) Goodbye to "one by one" genetics. Genome Biol 2:comment2004.1. https://doi.org/10.1186/gb-2001-2-4-comment2004

Tiwari JK, Rawat S, Luthra SK et al (2021) Genome sequence analysis provides insights on genomic variation and late blight resistance genes in potato somatic hybrid (parents and progeny). Mol Biol Rep 48:623–635. https://doi.org/10.1007/s11033-020-06106-x

Tuskan GA, DiFazio S, Jansson S et al (2006) The genome of black cottonwood, *Populus trichocarpa* (Torr. & Gray). Science 313:1596–1604. https://doi.org/10.1126/science.1128691

Varshney RK, Roorkiwal M, Sun S et al (2021) A chickpea genetic variation map based on the sequencing of 3,366 genomes. Nature 599:622–627. https://doi.org/10.1038/s41586-021-04066-1

Waese J, Provart NJ (2017) The bio-analytic resource for plant biology. In: Van Dijk ADJ (ed) Plant genomics databases. Springer New York, New York, pp 119–148

Wang W, Mauleon R, Hu Z et al (2018) Genomic variation in 3,010 diverse accessions of Asian cultivated rice. Nature 557:43–49. https://doi.org/10.1038/s41586-018-0063-9

Wang P, Luo Y, Huang J et al (2020) The genome evolution and domestication of tropical fruit mango. Genome Biol 21:60. https://doi.org/10.1186/s13059-020-01959-8

Wang B, Yang X, Jia Y et al (2022) High-Quality *Arabidopsis thaliana* genome assembly with Nanopore and HiFi long reads. Genomics Proteomics Bioinformatics 20:4–13. https://doi.org/10.1016/j.gpb.2021.08.003

Wang S, Qian Y-Q, Zhao R-P et al (2023a) Graph-based pan-genomes: increased opportunities in plant genomics. J Exp Bot 74:24–39. https://doi.org/10.1093/jxb/erac412

Wang K, Abid MA, Rasheed A et al (2023b) DNNGP, a deep neural network-based method for genomic prediction using multi-omics data in plants. Mol Plant 16:279–293. https://doi.org/10.1016/j.molp.2022.11.004

Wanunu M (2022) Back and forth with nanopore peptide sequencing. Nat Biotechnol 40:172–173. https://doi.org/10.1038/s41587-021-01205-x

Ware DH, Jaiswal P, Ni J et al (2002) Gramene, a tool for grass genomics. Plant Physiol 130:1606–1613. https://doi.org/10.1104/pp.015248

Wijesundara UK, Masouleh AK, Furtado A et al (2024) A chromosome-level genome of mango exclusively from long-read sequence data. Plant Genome 17:e20441. https://doi.org/10.1002/tpg2.20441

Wilkinson MD, Dumontier M, Aalbersberg IJ et al (2016) The FAIR Guiding Principles for scientific data management and stewardship. Sci Data 3:160018. https://doi.org/10.1038/sdata.2016.18

Woodhouse MR, Cannon EK, Portwood JL et al (2021) A pan-genomic approach to genome databases using maize as a model system. BMC Plant Biol 21:385. https://doi.org/10.1186/s12870-021-03173-5

Xia E-H, Zhang H-B, Sheng J et al (2017) The tea tree genome provides insights into tea flavor and independent evolution of caffeine biosynthesis. Mol Plant 10:866–877. https://doi.org/10.1016/j.molp.2017.04.002

Xia K, Sun H-X, Li J et al (2022) The single-cell stereo-seq reveals region-specific cell subtypes and transcriptome profiling in Arabidopsis leaves. Dev Cell 57:1299–1310.e4. https://doi.org/10.1016/j.devcel.2022.04.011

Xie L, Gong X, Yang K et al (2024) Technology-enabled great leap in deciphering plant genomes. Nat Plants 10:551–566. https://doi.org/10.1038/s41477-024-01655-6

Yang T, Cai Y, Huang T et al (2024) A telomere-to-telomere gap-free reference genome assembly of avocado provides useful resources for identifying genes related to fatty acid biosynthesis and disease resistance. Hortic Res 11:uhae119. https://doi.org/10.1093/hr/uhae119

Yoosefzadeh-Najafabadi M, Tulpan D, Eskandari M (2021) Using hybrid artificial intelligence and evolutionary optimization algorithms for estimating soybean yield and fresh biomass using hyperspectral vegetation indices. Remote Sens 13:2555. https://doi.org/10.3390/rs13132555

Yu J, Hu S, Wang J et al (2002) A draft sequence of the rice genome (*Oryza sativa* L. ssp. *indica*). Science 296:79–92. https://doi.org/10.1126/science.1068037

Yue J, Chen Q, Wang Y et al (2023) Telomere-to-telomere and gap-free reference genome assembly of the kiwifruit *Actinidia chinensis*. Hortic Res 10:uhac264. https://doi.org/10.1093/hr/uhac264

Zhang J, Chen L-L, Sun S et al (2016) Building two indica rice reference genomes with PacBio long-read and Illumina paired-end sequencing data. Sci Data 3:160076. https://doi.org/10.1038/sdata.2016.76

Zhang C, Xie L, Yu H et al (2023) The T2T genome assembly of soybean cultivar ZH13 and its epigenetic landscapes. Mol Plant 16:1715–1718. https://doi.org/10.1016/j.molp.2023.10.003

Zhou Q, Tang D, Huang W et al (2020) Haplotype-resolved genome analyses of a heterozygous diploid potato. Nat Genet 52:1018–1023. https://doi.org/10.1038/s41588-020-0699-x

Zhou Y, Zhang Z, Bao Z et al (2022) Graph pangenome captures missing heritability and empowers tomato breeding. Nature 606:527–534. https://doi.org/10.1038/s41586-022-04808-9

Zhou Y, Xiong J, Shu Z et al (2023) The telomere-to-telomere genome of *Fragaria vesca* reveals the genomic evolution of *Fragaria* and the origin of cultivated octoploid strawberry. Hortic Res 10:uhad027. https://doi.org/10.1093/hr/uhad027

Zong J, Wang L, Zhu L et al (2022) A rice single cell transcriptomic atlas defines the developmental trajectories of rice floret and inflorescence meristems. New Phytol 234:494–512. https://doi.org/10.1111/nph.18008

Chapter 6
Development of Low-Cost Genotyping Platforms

Reshmi Rani Das, Shaina Sharma, Pallavi Sinha, and Vikas Kumar Singh

Abstract The advent of low-cost genotyping platforms has revolutionised modern plant breeding, addressing critical global food security challenges amid climate change and a growing population. Genomic-assisted breeding approaches, leveraging technologies such as single nucleotide polymorphism (SNP), enable the precise selection of complex traits, accelerating the development of climate-resilient and high-yielding crop varieties. While high-throughput genotyping techniques are widely adopted in the private and public sector breeding programmes, they often face significant financial, technical, and infrastructural constraints. This chapter reviews advances in SNP marker technologies, spanning low- to high-density platforms, and their applications in genomic selection (GS), marker-assisted selection (MAS), and trait mapping. It also explores innovative solutions, such as shared genotyping resources including the KASP low-density platform and mid-density panel—that expand access to advanced molecular breeding tools. Special emphasis is placed on the role of public sector breeding programmes, CGIAR initiatives, and bioinformatics integration in enhancing the utility of genotyping-by-sequencing (GBS) and SNP arrays. By synthesising recent advancements and identifying future prospects, this chapter highlights the transformative potential of low-cost genotyping in empowering resource-limited breeding programmes. Ultimately, it underscores its role in accelerating genetic gains, fostering sustainable agricultural development, and strengthening global food security.

Keywords SNP markers · Mid-density array · KASP assay · 4K HaploPanel · Low-cost genotyping

---

R. R. Das · S. Sharma · P. Sinha (✉) · V. K. Singh (✉)
IRRI-South Asia Hub, Hyderabad, India
e-mail: P.Sinha@cgiar.org; V.K.Singh@cgiar.org

V. Garg et al. (eds.), *DNA of Sustainability*, Sustainability Sciences in Asia and Africa,
https://doi.org/10.1007/978-981-95-5109-5_6

Abbreviations and Symbols

BACs	Bacterial Artificial Chromosomes
CGIAR	Consultative Group for International Agricultural Research
cM	centiMorgan
DArTag	Diversity Arrays Technology
EiB	Excellence in Breeding
GBS	Genotyping-by-Sequencing
GEBV	Genomic Estimated Breeding Value
GS	Genomic Selection
GWAS	Genome-Wide Association Study
HDSG	High-Density SNP Genotyping
HTPG	High-Throughput Genotyping
IRGSP	International Rice Genome Sequencing Project
IRRI	International Rice Research Institute.
IWGSC	International Wheat Genome Sequencing Consortium
KASP	Kompetitive Allele-Specific PCR
LD	Linkage disequilibrium
LDSG	Low-Density SNP Genotyping
MABC	Marker-Assisted Backcrossing
MAS	Marker-Assisted Selection
MDSG	Mid-Density SNP Genotyping
NGS	Next-Generation Sequencing
PACs	P1-derived Artificial Chromosomes
PCR	Polymerase Chain Reaction
QC	Quality Control
QTL	Quantitative Trait Loci
SNP	Single Nucleotide Polymorphism
WGA	Whole-Genome Assembly

6.1 Introduction

The increasing urgency of global food security, driven by rapid population growth and the intensifying effects of climate change, necessitates transformative advancements in agricultural production. Feeding a projected global population of 9.8 billion by 2050 sustainably, according to the United Nations, requires a shift in traditional approaches to crop production, especially in stress-prone and resource-limited environments. This challenge is compounded by the imperative to conserve land and natural resources while optimising agricultural output. Historically, plant breeding has transitioned from art to science by integrating genetics, genomics, and data-driven methodologies to address modern agricultural challenges (Xu 2010). Conventional breeding approaches have been instrumental in developing high-yielding, stress-resilient varieties of staple crops such as rice (Khush 2001). However, these methods are limited by prolonged breeding cycles, low predictability, and weak correlations between traits of interest.

Advances in molecular biology and plant genomics have introduced a new paradigm in plant breeding, enabling precision and speed in trait selection (Cobb et al. 2019; Crossa et al. 2017; Bohar et al. 2020). The integration of genotyping technologies, mainly single nucleotide polymorphism (SNP) markers, and next-generation sequencing (NGS) platforms, including genotyping-by-sequencing (GBS) and whole-genome resequencing, has facilitated the development of genomics-based breeding strategies (Elshire et al. 2011; Edwards et al. 2012). These innovations address the limitations of traditional breeding approaches by enabling the accurate selection of genotypes, even in complex traits governed by multiple genes and genotype-environment interactions (Lamichhane and Thapa 2022). For instance, genomic selection (GS), a robust approach that employs genome-wide markers to estimate genomic estimated breeding values (GEBVs), has significantly advanced the capacity to improve complex traits, such as yield. This approach circumvents the need for extensive phenotyping and simplifies trait prediction in the early breeding stages (Meuwissen et al. 2001; Nakaya and Isobe 2012).

Recent advancements in low-, mid-, and high-density genotyping platforms, such as Kompetitive Allele-Specific PCR (KASP™) and genotyping-by-sequencing (GBS), have rendered GS more accessible and cost-effective in resource-limited breeding programmes. A prerequisite for developing SNP genotyping panels is SNP discovery/identification, which traditionally requires separate processes for marker development and genotyping. To address this limitation, genotyping-by-sequencing (GBS) and its modified forms have been optimised to enable simultaneous SNP discovery and genotyping in a single step. Thereby saving both time and costs associated with genetic analysis. This method is particularly advantageous for plant breeders and researchers, as it allows for rapid and reproducible detection of genetic loci without the ascertainment bias typically found in traditional SNP selection methods (Elshire et al. 2011; He et al. 2014). For SNP validation, the most frequently used high-throughput (HTP) genotyping platforms and assays include Life Technologies' TaqMan assay (Livak et al. 1995) integrated with the OpenArray platform (TaqMan OpenArray Genotyping System), KBiosciences' Competitive Allele-Specific PCR (KASPar) combined with the SNP Line platform (SNP Line XL; http://www.kbioscience.co.uk), and Illumina's BeadArray technology-based assays, such as GoldenGate (GG) (Fan et al. 2003) and Infinium (Steemers and Gunderson 2007). Recent studies have highlighted the advantages of these platforms. For instance, KASP markers have shown high assay design flexibility and success rates compared to TaqMan assays, making them suitable for various applications in crop improvement (Ayalew et al. 2019). Modern genotyping assays and platforms differ in their chemistry, costs, sample throughput, and the number of validated SNPs. The selection of an appropriate chemistry and genotyping platform is influenced by several factors, including the length of the SNP context sequence, the total number of SNPs to be genotyped, and the available research funding. These panels have been successfully developed and implemented across multiple crops, including rice, wheat, chickpea, and pigeon pea, for applications such as genetic mapping, diversity analysis, and marker-assisted selection (Roorkiwal et al. 2020) (Table 6.1). Despite the remarkable potential of genomics-assisted breeding, public sector breeding programmes face substantial barriers to adopting these technologies, limited funding, technical expertise, and access to cutting-edge platforms impede their ability to achieve efficiency comparable to that of private sector programmes. These challenges highlight the importance of collaborative efforts, such as shared genotyping platforms and public–private partnerships, to democratise access to advanced breeding tools and promote their integration into public breeding pipelines (Bohar et al. 2020).

This chapter explores the evolution and application of forward-breeding genomic resources with a particular focus on low- to mid-density genotyping platforms. These platforms serve as pivotal tools for addressing food security challenges by accelerating genetic gains, enhancing breeding precision, and fostering sustainable agricultural practices. By synthesising the current advancements and their applications, this chapter emphasises the essential role of innovative genotyping technologies in transforming global agricultural research and development.

6.2 Overview of SNP Marker-Based Platforms

6.2.1 Importance of SNP Markers in Genetic Research and Breeding

Rapid advancements in molecular biology and plant genomics have catalysed the development of diverse single-nucleotide polymorphism (SNP) marker-based platforms tailored to meet the varied requirements of genetic research and breeding programmes. SNP markers represent the most abundant form of genetic variation and have transformed plant breeding by providing cost-effective and high-resolution tools for trait mapping, marker-assisted selection (MAS), and genomic selection (GS). These platforms can be classified into three main categories based on their marker density and intended use: high-density SNP genotyping platforms (HDSG), medium-density SNP genotyping platforms (MDSG), and low-density SNP genotyping platforms (LDSG), each serving a distinct purpose in modern crop improvement initiatives.

6.2.2 Classification of SNP Genotyping Platforms

6.2.2.1 High-Density SNP Genotyping Platforms (HDSG)

High-density SNP genotyping platforms are essential for applications that require extensive genome-wide data, such as high-resolution fine mapping of quantitative

Table 6.1 List of cereals & legumes with well-assembled and quality genome sequences

Sl. No.	Crop	Genome size (Mb)	No. of genes	Genome coverage	Sequencing technology	References
1	Rice	~389 Mb	37,544 non-transposable-element-related protein-coding genes	~95%	Hierarchical clone-by-clone sequencing method. This approach involved using bacterial and P1 artificial chromosome clones (BACs and PACs) for high-density genetic mapping and sequencing.	International Rice Genome Sequencing Project (2005); Hamilton et al. (2025)
2	Wheat	~14,000 Mb	107,891 high-confidence protein-coding genes	~94%	de novo whole-genome assembly (WGA). This assembly utilised Illumina short-read sequencing processed with the NRGene DeNovoMagic2 platform.	International Wheat Genome Sequencing Consortium (IWGSC), *Science* (2018)
3	Maize	~2300 Mb	~32,540 protein-coding genes	~95%	The sequencing technology used for the maize genome involved a minimum tiling path approach using bacterial artificial chromosomes (BACs) and fosmid clones. These were sequenced with shotgun sequencing to a coverage of 4–6x and then assembled into the B73 reference genome.	Schnable et al., B73 Maize Genome Project, *Science* (2009)
4	Soybean	~1115 Mb	~46,430 protein-coding genes	~85%	Whole-Genome Shotgun (WGS) sequencing using Sanger sequencing	Schmutz et al., *Nature* (2010)
5	Chickpea	~738 Mb	~28,269	~73.8%	Whole-Genome Shotgun (WGS) sequencing using Illumina sequencing with 11 genomic libraries and SOAPdenovo for assembly	Varshney et al., *Nature Biotechnology* (2013)
6	Common Bean	~587 Mb	~27,197	~98%	Whole-genome shotgun sequencing strategy combining Roche 454 (18.6× coverage), Illumina sequencing (24.1 Gb of reads), Sanger sequencing (fosmid and BAC libraries)	Schmutz et al., *Nature Genetics* (2014)

trait loci (QTLs), genome-wide association studies (GWAS), and advanced functional genomics. These platforms are particularly effective for precision breeding, enabling breeders to analyse genetic diversity, uncover trait loci, and implement GS with enhanced accuracy. Despite their transformative potential, the adoption of high-density platforms in breeding programmes is often constrained by cost and infrastructure requirements (McCouch et al. 2016).

Prominent technologies in this category include microarray-based systems, such as the Illumina Infinium and Affymetrix arrays, which enable the simultaneous genotyping of thousands of SNPs across large populations. These platforms deliver an unparalleled resolution, allowing for comprehensive genomic analysis. However, their cost-intensive nature limits their accessibility, particularly in resource-constrained public sector breeding programmes. Enhancing data management capabilities and reducing costs are critical for maximising the utility of high-density SNP platforms in breeding pipelines. High-density SNP genotyping platforms typically include an estimated range of *10,000 to over one million SNPs* (Wang et al. 2015).

6.2.2.2 Medium-Density SNP Genotyping Platforms (MDSG)

Medium-density SNP platforms provide an optimal balance between marker density and affordability, making them suitable for a wide range of breeding applications, including GS, MAS, diversity studies, and background recovery in marker-assisted backcrossing (MABC). These platforms are particularly advantageous in environments with budget constraints, as they deliver sufficient genomic information for decision-making while remaining cost-effective (Mid-density Genotyping Services n.d.; Semalaiyappan et al. 2023).

Medium-density SNP genotyping platforms are designed to provide an estimated range of approximately 800 to 10,000 SNPs. A notable example of such a platform is the 1k RiCA panel, developed for rice improvement, which exemplifies the potential of medium-density genotyping systems. It features approximately 800 genome-wide markers (Arbelaez et al. 2019a) along with trait-specific and quality-control markers. Such platforms are designed to optimise genotyping accuracy and resolution for specific crops, thereby facilitating the efficient integration of genomic predictions into breeding programmes.

6.2.2.3 Low-Density SNP Genotyping Platforms (LDSG)

Low-density SNP genotyping platforms cater to targeted breeding applications such as marker-assisted selection (MAS), marker-assisted backcrossing (MABC), and quality control (QC) analysis. These platforms are characterised by their affordability and ease of use, enabling widespread adoption across public and private breeding programmes. Kompetitive Allele-Specific PCR (KASP) is a widely utilised low-density platform, offering high specificity and throughput at a fraction of the cost of high-density systems. Low-density SNP genotyping platforms generally consist of 10–500 SNPs, making them suitable for targeted applications in breeding programmes, particularly effective for forward breeding applications, including the introgression of specific traits and validation of genetic material precision in trait identification and validation (Rasheed et al. 2017). This cost-effective approach allows breeders to focus on significant traits while maintaining efficiency in their programmes (Dipta et al. 2024; Steele et al. 2024).

The simplicity and efficiency of low-density platforms make them highly suitable for programmes aimed at introgressing specific traits into breeding lines while maintaining stringent QC. Moreover, their rapid turnaround times and reduced costs allow breeding programmes to focus on resources in other critical areas, such as phenotyping and field trials.

6.2.3 Applications and Impacts of SNP Marker-Based Platforms in Plant Breeding

The choice of an SNP marker platform significantly influences the efficiency and success of breeding programmes. High-density platforms provide extensive genomic coverage, enabling in-depth analysis of complex traits. Medium-density platforms strike a balance between thoroughness and practicality, offering sufficient marker resolution for genomic predictions while remaining cost-effective. Low-density platforms, such as KASP, excel in targeted applications, thereby enhancing the efficiency of trait introgression and validation.

Collectively, these platforms enable breeding programmes to address key challenges in crop improvement, including enhancing yield, improving resilience to biotic and abiotic stresses, and ensuring adaptability to diverse environmental conditions (Table 6.2).

6.3 Low-Density Genotyping Using Kompetitive Allele-Specific PCR (KASP)

Kompetitive allele-specific PCR (KASP) technology is a widely utilised genotyping technique that utilises allele-specific primers to amplify target DNA sequences. This high-throughput and cost-effective method is based on fluorescence resonance energy transfer (FRET) and allele-specific oligonucleotide extension. KASP technology facilitates precise bi-allelic genotyping, making it a valuable tool in modern breeding programmes across a variety of crops (Steele et al. 2018; He et al. 2014; Dipta et al. 2024; PCR-based Genotyping with KASP | LGC Biosearch Technologies n.d.).

Table 6.2 List of successfully developed and implemented SNP panels

Sl. No	Crop	SNP panel	Details	Reference
1	Rice	1K-Rice Custom Amplicon (1k-RiCA)	Medium-density panel (~1000 SNPs) developed for high-resolution genotyping and diversity analysis.	Arbelaez et al. (2019a)
2	Rice	Rice 6K SNP Chip	Designed for genotyping using Illumina's Infinium platform for genome-wide diversity analysis.	Thomson et al. (2017)
3	Wheat	Wheat 90K iSelect SNP Chip	High-density SNP panel with ~90,000 SNPs developed for breeding and genome-wide association studies.	Wang et al. (2014)
4	Wheat	Wheat 3.9K SNP Panel	High-density SNP panel is suitable for genetic diversity studies and genomic selection (GS) applications.	Sehgal et al. (2020)
5	Wheat	Wheat 35K Axiom Array	SNP array developed for targeted genotyping and breeding applications.	Allen et al. (2016), CerealsDB (n.d.)
6	Chickpea	Chickpea 1K SNP Panel	A cost-effective panel (~1000 SNPs) optimised for targeted genotyping and breeding.	Hiremath et al. (2012)
7	Pigeon pea	Pigeon Pea 62K SNP Array	High resolution fingerprinting, QTL mapping and genome wide as well as gene-based association studies in pigeonpea.	Singh et al. (2020)
8	Pigeon pea	Pigeon Pea 2K SNP Panel	Suitable for estimating background genome recovery, selection with traits associated markers in genomics-assisted breeding, genetic diversity and germplasm management applications	Varshney et al. (2017)
9	Maize	Maize 50K SNP Array	High-density panel with ~50,000 SNPs for genome-wide association studies and breeding applications.	Ganal et al. (2011)
10	Maize	Maize 600K Axiom Array	Ultra-high-density array (~600,000 SNPs) for fine mapping, genomic selection, and GWAS.	Unterseer et al. (2014)
11	Maize	Maize mid-density genotyping panel	The mid-density genotyping panel for maize consists of 3305 genome-wide markers. The panel enables genomic selection, QTL mapping and fingerprinting of elite material, QC, estimating background genome recovery in genomics-assisted breeding, genetic diversity, and germplasm management applications.	Bukowski et al. (2018), Chen et al. (2016)
12	Potato	Potato mid-density genotyping panel	Marker panel containing 2147 SNP markers and an average density of 2–3 SNP/cM on all 12 potato chromosomes. Suitable for potato genomic selection and diversity studies.	Hamilton et al. (2011), Vos et al. (2015)

6.3.1 Applications of Low-Density KASP Genotyping

KASP genotyping is characterised by its affordability, simplicity, and flexibility. It enables efficient analysis of up to 200 single nucleotide polymorphism (SNP) markers per sample, making it particularly useful for breeding applications such as:

- Marker-Assisted Selection (MAS): Facilitating the introgression of desired traits into breeding populations.
- Marker-Assisted Backcrossing (MABC): Supporting background recovery and trait integration.
- Quality Control (QC) Analysis: Ensuring the genetic integrity of breeding materials.
- Trait Validation: Confirmation of the presence of target genetic loci.
- Forward Breeding: KASP markers are used to track specific traits, such as disease resistance or abiotic stress tolerance, throughout the breeding pipeline. This has accelerated the development of improved varieties.
- Pre-Breeding and Trait Discovery: KASP markers can help identify novel alleles and traits for crop improvement by analysing genetic diversity within germplasm collections.
- Genetic Purity Testing: KASP facilitates the verification of hybrid purity and parental lines, thereby ensuring the quality of breeding programmes.
- Haplotyping and Line Selection: Using KASP, breeders can define haplotypes associated with desired traits, thereby supporting more informed selection decisions.

Pre-validated KASP markers are widely accepted by the scientific community, bolstering the reliability of results (Rehman et al. 2021; Liu et al. 2012). The technology is not only cost-effective but also has a shorter turnaround time and a lower error rate compared to other genotyping methods. It has proven valuable in genotypic screening and candidate allele identification and has been extensively used for quality control analysis of germplasm to ensure genetic integrity (Semagn et al. 2012). Additionally, KASP has facilitated marker-assisted backcrossing (MABC) for developing rice varieties with superior cooking and eating qualities (Kim et al. 2021). The technology has also been instrumental in quantitative trait loci (QTL) mapping, aiding in identifying candidate genes associated with key agronomic traits. Furthermore, it has been employed in genetic mapping and bulk segregant analysis, contributing to a deeper understanding of complex genetic architectures (Ramirez-Gonzalez 2015). Lastly, KASP has enhanced marker-assisted selection (MAS), enabling breeders to select desirable traits efficiently (Leal-Bertioli et al. 2015).

6.3.2 Optimal Marker Density

The optimal marker density for KASP genotyping is recommended for applications that require up to 200 single nucleotide polymorphism (SNP) markers. However, the effectiveness of these markers in predicting phenotypes is contingent upon their quality and relevance to the specific traits of interest. Studies have indicated that a higher density of markers can improve the accuracy of genomic predictions, but the best density often depends on the genetic architecture of the traits being studied (Norman et al. 2018; Wang et al. 2024).

6.3.3 Availability of KASP Markers

A comprehensive collection of validated KASP markers can be accessed through the Excellence in Breeding (EiB) website (Genotyping/Sequencing Tools and Services n.d.). Various organisations, including CGIAR, NARES, and advanced research institutes, have validated these markers. It should be noted that if the genetic background of the population differs from that of the original design population, validation of the KASP markers in their material is advisable. Although these markers are publicly available and free to use, proper attribution is encouraged.

6.3.4 Integration of KASP within Breeding Programmes

Integrating KASP genotyping into breeding pipelines has transformed how breeders approach trait selection and variety development. Key benefits include cost-effectiveness; the typical turnaround time for KASP genotyping is approximately 10 business days from the receipt of samples by the laboratory, making it a relatively quick option for researchers. The cost per sample generally ranges from $1 to $5 with 10+ markers, depending on the specific requirements and scale of the project, which is significantly cheaper than in-house solutions through the Excellence in Breeding (EiB) website (Genotyping/Sequencing Tools and Services n.d.).

KASP's high throughput supports the simultaneous genotyping of large populations, accelerating the identification of superior lines. Furthermore, combining KASP with phenotypic and environmental data enhances the accuracy of genomic predictions, ensuring that the selected lines are better adapted to specific growing conditions (Fig. 6.1).

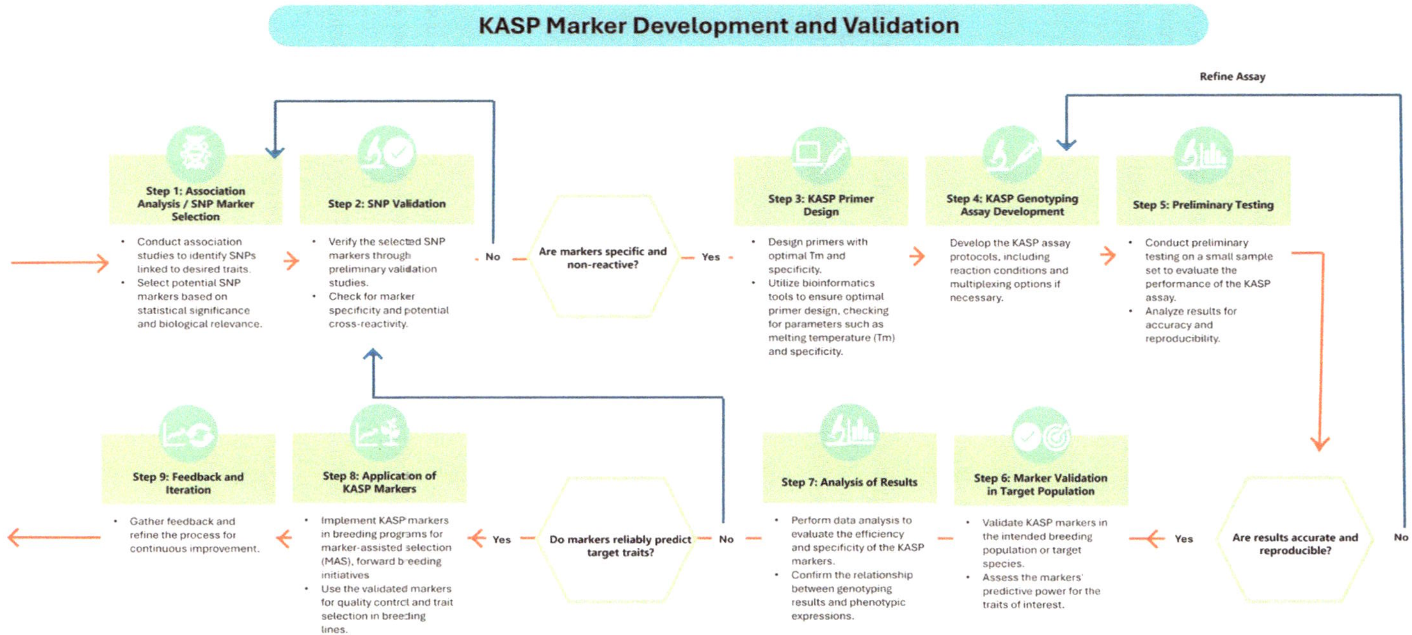

Fig. 6.1 KASP marker development and validation flow diagram. This flow diagram illustrates the stepwise process involved in developing and validating Kompetitive Allele-Specific PCR (KASP) markers for genetic studies and breeding applications. The process begins with SNP marker selection and validation, followed by primer design and assay development. Markers undergo preliminary testing to assess their accuracy and reproducibility. Validated markers are then tested in the target population and analysed for predictive accuracy. The final steps involve application in breeding programmes and continuous refinement based on feedback and iteration. The structured approach ensures that only reliable and effective KASP markers are integrated into molecular breeding programmes

6.4 Mid-Density Genotyping

The 1k RiCA mid-density genotyping panel for rice is a highly efficient and cost-effective tool designed to support the genomic selection, variety typing, QTL mapping, and fingerprinting of elite rice breeding materials. This panel was developed at the International Rice Research Institute (IRRI-ISL—External—RICA n.d.) to address the growing need for affordable mid-density genotyping solutions in rice breeding programmes. Currently offered by the Agriplex and DArT platforms, the 1k RiCA panel utilises custom amplicon workflows to reliably deliver nearly all markers across all samples, ensuring robust and consistent genotyping results (Fig. 6.2).

6.4.1 Applications of 1k RiCA Panel

The 1k RiCA panel includes approximately 800 genome-wide markers, 22 quality control (QC) markers, and up to 200 trait-specific markers that target 87 high-value genes and quantitative trait loci (QTLs). These markers collectively enable key applications, including:

1. Genomic Selection (GS): Estimating genomics-based breeding values, including integrating significant genes as fixed effects.
2. Variety Typing: Identification and differentiation of elite varieties.
3. QTL Mapping and Profiling: This involves identifying loci associated with desirable agronomic traits and fingerprinting breeding materials.

6.4.2 Genome-Wide Coverage

The genome-wide markers in the 1k RiCA panel were initially derived from a set of approximately 4700 SNPs included in the Illumina Infinium panel (Arbelaez et al. 2019a, b). The selection process focused on identifying the most informative markers, further supplemented with insights from the ("The 3,000 Rice Genomes Project" 2014; Wang et al. 2018). The panel primarily targeted the indica 1B group, which encompasses the majority of modern tropical breeding materials. Agriplex (Agriplex Genomics n.d.) and DArT platforms offer panels containing 797 and 504 genome-wide markers, respectively.

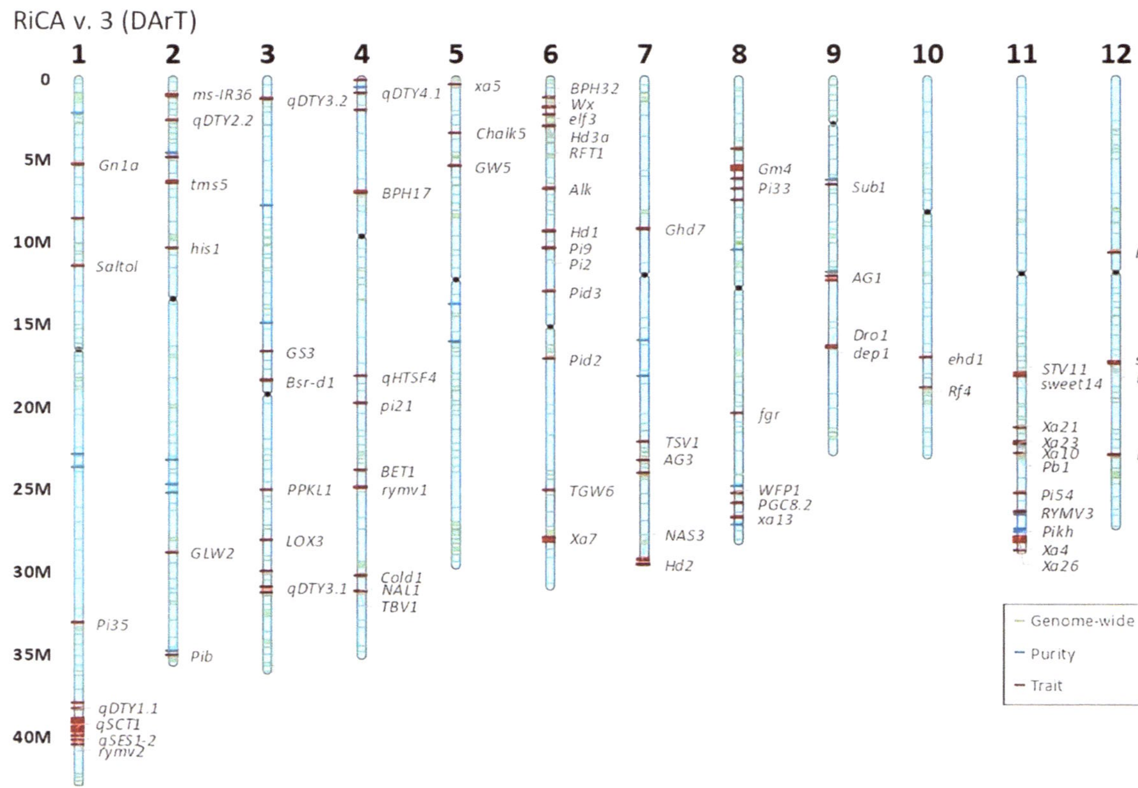

Fig. 6.2 SNP marker position in existing RiCA v.4 panels. This distribution ensures genome-wide coverage, which is critical for accurate estimation of genomic breeding values and effective integration into breeding workflows. In elite indica breeding materials, assessments have shown an average of 375 polymorphic markers in any two randomly chosen indica parents. While japonica material exhibits a lower average of 180 polymorphic markers, modelling studies indicate that this density is sufficient for genomic selection in IRRI breeding programmes. (IRRI-ISL—External—RICA. (n.d.). https://isl.irri.org/services/genotyping/rica)

6.4.3 Purity and Trait SNPs

The 22 QC markers incorporated into the 1k RiCA panel were routinely used for parental purity verification and F1 hybridity testing. These markers provide consistency and continuity across datasets generated on different platforms such as Intertek and Agriplex. Additionally, the panel featured 205 trait-specific markers in version 4 (Agriplex), distinguishing 237 alleles across 87 genes and QTLs. These markers cover traits critical to rice breeding, including abiotic stress tolerance, pest and disease resistance: Genes conferring resistance to blast, bacterial leaf blight (BLB), rice yellow mottle virus (RYMV), and brown planthopper (BPH). Grain Quality: Loci associated with grain yield and quality attributes. Notably, these trait markers are effective across the Oryza sativa genetic background, eliminating the need for prior genetic information about the material being fingerprinted.

6.5 4K HaploPanel

Developing SNP arrays imputed with trait-specific and diverse genomic knowledge has significantly enhanced plant breeding efforts, particularly for rice. While high-density arrays provide extensive data at high cost and low-density arrays lack resolution, the newly developed 4K HaploPanel strikes a balance by offering a mid-density genotyping platform. Designed with 4000+ SNPs and indels derived from 2268 functionally characterised genes, this innovative array leverages information from the 3000 Rice Genome (3K-RG) panel. It provides an improved SNP distribution over previous arrays, enabling the generation of genetic maps with more distinct bins, thus enhancing diversity analysis, trait mapping, haplotype identification, and genomic prediction.

4K HaploPanel serves as a next-generation genotyping platform, setting a benchmark for advanced SNP array technologies by providing high-resolution genetic data for breeding programmes. One of its key contributions is developing a haplotype database that enhances resources for rice breeding, genetic analysis, and functional genomics by cataloguing critical genetic variations. Additionally, the platform offers a cost-effective breeding strategy and significantly reduces genotyping costs while improving efficiency, making it accessible to resource-limited breeding programmes. Moreover, the 4K HaploPanel enhances breeding accuracy by facilitating genomic selection, trait mapping, and foreground and background selection, ultimately aiding breeders in making informed decisions to accelerate genetic gains. The 4K HaploPanel represents a transformative tool for rice improvement, enabling cost-efficient and precise genotyping for large-scale breeding applications (Unpublished).

6.6 CGIAR Strategic Support for Genotyping Services

The Consultative Group for International Agricultural Research (CGIAR n.d.). Implemented strategic initiatives to enhance global agricultural research and accelerate crop improvement efforts. A central component of this approach is the development of world-class shared genotyping services through key initiatives, including the High Throughput Genotyping Project (HTPG) (http://cegsb.icrisat.org/high-throughput-genotyping-project-htpg/), the Genomics Open-Source Breeding Informatics Initiative (GOBii) (http://gobiiproject.org/), and the Excellence in Breeding (EiB) platform (https://excellenceinbreeding.org/). Additionally, these initiatives aim to provide accessible, high-quality genotyping and sequencing resources, tools, and services of CGIAR through service providers (Intertek AgriTech n.d.) to researchers worldwide, particularly in developing countries, facilitating advancements in crop improvement and food security.

6.6.1 High Throughput Genotyping Project (HTPG) and Genomics Open-Source Breeding Informatics Initiative (GOBii)

The Genomics Open-Source Breeding Informatics Initiative (GOBii), funded by the Bill and Melinda Gates Foundation and launched in 2015, is a large-scale public initiative designed to expedite the breeding of staple crops through the use of high-density genotypic information. By leveraging open-source software and fostering global collaboration among researchers, GOBii aims to bridge the gap between plant breeders and genomic resources, facilitating more precise and data-driven breeding decisions (Genomic and Open-Source Breeding Informatics Initiative (GOBII) n.d.). The High Throughput Genotyping Project (HTPG) was established to provide cost-effective, high-throughput genotyping solutions for CGIAR and National Agricultural Research and Extension Systems (NARES). Led by ICRISAT from 2016 to 2020, the initiative later transitioned into the genotyping/sequencing tools and services module of the Excellence in Breeding (EiB) platform, with Intertek AgriTech serving as the service provider (Bohar et al. 2020; ICRISAT n.d.). This transition underscores the ongoing efforts to enhance breeding efficiency and accelerate genetic gains in diverse agricultural systems, ensuring broader accessibility to advanced genomic tools. These initiatives have improved agricultural productivity and addressed food security challenges in resource-limited regions (Bohar et al. 2020).

6.6.2 *Excellence in Breeding (EiB)*

EiB is a comprehensive platform designed to increase the efficiency of breeding programmes within CGIAR and National Agricultural Research and Extension Systems (NARES). EiB emphasises increasing genetic gain and reducing the age of crop varieties in farmers' fields. Supported by organisations like the Bill and Melinda Gates Foundation, EiB offers tools, advisory services, and capacity-building programmes to strengthen breeding pipelines, particularly in developing regions (Excellence in Breeding—CGIAR 2022). The platform fosters public–private partnerships, facilitates access to advanced technologies, and promotes data-driven decision-making for crop improvement (Table 6.3).

6.7 Navigating Challenges: Access to SNP Genotyping Platforms in Public Sector Breeding Programmes

Despite the availability of diverse single nucleotide polymorphism (SNP) marker-based platforms, public sector breeding programmes face significant challenges in adopting these advanced technologies compared to their private sector counterparts. These challenges include limited access to cutting-edge technologies, reliance on conventional phenotypic selection methods, inadequate technical expertise, insufficient infrastructure, inefficient genotyping protocols, restricted availability of high-quality diagnostic markers, and an inability to achieve the economies of scale required for cost-effective genotyping (Crossa et al. 2017; Varshney et al. 2017). Consequently, public breeding programmes are often unable to fully exploit marker-based platforms for modern breeding approaches, such as marker-assisted selection (MAS), F_1 hybridity testing, quantitative trait loci (QTL) profiling, DNA fingerprinting, and genomic selection. Furthermore, limited financial resources hinder the ability of public institutions to invest in the infrastructure and technologies necessary for efficient crop improvement, thereby slowing the development of climate-resilient and high-yielding crop varieties and posing a significant challenge to global food security.

To address these limitations, coordinated efforts within the scientific community are necessary, particularly in developing countries, which often lack adequate resources and infrastructure. Moreddu (2016) emphasised the importance of fostering public–private partnerships (PPPs) to bridge the gap between the public and private sectors. Such partnerships leverage both sectors' strengths and resources, promoting innovation in plant breeding (Xu et al. 2017; Cobb et al. 2019). Collaborations between public research institutions and private companies can enhance access to cutting-edge technologies, expertise, and funding opportunities while facilitating the transfer of knowledge and best practices. These partnerships have the potential to improve the capacity for joint crop improvement efforts significantly, enabling breeding programmes to respond more effectively to genetic gain.

Table 6.3 Highlighting success stories and publications that incorporate High Throughput Phenotyping and Genotyping (HTPG), Excellence in Breeding Low Density (EiB-LDSG), and Medium-Density SNP Genotyping (EiB-MDSG) platforms across various crops

	Crop	Platform	Application	Success story/Outcome	References
Cereals	Rice	HTPG / 1k-RiCA	Develop, validate, and deploy a single seed sampling strategy for marker-assisted selection of fixed lines in rice	Developed and validated a single-seed-based genotyping method for MAS, achieved 99% genotypic concordance between single-seed and leaf-based genotyping. Increased frequency of favourable alleles and improved cost-effectiveness in parental selection and varietal development	Arbelaez et al. (2019a, b)
	Rice	1k-RiCA	Genetic mapping to detect stringent QTLs using the 1k-RiCA SNP genotyping platform for salt tolerance	Identified 15 QTLs for nine traits related to salt stress across seven chromosomes. Four important genomic loci (qSES1, qSL1, qSUR1, qRL1) were identified on chromosome 1. Accelerated development of salt-tolerant rice varieties resistant to salt stress through marker-enabled breeding	Maniruzzaman et al. (2022)
	Sorghum	LDSG (Low-Density SNP Genotyping) using Kompetitive Allele Specific PCR (KASP)	Marker-Assisted Backcrossing (MABC) for introgression of stay-green (STG) QTLs associated with post-flowering drought tolerance into farmer-preferred sorghum varieties	Successfully introgressed STG 3A and STG 3B QTLs from donor parent B35 into two Tanzanian sorghum varieties (NACO Mtama 1 and Seguifa). Improved farmer-preferred varieties with enhanced drought tolerance	Mwamahonje et al. (2021)
Roots, tubers, and	Potato	LDSG (Low-Density SNP Genotyping) using Kompetitive Allele Specific PCR (KASP)	Development and validation of KASP markers for late blight	Two KASP markers for LB resistance and two for PVY (Ryadg gene) were validated with assay	Kante et al. (2021)

(continued)

Table 6.3 (continued)

	Crop	Platform	Application	Success story/Outcome	References
bananas (RTB)			(LB) and Potato virus Y (PVY) resistance	power between 0.65 and 0.88. The markers were successfully implemented in the International Potato Center (CIP) breeding programme. Improved genetic gain and selection accuracy by reducing mislabeling errors. Allowed faster, cost-effective genotyping compared to previous molecular marker systems	
	Cassava	LDSG (Low-Density SNP Genotyping) using Intertek platform	Identification of new resistance to Cassava Mosaic Disease (CMD). Validation of CMD resistance markers (S12_7926132 and S14_4626854). Marker-Assisted Selection (MAS) for CMD resistance in cassava breeding programmes	Identified eight cassava clones with high CMD resistance from the Vietnam farm collection (VNM142). Validated CMD resistance in two clones (CR63/PER262/TAI9 and KM57/VNM8/Xanh Vinh Phu). Found that the CMD2 markers are effective for segregating populations but less predictive in diverse populations. Suggested new sources of CMD resistance beyond CMD2 for breeding programmes in Southeast Asia. Provided the first report of CMD resistance from the CIAT genebank	Thuy et al. (2021)
	Banana	EiB-LDSG	Marker-assisted breeding for disease resistance and yield improvement	SNP marker linked to Fusarium wilt TR4 resistance validated in Musa germplasm. Developed resistance to pests and diseases, facilitating effective crosses	Garcia Oliveira et al. (2021)

Grain Legumes	Groundnut	HTPG	Developed single seed-based chipping genotyping (SSCGB) for early generation selection (MEGS), Used Kompetitive Allele Specific PCR (KASP) markers for genotyping, Improved marker-assisted selection (MAS) efficiency for traits like high oleic acid content and foliar disease resistance	Successfully implemented seed chipping for DNA extraction and genotyping, reducing reliance on leaf samples. Achieved 95–99% germination rate from chipped seeds, proving viability after chipping. Advanced 3.5 generations per year using a low-cost rapid generation turnover facility, compared to only two generations in field conditions. Reduced genotyping costs and time by 3–4 times compared to traditional leaf-based methods. Applied the method to >3000 groundnut samples in an active breeding programme, improving selection efficiency and genetic gain	Parmar et al. (2021)

The Bill and Melinda Gates Foundation (BMGF) has supported these efforts, particularly in regions facing critical food security and agricultural challenges. By investing in modernisation initiatives for plant breeding and supporting infrastructure development, the BMGF and similar donor organisations have substantially contributed to enhancing the effectiveness and efficiency of international crop improvement systems (https://www.gatesfoundation.org/). Their investments have helped bolster research capabilities, improve access to innovative technologies, and provide training opportunities for researchers and practitioners in developing countries. These initiatives highlight the importance of strengthening public–private collaborations and securing support from major donors to address the technology gap, advance global agricultural research, and improve food security worldwide.

6.8 Overcoming Challenges in Genotyping: Strategies for Optimisation and Efficient Implementation

6.8.1 Haplotype Variation

Haplotype variation can lead to discrepancies in genotyping results due to differences in linkage disequilibrium (LD) across the genome. This variation affects the number of frequent haplotypes (combinations of alleles inherited together) observed in small regions, which can lead to inconsistencies in the selection of SNPs for genotyping. Using haplotype information can optimise SNP selection; researchers can choose a smaller set of SNPs that still capture the relevant genetic variation within a population, making genotyping more efficient while maintaining data quality (Singh et al. 2024).

6.8.2 SNP Marker Origin

The source of single nucleotide polymorphism (SNP) markers, whether derived from existing literature or previous studies, can influence their applicability in genetic research. To minimise potential biases in genotyping results, it is critical to ensure that the selected SNPs accurately reflect the genetic composition of the target population. While commercial genome-wide genotyping platforms are generally effective at capturing common SNP variations, their performance can differ across populations, underscoring the necessity of selecting markers appropriate for the specific study context.

6.8.3 Ploidy Levels

Polyploid genetics introduces complexity in genotyping due to multiple sets of chromosomes, which can lead to challenges in accurately calling genotypes and inferring haplotypes. High-density SNP arrays and advanced clustering algorithms can help manage the complexity of polyploid data by providing accurate genotype calling even in the presence of multiple clusters (Wang et al. 2014).

6.8.4 Primer Design

Inefficient or poorly designed primers can result in amplification failures, which can lead to failed amplification during PCR, resulting in unreliable genotype calls due to non-specific amplification or inability to detect the target SNP variation, making it crucial to select primers that specifically target the desired SNP site and are compatible with the chosen genotyping platform.

6.8.5 Technical Issues

Laboratory-related problems, such as errors in automated SNP scoring, can affect the quality of genotyping results. Implementing comprehensive workflows that include error correction and validation steps can help ensure the accuracy of genotypic data.

6.9 Conclusion

The development and integration of low-cost genotyping platforms represent a significant milestone in modern plant breeding, offering a pathway to enhanced genetic gains. Advances in single nucleotide polymorphism (SNP) marker technologies, including low-, mid-, and high-density genotyping platforms, have transformed breeding programmes by facilitating the rapid identification and selection of desirable traits. These technologies have enabled the efficient deployment of genomic-assisted breeding approaches, reducing reliance on traditional phenotypic selection and accelerating the development of climate-resilient, high-yielding crop varieties. Despite these advancements, challenges remain, particularly for public-sector breeding programmes. Limited financial resources, restricted access to cutting-edge technologies, and the need for technical expertise continue to hinder the widespread adoption of genomic selection methodologies. Addressing these constraints requires concerted efforts through public–private partnerships, the establishment of shared genotyping platforms, and strategic investments in bioinformatics tools for data management and analysis.

The success of initiatives such as the High Throughput Genotyping Project (HTPG), the Genomics Open-Source Breeding Informatics Initiative (GOBii), and the Excellence in Breeding (EiB n.d.) platform underscores the importance of collaborative approaches in democratising access to advanced genotyping technologies. These initiatives have played a pivotal role in strengthening breeding pipelines, particularly in developing regions, by providing cost-effective and scalable genotyping solutions. Looking forward, the continued advancement of genotyping platforms, along with innovations in next-generation sequencing (NGS) and predictive breeding analytics, will further refine selection accuracy and breeding efficiency. Integrating phenotypic, genotypic, and environmental data will enhance genomic prediction models, paving the way for more precise trait selection and accelerated crop improvement. Additionally, investments in capacity-building and infrastructure development will be essential to utilise these technologies across diverse agricultural systems effectively.

Ultimately, the widespread adoption of low-cost genotyping platforms holds the potential to bridge the gap between resource-limited and well-funded breeding programmes. By fostering global collaborations and promoting equitable access to genomic resources, the agricultural community can drive sustainable advancements in plant breeding, contributing to a more resilient and food-secure future.

References

Agriplex Genomics (n.d.) IRRI RiCA V4 Panel. Retrieved from https://www.agriplexgenomics.com/irri-v4

Allen AM, Winfield MO, Burridge AJ, et al (2016) Characterization of a Wheat Breeders' Array suitable for high-throughput SNP genotyping of global accessions of hexaploid bread wheat (Triticum aestivum). Plant Biotechnol J 15:390–401. https://doi.org/10.1111/pbi.12635

Arbelaez JD, Dwiyanti MS, Tandayu E, Llantada K, Jarana A, Ignacio JC, Platten JD, Cobb J, Rutkoski JE, Thomson MJ, Kretzschmar T (2019a) 1k-RiCA (1K-Rice Custom Amplicon) a novel genotyping amplicon-based SNP assay for genetics and breeding applications in rice. Rice 12(1). https://doi.org/10.1186/s12284-019-0311-0

Arbelaez JD, Tandayu E, Reveche MY, Jarana A, van Rogen P, Sandager L et al (2019b) Methodology: ssb-MASS: a single seed-based sampling strategy for marker-assisted selection in rice. Plant Methods 15(1):78–11. https://doi.org/10.1186/s13007-019-0464-2

Ayalew H, Tsang PW, Chu C, Wang J, Liu S, Chen C, Ma X (2019) Comparison of TaqMan, KASP and rhAmp SNP genotyping platforms in hexaploid wheat. PLoS One 14(5):e0217222. https://doi.org/10.1371/journal.pone.0217222

Bohar R, Chitkineni A, Varshney RK (2020) Genetic molecular markers to accelerate genetic gains in crops. BioTechniques 69(3):158–160. https://doi.org/10.2144/btn-2020-0066

Bukowski R, Guo X, Lu Y, Zou C, He B, Rong Z, Wang B, Xu D, Yang B, Xie C, Fan L, Gao S, Xu X, Zhang G, Li Y, Jiao Y, Doebley JF, Ross-Ibarra J, Lorant A et al (2018) Construction of the third-generation Zea mays haplotype map. Gigascience 7(4). https://doi.org/10.1093/gigascience/gix134

CerealsDB (n.d.). https://www.cerealsdb.uk.net/cerealgenomics/CerealsDB/axiom_download.php

CGIAR (n.d.) Genotyping services for global crop improvement. Retrieved from https://www.cgiar.org

Chen J, Zavala C, Ortega N, Petroli C, Franco J, Burgueño J, Costich DE, Hearne SJ (2016) The development of quality Control Genotyping Approaches: a case study using Elite Maize Lines. PLoS One 11(6):e0157236. https://doi.org/10.1371/journal.pone.0157236

Cobb JN, Biswas PS, Platten JD (2019) Back to the future: revisiting MAS as a tool for modern plant breeding. Theor Appl Genet 132(3):647–667. https://doi.org/10.1007/s00122-018-3205-5

Crossa J, Pérez-Rodríguez P, Cuevas J, Montesinos-López O, Jarquín D, De Los Campos G, Burgueño J, González-Camacho JM, Pérez-Elizalde S, Beyene Y, Dreisigacker S, Singh R, Zhang X, Gowda M, Roorkiwal M, Rutkoski J, Varshney RK (2017) Genomic selection in plant breeding: methods, models, and perspectives. Trends Plant Sci 22(11):961–975. https://doi.org/10.1016/j.tplants.2017.08.011

Dipta B, Sood S, Mangal V, Bhardwaj V, Thakur AK, Kumar V, Singh B (2024) KASP: a high-throughput genotyping system and its applications in major crop plants for biotic and abiotic stress tolerance. Mol Biol Rep 51(1). https://doi.org/10.1007/s11033-024-09455-z

Edwards D, Batley J, Snowdon RJ (2012) Accessing complex crop genomes with next-generation sequencing. Theor Appl Genet 126(1):1–11. https://doi.org/10.1007/s00122-012-1964-x

EiB (n.d.) Excellence in breeding platform. Retrieved from https://excellenceinbreeding.org/

Elshire RJ, Glaubitz JC, Sun Q, Poland JA, Kawamoto K, Buckler ES, Mitchell SE (2011) A robust, simple Genotyping-by-Sequencing (GBS) approach for high diversity species. PLoS One 6(5): e19379. https://doi.org/10.1371/journal.pone.0019379

Excellence in Breeding - CGIAR (2022, July 27) CGIAR. https://www.cgiar.org/annual-report/performance-report-2021/excellence-in-breeding/

Fan J, Oliphant A, Shen R, Kermani B, Garcia F, Gunderson K, Hansen M, Steemers F, Butler S, Deloukas P, Galver L, Hunt S, Mcbride C, Bibikova M, Rubano T, Chen J, Wickham E, Doucet D, Chang W et al (2003) Highly parallel SNP genotyping. Cold Spring Harb Symp Quant Biol 68(0):69–78. https://doi.org/10.1101/sqb.2003.68.69

Ganal MW, Durstewitz G, Polley A, Bérard A, Buckler ES, Charcosset A, Clarke JD, Graner E, Hansen M, Joets J, Paslier ML, McMullen MD, Montalent P, Rose M, Schön C, Sun Q, Walter H, Martin OC, Falque M (2011) A large maize (Zea mays L.) SNP genotyping array: development and germplasm genotyping, and genetic mapping to compare with the B73 reference genome. PLoS One 6(12):e28334. https://doi.org/10.1371/journal.pone.0028334

Garcia Oliveira AL, Ayele L, Uwimana B, Storr S, Tadesse T, Ng EH, Swennen R, Bohar R, Debaene J, Quinn M (2021) Access to genetic analysis accelerates banana breeding in East Africa. Retrieved from https://excellenceinbreeding.org/news/access-genetic-analysis-acceleratesbanana-breeding-east-africa

Genomic and Open-Source Breeding Informatics Initiative (GOBII) (n.d.) (2019, November 21) CIMMYT. https://www.cimmyt.org/funder_partner/cornell-university/genomic-and-open-source-breeding-informatics-initiative-gobii/

Genotyping / sequencing tools and services (n.d.) Excellence in breeding. https://excellenceinbreeding.org/module3

Hamilton JP, Hansey CN, Whitty BR, Stoffel K, Massa AN, Van Deynze A, De Jong WS, Douches DS, Buell CR (2011) Single nucleotide polymorphism discovery in elite North American potato germplasm. BMC Genomics 12(1). https://doi.org/10.1186/1471-2164-12-302

Hamilton JP, Li C, Buell CR (2025) The rice genome annotation project: an updated database for mining the rice genome. Nucleic Acids Res 53(D1):D1614–D1622. https://doi.org/10.1093/nar/gkae1061

He C, Holme J, Anthony J (2014) SNP genotyping: the KASP assay. In: Methods in molecular biology, pp 75–86. https://doi.org/10.1007/978-1-4939-0446-4_7

Hiremath PJ, Kumar A, Penmetsa RV, Farmer A, Schlueter JA, Chamarthi SK, Whaley AM, Carrasquilla-Garcia N, Gaur PM, Upadhyaya HD, Kishor PBK, Shah TM, Cook DR, Varshney RK (2012) Large-scale development of cost-effective SNP marker assays for diversity assessment and genetic mapping in chickpea and comparative mapping in legumes. Plant Biotechnol J 10(6):716–732. https://doi.org/10.1111/j.1467-7652.2012.00710.x

ICRISAT (n.d.) High-throughput genotyping project (HTPG). Retrieved from http://cegsb.icrisat.org/high-throughput-genotyping-project-htpg/

International Rice Genome Sequencing Project., Sasaki, T (2005) The map-based sequence of the rice genome. Nature 436(7052):793–800. https://doi.org/10.1038/nature03895

International Wheat Genome Sequencing Consortium (IWGSC), Appels R, Eversole K, Stein N, Feuillet C, Keller B, Rogers J, Pozniak CJ, Choulet F, Distelfeld A, Poland J, Ronen G, Sharpe AG, Barad O, Baruch K, Keeble-Gagnère G, Mascher M, Ben-Zvi G, Josselin A, Himmelbach A et al (2018) Shifting the limits in wheat research and breeding using a fully annotated reference genome. Science 361(6403). https://doi.org/10.1126/science.aar7191

Intertek AgriTech (n.d.). http://www.intertek.com/agriculture/agritech/

IRRI-ISL - External - RICA (n.d.). https://isl.irri.org/services/genotyping/rica

Kante M, Lindqvist-Kreuze H, Portal L, David M, Gastelo M (2021) Kompetitive Allele Specific PCR (KASP) markers for potato: an effective tool for increased genetic gains. Agronomy 11(11):2315. https://doi.org/10.3390/agronomy11112315

Khush GS (2001) Green revolution: the way forward. Nat Rev Genet 2(10):815–822. https://doi.org/10.1038/35093585

Kim M, Yang J, Yu J, Lee Y, Park Y, Kang K, Cho Y (2021) Breeding of high cooking and eating quality in rice by Marker-Assisted Backcrossing (MABC) using KASP markers. Plants 10(4): 804. https://doi.org/10.3390/plants10040804

Lamichhane S, Thapa S (2022) Advances from conventional to modern plant breeding methodologies. Plant Breed Biotechnol 10(1):1–14. https://doi.org/10.9787/pbb.2022.10.1.1

Leal-Bertioli SCM, Cavalcante U, Gouvea EG, Ballén-Taborda C, Shirasawa K, Guimarães PM, Jackson SA, Bertioli DJ, Moretzsohn MC (2015) Identification of QTLs for rust resistance in the peanut wild species Arachis magna and the development of KASP markers for marker-assisted selection. G3 (Bethesda) 5(7):1403–1413. https://doi.org/10.1534/g3.115.018796

Liu J, Huang S, Sun M, Liu S, Liu Y, Wang W, Zhang X, Wang H, Hua W (2012) An improved allele-specific PCR primer design method for SNP marker analysis and its application. Plant Methods 8(1). https://doi.org/10.1186/1746-4811-8-34

Livak KJ, Flood SJ, Marmaro J, Giusti W, Deetz K (1995) Oligonucleotides with fluorescent dyes at opposite ends provide a quenched probe system useful for detecting PCR product and nucleic acid hybridization. Genome Res 4(6):357–362. https://doi.org/10.1101/gr.4.6.357

Maniruzzaman S, Rahman MA, Hasan M, Rasul MG, Molla AH, Khatun H, Akter S (2022) Genetic mapping to detect stringent QTLs using 1k-RiCA SNP genotyping platform from the new landrace associated with salt tolerance at the seedling stage in rice. Plants (Basel) 11(11):1409. https://doi.org/10.3390/plants11111409

McCouch SR, Wright MH, Tung C, Maron LG, McNally KL, Fitzgerald M, Singh N, DeClerck G, Agosto-Perez F, Korniliev P, Greenberg AJ, Naredo MEB, Mercado SMQ, Harrington SE, Shi Y, Branchini DA, Kuser-Falcão PR, Leung H, Ebana K et al (2016) Open access resources for genome-wide association mapping in rice. Nat Commun 7(1). https://doi.org/10.1038/ncomms10532

Meuwissen THE, Hayes BJ, Goddard ME (2001) Prediction of total genetic value using Genome-Wide Dense Marker maps. Genetics 157(4):1819–1829. https://doi.org/10.1093/genetics/157.4.1819

Mid-density genotyping services (n.d.) Excellence in breeding. https://excellenceinbreeding.org/toolbox-collection/mid-density-genotyping-services

Moreddu C (2016) Public-Private partnerships for agricultural innovation. OECD Food, Agriculture and Fisheries Working Papers. https://doi.org/10.1787/5jm55j9p9rmx-en

Mwamahonje A, Eleblu JSY, Ofori K, Feyissa T, Deshpande S, Garcia-Oliveira AL et al (2021) Introgression of QTLs for drought tolerance into farmers' preferred sorghum varieties. Agriculture 11(9):883. https://doi.org/10.3390/agriculture11090883

Nakaya A, Isobe SN (2012) Will genomic selection be a practical method for plant breeding? Ann Bot 110(6):1303–1316. https://doi.org/10.1093/aob/mcs109

Norman A, Taylor J, Edwards J, Kuchel H (2018) Optimising genomic selection in wheat: effect of marker density, population size and population structure on prediction accuracy. G3 (Bethesda) 8(9):2889–2899. https://doi.org/10.1534/g3.118.200311

Parmar S, Deshmukh DB, Kumar R, Manohar SS, Joshi P, Sharma V, Chaudhari S, Variath MT, Gangurde SS, Bohar R, Singam P, Varshney RK, Janila P, Pandey MK (2021) Single Seed-Based High-Throughput genotyping and rapid generation advancement for accelerated groundnut genetics and breeding research. Agronomy 11(6):1226. https://doi.org/10.3390/agronomy11061226

PCR-based Genotyping with KASP | LGC Biosearch Technologies (n.d.) LGC biosearch technologies. https://www.biosearchtech.com/support/education/kasp-genotyping-reagents/kasp-overview

Ramirez-Gonzalez RH (2015) RNA-SEQ bulked segregant analysis enables the identification of High-Resolution genetic markers for breeding in hexaploid wheat. Plant and Animal Genome XXIII Conference. https://pag.confex.com/pag/xxiii/webprogram/Paper14374.html

Rasheed A, Hao Y, Xia X, Khan A, Xu Y, Varshney RK, He Z (2017) Crop breeding chips and genotyping platforms: progress, challenges, and perspectives. Mol Plant 10(8):1047–1064. https://doi.org/10.1016/j.molp.2017.06.008

Rehman SU, Sher MA, Saddique M a B, Ali Z, Khan MA, Mao X, Irshad A, Sajjad M, Ikram RM, Naeem M, Jing R (2021) Development and exploitation of KASP assays for genes underpinning drought tolerance among wheat cultivars from Pakistan. Front Genet 12. https://doi.org/10.3389/fgene.2021.684702

Roorkiwal M, Bharadwaj C, Barmukh R, Dixit GP, Thudi M, Gaur PM, Chaturvedi SK, Fikre A, Hamwieh A, Kumar S, Sachdeva S, Ojiewo CO, Tar'an B, Wordofa NG, Singh NP, Siddique KHM, Varshney RK (2020) Integrating genomics for chickpea improvement: achievements and opportunities. Theor Appl Genet 133(5):1703–1720. https://doi.org/10.1007/s00122-020-03584-2

Schmutz J, Cannon SB, Schlueter J, Ma J, Mitros T, Nelson W, Hyten DL, Song Q, Thelen JJ, Cheng J, Xu D, Hellsten U, May GD, Yu Y, Sakurai T, Umezawa T, Bhattacharyya MK, Sandhu D, Valliyodan B et al (2010) Genome sequence of the palaeopolyploid soybean. Nature 463(7278):178–183. https://doi.org/10.1038/nature08670

Schmutz J, McClean PE, Mamidi S, Wu GA, Cannon SB, Grimwood J, Jenkins J, Shu S, Song Q, Chavarro C, Torres-Torres M, Geffroy V, Moghaddam SM, Gao D, Abernathy B, Barry K, Blair M, Brick MA, Chovatia M et al (2014) A reference genome for common bean and genome-wide analysis of dual domestications. Nat Genet 46(7):707–713. https://doi.org/10.1038/ng.3008

Schnable PS, Ware D, Fulton RS, Stein JC, Wei F, Pasternak S, Liang C, Zhang J, Fulton L, Graves TA, Minx P, Reily AD, Courtney L, Kruchowski SS, Tomlinson C, Strong C, Delehaunty K, Fronick C, Courtney B et al (2009) The B73 maize genome: complexity, diversity, and dynamics. Science 326(5956):1112–1115. https://doi.org/10.1126/science.1178534

Sehgal D, Mondal S, Crespo-Herrera L, Velu G, Juliana P, Huerta-Espino J, Shrestha S, Poland J, Singh R, Dreisigacker S (2020) Haplotype-Based, Genome-Wide Association study reveals stable genomic regions for grain yield in CIMMYT spring bread wheat. Front Genet 11. https://doi.org/10.3389/fgene.2020.589490

Semagn K, Beyene Y, Makumbi D, Mugo S, Prasanna BM, Magorokosho C, Atlin G (2012) Quality control genotyping for assessment of genetic identity and purity in diverse tropical maize inbred lines. Theor Appl Genet 125(7):1487–1501. https://doi.org/10.1007/s00122-012-1928-1

Semalaiyappan J, Selvanayagam S, Rathore A, Gupta S, Chakraborty A, Gujjula KR, Haktan S, Viswanath A, Malipatil R, Shah P, Govindaraj M, Ignacio JC, Reddy S, Singh AK, Thirunavukkarasu N (2023) Development of a new AgriSeq 4K mid-density SNP genotyping panel and its utility in pearl millet breeding. Front Plant Sci 13. https://doi.org/10.3389/fpls.2022.1068883

Singh S, Mahato AK, Jayaswal PK, Singh N, Dheer M, Goel P, Raje RS, Yasin JK, Sreevathsa R, Rai V, Gaikwad K, Singh NK (2020) A 62K genic-SNP chip array for genetic studies and breeding applications in pigeonpea (Cajanus cajan L. Millsp.). Sci Rep 10(1). https://doi.org/10.1038/s41598-020-61889-0

Singh P, Sundaram KT, Vinukonda VP, Venkateshwarlu C, Paul PJ, Pahi B, Gurjar A, Singh UM, Kalia S, Kumar A, Singh VK, Sinha P (2024) Superior haplotypes of key drought-responsive genes reveal opportunities for the development of climate-resilient rice varieties. Commun Biol 7(1). https://doi.org/10.1038/s42003-024-05769-7

Steele KA, Quinton-Tulloch MJ, Amgai RB, Dhakal R, Khatiwada SP, Vyas D, Heine M, Witcombe JR (2018) Accelerating public sector rice breeding with high-density KASP markers derived from whole genome sequencing of indica rice. Mol Breed 38(4). https://doi.org/10.1007/s11032-018-0777-2

Steele K, Quinton-Tulloch M, Vyas D, Witcombe J (2024) Thousands of trait-specific KASP markers designed for diverse breeding applications in rice (Oryza sativa). G3 (Bethesda). https://doi.org/10.1093/g3journal/jkae251

Steemers FJ, Gunderson KL (2007) Whole genome genotyping technologies on the BeadArray™ platform. Biotechnol J 2(1):41–49. https://doi.org/10.1002/biot.200600213

The 3,000 rice genomes project (2014) Gigascience 3(1). https://doi.org/10.1186/2047-217x-3-7

Thomson MJ, Singh N, Dwiyanti MS, Wang DR, Wright MH, Perez FA, DeClerck G, Chin JH, Malitic-Layaoen GA, Juanillas VM, Dilla-Ermita CJ, Mauleon R, Kretzschmar T, McCouch SR (2017) Large-scale deployment of a rice 6 K SNP array for genetics and breeding applications. Rice 10(1). https://doi.org/10.1186/s12284-017-0181-2

Thuy CTL, Lopez-Lavalle LAB, Vu NA, Hy NH, Nhan PT, Ceballos H (2021) Identifying resistance to CMD and validating CMD2 markers. Agriculture 11(9):829. https://doi.org/10.3390/agriculture11090829

Unterseer S, Bauer E, Haberer G, Seidel M, Knaak C, Ouzunova M, Meitinger T, Strom TM, Fries R, Pausch H, Bertani C, Davassi A, Mayer KF, Schön C (2014) A powerful tool for genome analysis in maize: development and evaluation of the high density 600 k SNP genotyping array. BMC Genomics 15(1). https://doi.org/10.1186/1471-2164-15-823

Varshney RK, Song C, Saxena RK, Azam S, Yu S, Sharpe AG, Cannon S, Baek J, Rosen BD, Tar'an B, Millan T, Zhang X, Ramsay LD, Iwata A, Wang Y, Nelson W, Farmer AD, Gaur PM, Soderlund C et al (2013) Draft genome sequence of chickpea (Cicer arietinum) provides a resource for trait improvement. Nat Biotechnol 31(3):240–246. https://doi.org/10.1038/nbt.2491

Varshney RK, Saxena RK, Upadhyaya HD, Khan AW, Yu Y, Kim C, Rathore A, Kim D, Kim J, An S, Kumar V, Anuradha G, Yamini KN, Zhang W, Muniswamy S, Kim J, Penmetsa RV, Von Wettberg E, Datta SK (2017) Whole-genome resequencing of 292 pigeonpea accessions identifies genomic regions associated with domestication and agronomic traits. Nat Genet 49(7):1082–1088. https://doi.org/10.1038/ng.3872

Vos PG, Uitdewilligen JG a ML, Voorrips RE, Visser RGF, Van Eck HJ (2015) Development and analysis of a 20K SNP array for potato (Solanum tuberosum): an insight into the breeding history. Theor Appl Genet 128(12):2387–2401. https://doi.org/10.1007/s00122-015-2593-y

Wang S, Wong D, Forrest K, Allen A, Chao S, Huang BE, Maccaferri M, Salvi S, Milner SG, Cattivelli L, Mastrangelo AM, Whan A, Stephen S, Barker G, Wieseke R, Plieske J, Lillemo M, Mather D, Appels R et al (2014) Characterization of polyploid wheat genomic diversity using a high-density 90 000 single nucleotide polymorphism array. Plant Biotechnol J 12(6):787–796. https://doi.org/10.1111/pbi.12183

Wang X, Yu K, Li H, Peng Q, Chen F, Zhang W, Chen S, Hu M, Zhang J (2015) High-density SNP map construction and QTL identification for the apetalous character in Brassica napus L. Front Plant Sci 6. https://doi.org/10.3389/fpls.2015.01164

Wang W, Mauleon R, Hu Z, Chebotarov D, Tai S, Wu Z, Li M, Zheng T, Fuentes RR, Zhang F, Mansueto L, Copetti D, Sanciangco M, Palis KC, Xu J, Sun C, Fu B, Zhang H, Gao Y, Leung H (2018) Genomic variation in 3,010 diverse accessions of Asian cultivated rice. Nature 557(7703):43–49. https://doi.org/10.1038/s41586-018-0063-9

Wang H, Bai Y, Biligetu B (2024) Effects of SNP marker density and training population size on prediction accuracy in alfalfa (Medicago sativa L.) genomic selection. Plant Genome 17(1). https://doi.org/10.1002/tpg2.20431

Wensheng, Wang Ramil, Mauleon Zhiqiang, Hu Dmytro, Chebotarov Shuaishuai, Tai Zhichao, Wu Min, Li Tianqing, Zheng Roven Rommel, Fuentes Fan, Zhang Locedie, Mansueto Dario, Copetti Millicent, Sanciangco Kevin Christian, Palis Jianlong, Xu Chen, Sun Binying, Fu Hongliang, Zhang Yongming, Gao Xiuqin, Zhao Fei, Shen Xiao, Cui Hong, Yu Zichao, Li Miaolin, Chen Jeffrey, Detras Yongli, Zhou Xinyuan, Zhang Yue, Zhao Dave, Kudrna Chunchao, Wang Rui, Li Ben, Jia Jinyuan, Lu Xianchang, He Zhaotong, Dong Jiabao, Xu Yanhong, Li Miao, Wang Jianxin, Shi Jing, Li Dabing, Zhang Seunghee, Lee Wushu, Hu Alexander, Poliakov Inna, Dubchak Victor Jun, Ulat Frances Nikki, Borja John Robert, Mendoza Jauhar, Ali Jing, Li Qiang, Gao Yongchao, Niu Zhen, Yue Ma. Elizabeth B, Naredo Jayson, Talag Xueqiang, Wang Jinjie, Li Xiaodong, Fang Ye, Yin Jean-Christophe, Glaszmann Jianwei, Zhang Jiayang, Li Ruaraidh Sackville, Hamilton Rod A, Wing Jue, Ruan Gengyun, Zhang Chaochun, Wei Nickolai, Alexandrov Kenneth L, McNally Zhikang, Li Hei, Leung (2018) Genomic variation in 3010 diverse accessions of Asian cultivated rice. Nature 557 (7703):43–49. https://doi.org/10.1038/s41586-018-0063-9

Xu Y (2010) Molecular plant breeding. CABI International, Wallingford, Oxfordshire

Xu Y, Li P, Zou C, Lu Y, Xie C, Zhang X, Prasanna BM, Olsen MS (2017) Enhancing genetic gain in the era of molecular breeding. J Exp Bot 68(11):2641–2666. https://doi.org/10.1093/jxb/erx135

Chapter 7
Advances in Synthetic Genomics (SynGen) for Crop Improvement and Beyond

Ali Raza, Yiran Li, Chunli Guo, Mengkai Zhao, and Zhangli Hu

Abstract Synthetic genomics (SynGen) has emerged as a new game-changing platform for future crop improvement by allowing the design and construction of designer-made plant genomes for sustainable agriculture. This chapter first discussed the conceptual and technological foundations of SynGen for future genome design. We investigated how SynGen is being applied to improve yield, stress tolerance, nutritional profile, metabolic pathway engineering, plant–microbe interactions, and to design minimal or fully synthetic plant genomes. Integration with multi-omics, biofoundries, and gene drives further expands its capability to fast-track future crop design. Nevertheless, widespread application encounters major technical, ethical, and regulatory challenges, along with some bottlenecks and implications of redesigned crops, which demand transparent governance and inclusive public dialogue. To address these issues, we propose that innovations in synthetic biology and public–private collaborations could assist in the design of future climate-smart, high-

A. Raza (✉) · Y. Li · C. Guo · M. Zhao
Guangdong Key Laboratory of Plant Epigenetics, College of Life Sciences and Oceanography, Shenzhen University, Shenzhen, China

Shenzhen Engineering Laboratory for Marine Algal Biotechnology, Guangdong Technology Research Center for Marine Algal Biotechnology, Longhua Innovation Institute for Biotechnology, College of Life Sciences and Oceanography, Shenzhen University, Shenzhen, China
e-mail: alirazamughal143@gmail.com

Z. Hu (✉)
Guangdong Key Laboratory of Plant Epigenetics, College of Life Sciences and Oceanography, Shenzhen University, Shenzhen, China

Shenzhen Engineering Laboratory for Marine Algal Biotechnology, Guangdong Technology Research Center for Marine Algal Biotechnology, Longhua Innovation Institute for Biotechnology, College of Life Sciences and Oceanography, Shenzhen University, Shenzhen, China

Guangdong Provincial Key Laboratory of Functional Substances in Medicinal Resources and Healthcare Products, School of Life Sciences and Food Engineering, Hanshan Normal University, Chaozhou, China
e-mail: huzl@szu.edu.cn

V. Garg et al. (eds.), *DNA of Sustainability*, Sustainability Sciences in Asia and Africa,
https://doi.org/10.1007/978-981-95-5109-5_7

performance crops. In short, SynGen could serve as a game-changing approach that can help achieve global food security in the face of changing climate.

Keywords Artificial intelligence · Design-Build-Test-Learn cycle · Rational genome design · Stress tolerance · Synthetic gene circuits · Synthetic plant genomes

7.1 Introduction

Synthetic biology (SynBio) is a fast-growing field that aims to redesign and construct new biological chunks, designs, and systems for effective purposes. It integrates principles from engineering, biology, and computer science to develop standardized genetic circuits and modular biological systems (Konig et al. 2013; Goold et al. 2018; Pixley et al. 2019). A specialized subfield of SynBio, i.e., synthetic genomics (hereafter named as "SynGen"), focuses on writing and assembling entire genomes from scratch. Notably, it leverages whole-genome synthesis (or even the whole chromosome) and large-scale engineering (rather than single-gene edits) (Konig et al. 2013; Pixley et al. 2019; Zhang et al. 2020). For example, the literature suggests that scientists have chemically synthesized entire viral (Cello et al. 2002; Becker et al. 2008; Noyce et al. 2018), bacterial (Gibson et al. 2008; Fredens et al. 2019) genomes, and the first fully synthetic eukaryotic genome (yeast 2.0) (Richardson et al. 2017). The option to generate designer genomes delivers essentially new openings to probe and adjust biological functions that can have massive economic and societal impacts on different areas, including biomedicine, biomanufacturing, bioenergy, agriculture, bioconservation, and others.

Conventional breeding techniques have played a critical role in improving several agronomic and stress tolerance-related traits across major crops (Ahmar et al. 2020; Bhogireddy et al. 2021; Varshney et al. 2021). However, despite these successes, breeding efforts face limitations such as narrow genetic diversity, especially in elite germplasm pools (Varshney et al. 2021). Furthermore, challenges posed by climate change and emerging stresses demand more rapid and targeted innovations (Bhogireddy et al. 2021; Depuydt et al. 2023; Raza et al. 2025a, b). Though biotechnology has enhanced breeding efficiency by developing advanced cultivars, its potential is often constrained by the complexity of polygenic traits (Varshney et al. 2021). Therefore, advanced system-level tools such as SynBio are now being investigated to overcome these limitations and sustainably enhance crop performance in the climate change era.

Among different tools, SynGen is gaining attention in agriculture as a game-changing approach for designing high-performing, climate-smart crops. Synthetic approaches can drive new traits (e.g., higher genetic gain) in crops at a record scale (Liu and Stewart 2015; Goold et al. 2018; Pixley et al. 2019; Wurtzel et al. 2019; Roell and Zurbriggen 2020; Sargent et al. 2022). By permitting synchronized, large-scale modifications, SynGen can deliver new solutions for global food challenges (e.g., population growth and climate change) that go beyond established genetic engineering methods (Goold et al. 2018; Roell and Zurbriggen 2020). Traditional genetic engineering usually modifies one or a few genes, whereas SynGen involves designing entire chromosomes or genomes (Konig et al. 2013; Chen et al. 2024; Du et al. 2024). Notably, SynBio more broadly includes building new genetic circuits and parts (Goold et al. 2018; Andres et al. 2019; Wurtzel et al. 2019), and the application of SynGen specifically results in writing and assembling DNA on a genome-wide scale. In practice, the two fields overlap, but SynGen emphasizes "digital" genome design (e.g., de novo sequences and recoded genomes) by moving forward from conventional transgenes (Konig et al. 2013; Pixley et al. 2019; Du et al. 2024). During the last few years, a wide range of SynGen-based applications have emerged and are still evolving for global crop improvement, which are organized into different categories (Fig. 7.1).

SynGen is a broad domain and comprises many new directions, and in this chapter, we have provided an overview of key advances in SynGen for crop improvement, proposed practical considerations for its application in agriculture, and explored how it may contribute to the sustainability of food systems in the context of a changing climate.

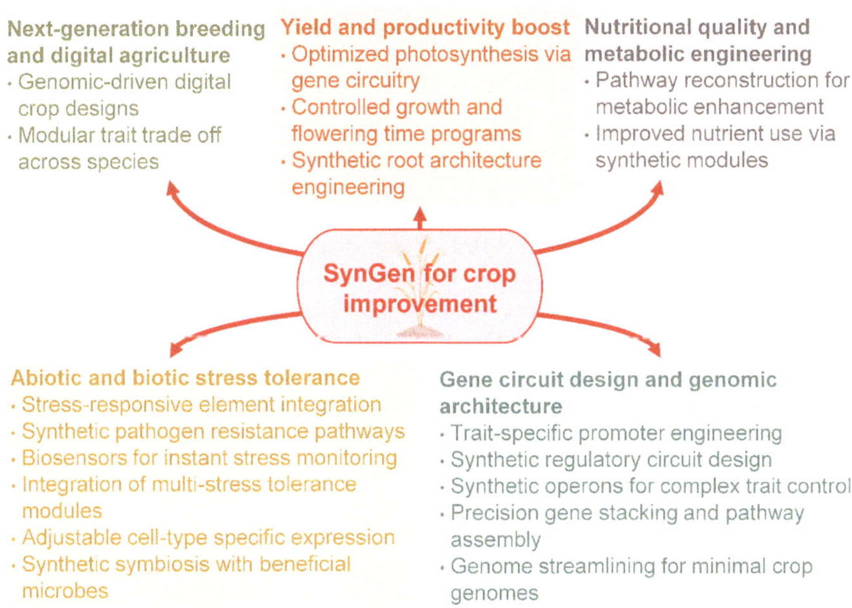

Fig. 7.1 Overview of the current and emerging major applications of synthetic genomics (SynGen) for crop improvement and global agriculture

7.2 Conceptual Fundamentals of Synthetic Genome Design

7.2.1 Rational Genome Design

SynGen builds upon the principles of modularity, minimalism, and rational engineering. Instead of altering individual genes, SynGen helps construct entire genomes/chromosomes based on a deep understanding of essential and non-essential genetic elements. One major milestone has been the design of minimal genomes, which are those stripped of non-essential genes and repetitive elements and that offer a stable and predictable chassis for bioengineering purposes (Hutchison III et al. 2016; Fredens et al. 2019; Xu et al. 2023). This oversimplified approach frees up metabolic capacity for synthetic pathways and guides scientists to study core biological functions in isolation that provide underlying insights into genome organization and evolution (Gibson et al. 2009).

7.2.2 Key Tools and Technologies

Four major powerful tools and technologies have been guiding the rapid development of SynGen. For example, (1) DNA synthesis: modern advances allow the construction of long DNA sequences with high fidelity and falling costs. Nowadays, entire gene clusters and bacterial chromosomes can be synthesized de novo (Gibson et al. 2009; Kosuri and Church 2014; Hutchison III et al. 2016). This capability highlights the power of SynGen for writing designer genomes from scratch. (2) Genome editing and integration: gene editing tools such as CRISPR/Cas9, TALENs, and ZFN provide site-specific insertion and editing capabilities, which enable the integration of synthetic DNA into native genomes (Barrangou and Doudna 2016; Jakhanwal et al. 2021; Yang et al. 2024). These techniques are often combined with DNA assembly methods such as Gibson Assembly or yeast-based homologous recombination, which support large-scale assembly and modular recombination (Gibson et al. 2009; Shao et al. 2009). (3) Computational design platforms: the design of synthetic constructs is progressively supported by in silico tools that optimize codon usage, minimize recombination hotspots, remove unwanted restriction sites, and simulate genetic behaviors (Medema et al. 2012; James et al. 2024). Advanced software now permits multi-objective optimization for complex metabolic pathways and whole-genome architectures (Rialle et al. 2010; Medema et al. 2012; Casini et al. 2014). (4) Chassis organisms: selecting the most-fitting host organism is essential for synthetic genome implementation. Regularly used chassis include *E. coli*, *Saccharomyces cerevisiae*, and *Mycoplasma mycoides* due to their well-characterized genetics and ease of manipulation (Fredens et al. 2019). For plant applications, chassis such as tobacco, Arabidopsis, and rice are being adapted for genome-scale adjustments (Wurtzel et al. 2019). Additionally, synthetic minimal chassis offer reduced background noise and improved control over engineered functions (Breuer et al. 2019).

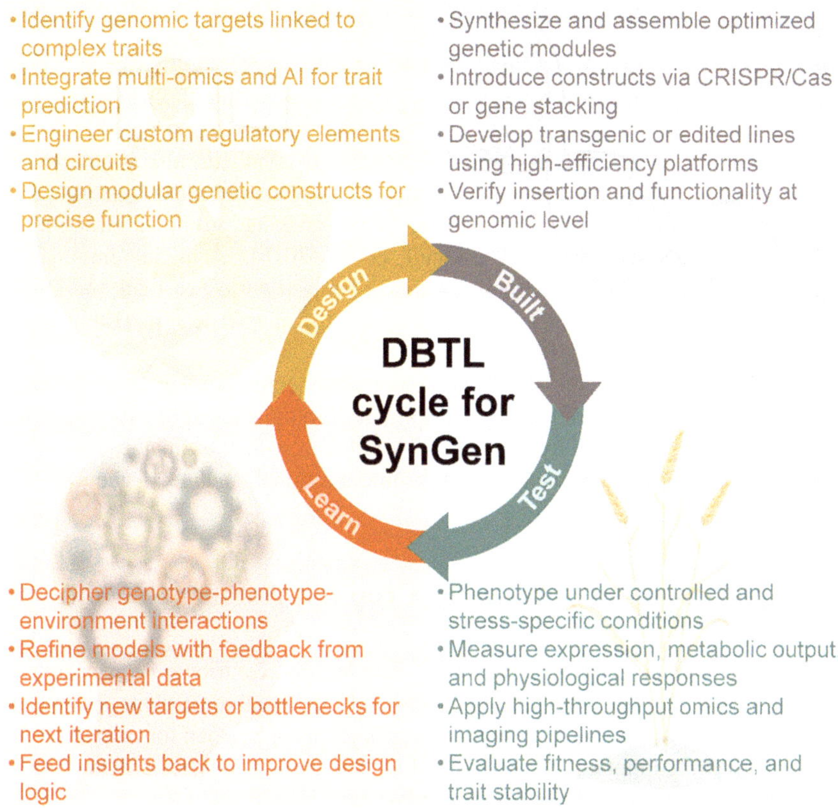

Fig. 7.2 The iterative Design-Build-Test-Learn (DBTL) cycle for crop synthetic genomics (SynGen). The insights from each round inform the next cycle, which rapidly converges on optimal engineered functions

7.2.3 Design-Build-Test-Learn Cycle: The "Engine of SynGen"

At the heart of SynGen is the Design-Build-Test-Learn (DBTL) cycle (Fig. 7.2), which is an iterative engineering structure adapted from SynBio and systems biology (Carbonell et al. 2018; Gurdo et al. 2023). In this cycle, the "design phase" is doubtfully the most critical of its four components, as flaws at this stage can propagate and upset the entire project. During the design phase, computational models and empirical data are used to formulate hypotheses about genetic function, network behavior, and regulatory logic. For genome-scale projects, this includes decisions about codon recoding, essentiality profiling, modularity, and potential off-target effects. Precision in design is vital, and even minor errors in regulatory element placement or unintended interactions can compromise viability or desired function (Ostrov et al. 2016; James et al. 2024). Once a synthetic genome is built using high-throughput DNA synthesis and assembly techniques, it is tested in appropriate chassis organisms. These tests assess phenotype, stability, gene expression, and metabolic output (Cameron et al. 2014; Carbonell et al. 2018). This step may include multi-omics analysis (e.g., transcriptomics, proteomics, and metabolomics) to assess the system-wide impacts of synthetic alterations. In the "learning phase," data from testing is evaluated to refine the next iteration. This learning may involve identifying design flaws, revising promoter strength, or re-balancing metabolic fluxes (Carbonell et al. 2018). Artificial intelligence (AI) and machine learning (ML) are nowadays being used to fast-track the DBTL cycle, which allows rapid convergence on optimal synthetic designs (Radivojević et al. 2020; Liao et al. 2022; Rai et al. 2024).

As SynGen continues, the DBTL cycle not only serves as a structure for engineering but also as a discovery engine for primary biological insights. Its power lies in closing the loop between hypothesis and validation, which ultimately fast-tracks genome writing from a provisional exercise into a rational, analytical science.

7.3 Applications of SynGen for Crop Improvement

SynGen, as a game-changing approach, can largely contribute to agricultural improvement by enabling the precise design and modification of plant genomes at scale. Subsequent sections highlight major applications of SynGen for crop improvement, which are supported by recent studies that demonstrate its efficacy across diverse traits (see Table 7.1 for key examples).

7.3.1 Improved Yield and Stress Tolerance

SynGen assists the rational engineering of plant traits to improve yield and tolerance against multiple stresses (Table 7.1). For instance, the synthetic regulation of *ZmDRO1*, a gene governing root growth angle in maize (Feng et al. 2022). While ancestral maize (*Zea mays ssp. mexicana*) showed drought-inducible *ZmDRO1* expression, this response was lost in modern cultivars. To restore it, Feng et al. (2022) designed an ABA-inducible synthetic promoter guiding *ZmDRO1*, which results in deeper rooting and an increase of over 40% in grain yield under drought stress. This represents how "synthetic promoters" can reprogram stress-responsive gene expression to reshape root architecture and improve drought adaptation.

Another frontier in SynGen is the design of Crassulacean Acid Metabolism (CAM) traits for improved water-use efficiency (WUE) in C_3 crops. For instance, Lim et al. (2019) introduced 13 CAM-related genes from *Mesembryanthemum crystallinum* into *Arabidopsis*, and this modular synthetic expression showed significant effects on stomatal behavior, organic acid accumulation, and vegetative

Table 7.1 Representative applications of synthetic genomics across plant species

Crop/ Organism	Targeted trait	Synthetic approach	Key outcome	References
Yield and stress				
Rice	Engineered trehalose biosynthesis under ABA promoter and abiotic stresses	Synthetic *TPSP* gene construct under an ABA-inducible promoter	• Improved drought, salinity, and sodicity tolerance • Enhanced grain yield under all stresses	Joshi et al. (2020)
Rice	Synthetic apomixis with high fertility for clonal seed production	Multiplex CRISPR/ Cas9 editing of *OsPLDα2, OsOSD1, OsPAIR1, OsREC8* (*Fix4* system)	• Establishment of transgenerational, fertile synthetic apomixis lines with diploid and clonal progeny	Hu et al. (2025)
Maize	Drought tolerance	Synthetic ABA/ drought-inducible promoter driving *ZmDRO1* expression	• Deeper roots • Improved yield under water stress	Feng et al. (2022)
Cotton	Drought-inducible cytokinin biosynthesis	*ipt* gene under SARK promoter (drought-inducible); stress-activated cytokinin synthesis	• Maintained green leaves • Higher photosynthesis and yield under drought stress	Kuppu et al. (2013)
Tobacco	Enhanced photosynthetic efficiency and growth	Introduction of synthetic glycolate metabolism pathways using genes from *E. coli*, algae, and plants; RNAi-mediated suppression of native glycolate transporter	• Increased photosynthesis, light-use efficiency, and field biomass under photorespiratory stress	South et al. (2019)
Tobacco	Heat-smart photosynthesis and biomass yield	Chloroplast-targeted synthetic photorespiration bypass pathway	• Increased biomass under heat stress • Less yield loss vs WT	Cavanagh et al. (2022)
Arabidopsis	Improved WUE and biomass	Modular overexpression of 13 core CAM (C_4 metabolism cycle) genes from *Mesembryanthemum crystallinum*	• Altered stomatal conductance • Increased titratable acidity • Enhanced rosette diameter and leaf fresh weight	Lim et al. (2019)
Arabidopsis	Root development	Synthetic gene circuits using bacterial TFs and synthetic promoters to reprogram root branching	• Predictable increase/ decrease in lateral roots • A platform for growth engineering (no yield data reported)	Brophy et al. (2022)

(continued)

Table 7.1 (continued)

Crop/ Organism	Targeted trait	Synthetic approach	Key outcome	References
Nutritional profile				
Rice	Pro-vitamin A (β-carotene)	Introduction of the synthetic β-carotene pathway	• Enhanced vitamin A content in rice endosperm	Ye et al. (2000), Datta et al. (2003), and Paine et al. (2005)
Rice	Vitamin B_1 (thiamine) biofortification in endosperm	Transgenic expression of codon-optimized *SiTBP* under the endosperm-specific GluD-1 promoter	• Increased thiamine content (3–4 times) in polished grains without affecting yield or agronomic traits	Fitzpatrick et al. (2024)
Potato	Pro-vitamin A (-β-carotene) in tubers	Induction of bacterial mini-pathway (*CrtB*, *CrtI*, and *CrtY*) in tuber cells	• Enhanced β-carotene (47 μg/g tuber) • ~3600× in WT	Diretto et al. (2007)
Maize	β-carotene, vitamin C (ascorbate), and folate	Multi-pathway stacking in the endosperm (genes for carotenoids, ascorbate, and folate)	• Enhanced nutritional profile • Kernels had 169× β-carotene, 6× ascorbate, and 2× folate	Naqvi et al. (2009)
Tomato	Anthocyanin pigments in fruit	Transgenic expression of two TFs (Delila + Rosea1)	• Enhanced pigments by ~5.0 g/kg dry fruit anthocyanins (vs ~0 in wild type)	Su et al. (2016)
Tomato	GABA in leaves/ fruits	Multiplex CRISPR/ Cas9 knockout of five GABA-shunt genes	• Increased GABA levels (~19× GABA vs wild type) in leaves	Li et al. (2018)
Arabidopsis	Pro-vitamin A (-β-carotene) in seeds	Multi-gene stacking (seed *PSY*, OR^{His}, and *HGGT*) + CRISPR knockout (*BCH2*)	• Enhanced nutritional profile up to ~65× β-carotene; 63% of carotenoids as pro-vitamin A	Sun et al. (2021)
Metabolic pathway engineering				
Rice	C and N assimilation under low N	Transgenic TF (maize Dof1 under constitutive promoter)	• Upregulation of PEPC • Increased C/N flux • Higher photosynthesis and growth under N-deficit	Kurai et al. (2011)
Camelina	ω-3 long-chain-PUFA (EPA and DHA)	Multi-gene algal pathway (desaturases/ elongases) stacked in a single construct, and seed-specific expression	• Higher and stable EPA/DHA accumulation in seed TAG	Han et al. (2020)

(continued)

Table 7.1 (continued)

Crop/ Organism	Targeted trait	Synthetic approach	Key outcome	References
Arabidopsis	Conjugated fatty acid (eleostearic acid) and oil	Gene stacking (tung DGAT1/2, oleosin, plus FADX desaturase) in mutants and leaf expression	• Synergistic increase of leaf TAG • Higher accumulation of eleostearic acid	Yurchenko et al. (2017)
Arabidopsis	Triterpenes (squalene and botryococcene)	Gene stacking with seed-specific promoters (triterpene synthase + FPS + DXS) and plastid targeting	• High triterpene yields (botryococcene ~0.53 mg/g, and squalene ~0.23 mg/g seed) • >1 mg/g botryococcene in the cytosol	Kempinski and Chappell (2019)
Tobacco	Photorespiration bypass (C fixation)	Multi-gene pathway (bacterial/algal glycolate pathway enzymes) + RNAi of transporter	• Enhanced photosynthesis and biomass via CO_2-conserving routes	South et al. (2019)
Tobacco	TAG oil content	Transient expression of *camelina WRI1* TFs under 35S promoter	• Higher TAG levels in infiltrated leaves (~2.5–4×)	An et al. (2017)
Tobacco	Terpenoid volatiles (linalool, and costunolide)	Combinatorial MoClo expression of MEP/MVA pathway genes and terpene synthases	• Increased linalool (~50–90 ng/mg FW) and costunolide (~70–90 ng/mg FW)	Park et al. (2022)
Tobacco	TAG oil content	Gene stacking + TF/RNAi (overexpression of *WRI1/LEC2*, and silencing of lipase "SDP1")	• Remarkable increase in leaf TAG (up to ~30–33% DW)	Vanhercke et al. (2017)

Abbreviations: ABA abscisic acid, *C* carbon, *CRISPR* clustered regularly interspaced short palindromic repeats, *DW* dry weight, *DHA* docosahexaenoic acid, *EPA* eicosapentaenoic acid, *FW* fresh weight, *GABA* γ-aminobutyric acid, *MoClo* modular cloning, *N* nitrogen, *PEPC* phosphoenolpyruvate carboxylase, *PUFA* polyunsaturated fatty acids, *TPSP* trehalose-6-phosphate synthase/phosphatase, *TAG* triacylglycerol, *TF* transcription factor, *WT* wild-type, *WUE* water-use efficiency

growth. These discoveries signify foundational steps toward full CAM pathway engineering in major crops for enhanced photosynthesis and productivity under arid conditions (Lim et al. 2019). Moreover, to address inefficiencies in C_3 photosynthesis, South et al. (2019) introduced synthetic photorespiratory bypasses into tobacco, incorporating enzymes from *E. coli*, algae, and plants. Combined with suppression of the native glycolate transporter, the modified pathway boosted light-use efficiency and increased field biomass by up to 24% (South et al. 2019). Similarly, Cavanagh et al. (2022) evaluated synthetic photorespiratory bypasses in tobacco under heat stress. Engineered lines showed a 26% biomass increase and 19% yield loss reduction under +5 °C conditions compared to wild-type plants. These results highlight the power of SynGen to reconfigure native metabolic bottlenecks and boost tolerance against extreme temperatures.

In rice, Joshi et al. (2020) engineered a synthetic trehalose biosynthesis pathway by expressing a fusion enzyme (TPSP from *E. coli*) under an ABA-inducible promoter. This modification significantly enhanced tolerance against drought, salinity, and sodicity by maintaining higher water content, photosynthesis, and grain yield. Trehalose levels positively correlated with yield traits, which highlights the effectiveness of synthetic metabolic rewiring for multi-stress management in plants.

7.3.2 Enhanced Nutritional Profile (Biofortification)

SynGen enables the biofortification of crops to address micronutrient deficiencies (Table 7.1). For instance, golden rice is a landmark example of SynGen enabling nutritional biofortification in staple crops. By introducing key carotenoid biosynthesis genes (i.e., *psy*, *crtI*, and later *lcy*) into rice endosperm, previous studies have successfully reconstituted the provitamin A pathway in a tissue that naturally lacks carotenoids (Ye et al. 2000; Datta et al. 2003; Paine et al. 2005). For instance, Ye et al. (2000) first achieved β-carotene production using just two transgenes, and then Datta et al. (2003) advanced the approach with a multi-gene construct in indica rice. In these experiments, key innovations included tissue-specific promoter engineering, plastid-targeted enzyme localization, and multi-gene cassette assembly. The key outcomes include bright yellow grains rich in β-carotene that retain their nutritional value after cooking. Hence, the resulting golden rice provides a scalable, food-based solution to combat vitamin A deficiency in vulnerable populations that proves the practical impact of SynBio on global nutrition (Datta et al. 2003; Ye et al. 2000; Paine et al. 2005). Interestingly, modulating pathways like carotenoids not only enhances nutrition but can also affect plant architecture and yield (Stra et al. 2023). SynBio/SynGen enables such multi-gene trait enhancements simultaneously to fast-track future breeding efforts as well.

Analogous multi-gene pathway insertions have fortified other staples. For example, Diretto et al. (2007) expressed three Erwinia genes (*CrtB*, *CrtI*, and *CrtY*) under a tuber-specific promoter in potatoes that yielded deep-yellow "golden tubers" with ~114 μg/g total carotenoids and 47 μg/g β-carotene, and this is ~20-fold and ~3600-fold increase over wild-type levels, respectively (Diretto et al. 2007). In maize, Naqvi et al. (2009) assembled transgenes from three distinct vitamin pathways in the endosperm. The resulting "multi-vitamin maize kernels" contained ~169-fold the normal β-carotene, six-fold the normal ascorbate (vitamin C) and 2 times the normal folate (Naqvi et al. 2009). These examples demonstrate multi-gene pathway assembly in crops to raise vitamin and provitamin levels far beyond what conventional breeding achieves.

Similar to other traits, Arabidopsis has also served as a model platform for advanced nutritional stacking. For instance, a recent study used Arabidopsis as a testbed to stack multiple traits, i.e., seed-specific PSY expression, an "Orange" (OR^His) plastid biogenesis factor, and a tocotrienol cyclase (HGGT) for antioxidant protection, plus CRISPR knockout of a β-carotene hydroxylase. The engineered seeds accumulated up to ~65× more β-carotene and ~10× more total carotenoids (with over 63% provitamin A) (Sun et al. 2021). In this way, multi-strategy engineering (gene stacking combined with genome editing) greatly improved both the level and stability of nutrients in plant tissues.

SynGen has also targeted non-carotenoid nutrients. A notable case is tomato, where the authors utilized a multiplex CRISPR/Cas9 system to knock out five genes in the tomato GABA shunt. The resulting lines showed dramatically higher GABA levels, up to a 19-fold rise in leaf GABA than wild-type (with similarly elevated GABA in fruit) (Li et al. 2018). In another tomato study, Su et al. (2016) have biofortified tomatoes with health-promoting polyphenols. By expressing two snapdragon transcription factors (TFs; Delila and Rosea1) under a fruit-specific promoter, a "purple" tomato line was created that synthesizes high anthocyanin levels. This engineered tomato produced about 5 g/kg dry weight of anthocyanins (flavonoid pigments) in the fruit (whereas wild-type fruit had essentially none) (Su et al. 2016). These cases illustrate both genome editing (inactivating multiple endogenous enzymes) and heterologous pathway induction (adding transcriptional regulators) to rewire the diversity of SynGen-enabled biofortification strategies across metabolic classes and plant species.

7.3.3 Metabolic Pathway Engineering

SynGen enables a comprehensive redesign of metabolic networks by stacking genes and modulating regulatory circuits to achieve novel traits (Table 7.1). For example, to engineer C_4 photosynthesis into rice, several previous studies have used Golden Gate cloning and synthetic promoters to assemble multi-gene constructs expressing multiple C_4-cycle enzymes and transporters in bundle sheath cells, as reviewed by Ermakova et al. (2020). These "gene-building-block" approaches illustrate how SynGen can introduce entire pathways into plant hosts.

Photorespiration has been another metabolic target. For instance, South et al. (2019) introduced synthetic photorespiratory bypass routes into tobacco, inserting bacterial and algal enzymes (e.g., *Chlamydomonas* glycolate dehydrogenase and Cucurbita malate synthase) together with RNAi knockdown of the glycolate transporter. These synthetic routes recycle glycolate more efficiently, which results in enhanced photosynthesis and biomass (South et al. 2019). Likewise, the introduction of a maize *Dof1* TF in rice upregulated phosphoenolpyruvate carboxylase and other anaplerotic enzymes, which boosted carbon flow into amino acids and increased nitrogen/carbon assimilation under low nitrogen conditions (Kurai et al. 2011). In each case, the synthetic strategy (gene stacking or TF engineering) reprogrammed core metabolic flux to improve growth and nutrient use.

Other studies have focused on increasing carbon allocation to energy-rich storage compounds. In tobacco, Vanhercke et al. (2017) overexpressed a seed developmental TF (*LEC2*) and silenced a lipase (*SDP1*) in a previously engineered high-oil background, which increases leaf triacylglycerol (TAG) from ~15 to over 30% dry weight (Vanhercke et al. 2017). Similarly, An et al. (2017) showed that expressing *Camelina* WRINKLED1 TFs in tobacco leaves induced a 2.5–4× increase in TAG content. Another study took a complementary strategy, and they first generated Arabidopsis lines accumulating high leaf oil (via tung tree DGAT and oleosin genes and lipid-turnover mutants) and then expressed the tung FAT8 desaturase/conjugase (*FADX*). Co-expression of *FADX* with *DGAT2* led to synergistic increases in both leaf oil content and production of the unusual conjugated fatty acid eleostearic acid (ESA) (Yurchenko et al. 2017). These examples use multigene stacking (combining biosynthetic enzymes, oleosins, or TFs) to reroute carbon into oils and specialty fatty acids in non-seed tissues.

Moreover, synthetic designs have enabled plant synthesis of valuable compounds. For instance, a study constructed a multigene algal pathway in *Camelina sativa* seeds to produce the fish oil ω-3 fatty acids, EPA and DHA. A single "DHA2015.1" construct contained an optimized gene set of algal desaturases/elongases that trigger the vigorous accumulation of EPA and DHA in seed triacylglycerols across field trials (Han et al. 2020). In Arabidopsis seeds, triterpene pathways were engineered by combining a plant triterpene synthase with upstream isoprenoid enzymes (DXS and FPS) in a seed-specific gene stack. By targeting the pathway to plastids, they achieved up to ~0.53 mg/g botryococcene and 0.23 mg/g squalene in seeds (and >1 mg/g botryococcene when targeted to cytosol) (Kempinski and Chappell 2019). These results demonstrate how subcellular targeting and multigene stacking can harness seed oil bodies to produce non-native terpenes. In tobacco leaf biomass, a recent study used transient MoClo assembly of MEP- and MVA-pathway genes to boost terpenoid volatiles. Combinatorial expression of linalool and costunolide pathway genes in tobacco leaves increased linalool (a monoterpene) to ~50–90 ng/mg FW and costunolide (a sesquiterpene lactone) to ~70–90 ng/mg FW (Park et al. 2022). Based on the above examples, it can be concluded that synthetic rewiring of plant metabolism by stacking heterologous biosynthetic genes (often under engineered promoters) and redirecting flux, plants can be programmed or redesigned to accumulate new or enhanced metabolic traits.

7.3.4 Synthetic Chromosomes and Genome Redesign

Advances in SynBio are pushing the boundaries of what is possible in genome engineering. For instance, the landmark "yeast 2.0" project demonstrated that entire eukaryotic genomes can be synthetically rewritten to optimize multiple traits (Richardson et al. 2017). Translating this vision to plants can serve as a game-changing approach for agriculture, where designing synthetic chromosomes/

genomes or large chromosomal segments could accelerate breeding and trait stacking at unprecedented scales.

A major breakthrough in this direction was recently achieved in *Physcomitrium patens*, a model moss (Chen et al. 2024). The authors replaced one-third of a chromosomal arm with chemically synthesized, redesigned DNA that removes over half of the native sequence, including numerous transposable elements, without compromising plant growth (Chen et al. 2024; Yu et al. 2024). Despite extensive sequence and structural modifications, the engineered lines preserved their complex epigenetic landscape, which establishes the feasibility of genome-scale redesign in multicellular plants. This groundbreaking work lays the foundation for the SynMoss project and signals a new era of SynGen for plant systems (Chen et al. 2024). We anticipate that the integration of tools and concepts from projects like "synthetic yeast" could inform efforts to refactor polyploid crop genomes. SynGen can guide the building of optimized, modular plant genomes capable of responding to global challenges such as stress tolerance and nutritional security to sustainable productivity.

7.3.5 Engineering Plant–Microbe Interactions

Recent advances in SynBio have opened other possibilities for engineering plant–microbe interactions to improve crop performance under adverse conditions. One prominent goal has been the reconstitution of biological nitrogen fixation in non-leguminous crops, which aims to reduce reliance on synthetic nitrogen fertilizers. Nitrogenase, the bacterial enzyme complex responsible for nitrogen fixation, is naturally absent in cereals. However, recent breakthroughs have demonstrated the successful expression and mitochondrial targeting of multiple nitrogenase (Nif) proteins in plant systems, which opened the critical groundwork to enable *planta* nitrogen fixation (Allen et al. 2017). These developments support the long-term vision of engineering cereals either to express a fully functional nitrogenase system or to establish strong synthetic symbioses with diazotrophic microbes. This once-ambitious goal is now increasingly perceptible thanks to SynBio innovations (Rogers and Oldroyd 2014).

Among the most promising directions within SynBio is the design and deployment of synthetic microbial communities (SynComs), rationally constructed microbial consortia that enhance plant tolerance, survival, and performance under stress. These SynComs are adapted to mimic, complement, or enhance native microbial interactions, which often results in improved WUE, nutrient cycling, and stress mitigation (De Souza et al. 2020; Durán et al. 2025; Delgado-Baquerizo et al. 2025). For instance, Armanhi et al. (2021) developed a SynCom of beneficial bacteria that significantly enhanced drought tolerance in commercial maize hybrids. Using a simultaneous phenotyping platform, they reported improved WUE, lower leaf temperatures, faster post-drought recovery, and enhanced recruitment of endogenous beneficial microbes (Armanhi et al. 2021).

In saline soils, a recent study isolated endophytes from the halophyte *Kalidium schrenkianum* and constructed a salt-tolerant SynCom that improved wheat growth by boosting seed germination, root vigor, antioxidant enzyme activity, and chlorophyll content. This study demonstrated the ecological potential of microbiomes adapted to extreme environments as SynGen platforms (Zhu et al. 2025). Meanwhile, Kaur et al. (2022) emphasized the importance of SynCom delivery strategies. Their study compared seed-dressing and soil application of a four-strain SynCom across five cotton cultivars. Seed-based applications yielded the strongest benefits, including higher germination, biomass, flower number, and yield, alongside enrichment of beneficial bacterial (*Actinobacteria* and *Firmicutes*) and fungal (*Basidiomycota* and *Chytridiomycota*) taxa in the rhizosphere, and improved soil nitrate availability. Their findings highlight that plant genotype and delivery mode critically influence SynCom effectiveness, which offers insights for field-scale optimization.

To further expand the SynCom application across stressed ecosystems, Flores-Duarte et al. (2023) constructed a SynCom from *Medicago* nodule endophytes and showed that it improved *M. sativa* growth, nodulation, photosynthesis, and nitrogen content under multiple abiotic stresses in degraded estuarine soils. Notably, SynCom enhanced antioxidant responses and sequestered metals in roots that reduce translocation to shoots, a critical biosafety consideration for sustainable soil remediation. The above studies highlight the growing potential of SynComs as scalable, programmable, and eco-friendly bio-tools capable of reprogramming plant–microbe interactions, modulating host stress physiology, and reshaping the rhizosphere microbiome for improved plant tolerance across diverse agroecosystems.

To fully harness the potential of SynGen in agricultural systems, future research must shift from empirical assembly to function-driven and genome-informed design of microbial consortia by leveraging tools such as metagenomics, single-cell genomics, metabolic modeling, and CRISPR-based genome editing. Standardizing SynCom construction, delivery methods (e.g., seed coatings or root colonization enhancers), and host genotype compatibility will be essential for consistent field-level outcomes. Moreover, integrating biosensors, chassis engineering, and synthetic signaling circuits could enable SynComs to respond to environmental signals and host needs in natural conditions.

7.4 Integration of SynGen with Other Modern Technologies

7.4.1 Artificial Intelligence and Machine Learning: The Future Is Here

Integration of AI and ML with SynGen is transforming the landscape of crop improvement. Over the last few years, AI-driven models have been gradually employed to predict, design, and optimize genetic constructs that further advance the DBTL cycle in SynBio (Radivojević et al. 2020; Liao et al. 2022; van Lent et al. 2023; Goshisht 2024; Rai et al. 2024). Generative models (including deep learning) can design synthetic promoters, optimize enzyme variants, and even construct complete metabolic pathways underlying the improvement of specific agronomic traits (Zhang et al. 2024).

Within the context of crop engineering programs, AI/ML can process multi-omics datasets to discover novel targets and inform SynGen designs for enhanced stress tolerance, yield potential, or nutrient efficiency (Depuydt et al. 2023; Zhang et al. 2024; Raza et al. 2025c). Ultimately, these outcomes enable data-driven trait discovery, particularly under complex environmental interactions.

Moreover, recent advances in large language models (LLMs) such as ChatGPT (https://chatgpt.com/) (Van Dis et al. 2023), DeepSeek (https://chat.deepseek.com/) (Liu et al. 2024; Guo et al. 2024), and CRISPR-GPT (https://www.crispr-gpt.com/auth/signin; https://github.com/cong-lab/crispr-gpt-pub) (Huang et al. 2024) are guiding how we manipulate genomes. For instance, CRISPR-GPT integrates domain-specific tools to automate CRISPR-based editing by helping with guided RNA design, delivery method selection, protocol drafting, and experimental planning. These LLMs provide automated thoughts and decision-making, hence reducing time and human errors in the experimental workflow.

AI also facilitates the development of a global biological knowledge atlas by mining vast literature datasets and biological records. Moreover, iterative learning models can refine breeding strategies over time and help improve predictive power and design accuracy continuously. Notably, tools like G2PDeep (https://g2pdeep.org/) (Zeng et al. 2021) and DeepG2P (Sharma et al. 2022) utilize deep learning to predict genotype-phenotype relationships using multi-modal data such as genomic sequences, environmental conditions, and field practices. Recently developed new platforms such as CropGPT (https://cropgpt.ai/) (Zhu et al. 2024) and PlantGPT (http://www.plantgpt.icu) (Zhang et al. 2025) are now aiming to integrate genome editing, omics data, and intelligent algorithms to fast-track precision breeding. CropGPT leverages double-haploid technology and LLMs for germplasm enhancement. Whereas PlantGPT provides a user-friendly agent trained on Arabidopsis data to support plant functional genomics research in model plants that can help translate these findings to the major food crops.

We assume that integrating AI with SynGen will be vital for addressing the complexity of polygenic traits, gene regulatory networks, and crop stress adaptation. AI enables the predictive design of synthetic regulatory circuits, intelligent screening of chassis organisms, and prioritization of gene targets using pan-genomic and epigenomic data. Emerging closed-loop AI systems can self-optimize synthetic designs through iterative learning. To fully understand these potentials, future efforts must focus on developing standardized AI-ready datasets that can enhance ML models' interpretability, integrate AI into actual cloud-based SynGen platforms, and advance interdisciplinary collaboration.

7.4.2 Big Data and Omics Integration

The convergence of omics technologies, such as genomics, transcriptomics, proteomics, metabolomics, and epigenomics, has unlocked vast datasets that are critical for informing synthetic genome design. When coupled with ML, these data-rich platforms can discover complex biological patterns and regulatory logic, thereby fast-tracking the development of more precise and predictive SynGen strategies. For instance, single-cell transcriptomics in key crops such as maize (Ortiz-Ramírez et al. 2021; Marand et al. 2021; Xu et al. 2021; Cao et al. 2023; Chen et al. 2025), wheat (Zhang et al. 2023; Du et al. 2025; Guo et al. 2025) and rice (Liu et al. 2021; Zhang et al. 2021b; Zong et al. 2022; Wang et al. 2025) are delivering insights into cell-type-specific gene expression dynamics. These insights enable the engineering of synthetic promoters or regulatory circuits designed for specific tissues, such as roots, leaves, or meristems, thereby enhancing spatiotemporal control of gene expression (Andres et al. 2019; Brophy et al. 2022). In parallel, advances in spatial transcriptomics provide additional context by mapping gene activity across tissue architecture, further refining SynGen designs (Yin et al. 2023; Wang et al. 2025; Zhong et al. 2025).

In addition to regular transcriptomics, integrative multi-omics (also called "panomics" approaches can help discover key metabolic bottlenecks, stress-responsive pathways, and gene-phenotype links that inform synthetic circuit design and chassis optimization. In this context, big data integration across omics layers also assists in the identification of new genetic markers and dynamic networks involved in complex traits (Zhang et al. 2024). As such, the synergy between SynGen and omics enables more innovative synthetic genome construction and alters SynGen and genomic-assisted breeding (GAB) into highly targeted, systems-level crop improvement pipelines. We also predict that future progress will depend on standardizing data formats, enhancing platform interoperability, and developing accessible, cloud-based tools that can harness multi-omics data for novel design and validation of synthetic components.

7.4.3 Biofoundries and Automation

Biofoundries are highly automated facilities that integrate robotics, AI, and SynBio, which can redesign the scale and speed of biological engineering. Their key role includes automation of the DBTL cycle, enabling rapid prototyping, high-throughput screening, and novel optimization of genetic constructs (Gurdo et al. 2023). In the context of SynGen, biofoundries allow systematic testing of thousands of synthetic circuits in parallel across model plants or cell systems, which vastly accelerates trait discovery and circuit refinement. Recent advances in chassis engineering, automated DNA assembly, and phenotyping platforms have also guided what was once artisanal benchwork into a streamlined and programmable pipeline (Zhang et al. 2021a).

However, understanding the full potential of biofoundries in agriculture requires close collaboration between synthetic biologists, software engineers, and robotics specialists. Beyond technological integration, standardized protocols, adaptable hardware, and user-friendly software interfaces are essential to ensure reproducibility and scalability (Zhang et al. 2021a). Initiatives like the Global Biofoundry Alliance (Hillson et al. 2019) feature the need for harmonized methods and open data infrastructures. By embedding SynGen within the operational fabric of biofoundries, we can move toward a future where crop improvement is not only faster and more precise but also predictive, modular, and field-relevant.

7.4.4 Gene Drives and Synthetic Ecology: Next-Frontier Tools of SynGen

Gene drives and synthetic ecology are considered the rapidly evolving branches of SynGen that offer unique tools for pest management and ecosystem engineering (Pixley et al. 2019; Schwille 2011). Though still largely in experimental stages, these technologies are ready to complement core SynGen strategies by expanding the functional reach of genomic manipulation beyond individual crop species (Schwille 2011; Pixley et al. 2019).

Gene drives, based on CRISPR/Cas systems, enable the biased inheritance of engineered traits, making them a promising tool for controlling insect vectors that transmit plant pathogens (Burt and Crisanti 2018; Pixley et al. 2019). Recent advances have demonstrated proof-of-concept for targeting diploid, sexually reproducing insects such as planthoppers and leafhoppers, which are major crop pests (Jairin et al. 2013; Gomez et al. 2019). However, gene drive strategies remain impractical in species with parthenogenesis or complex reproductive systems (Byrne and Devonshire 1996; Jacobson et al. 2016). Despite these biological constraints, their potential for population suppression or the disruption of pathogen replication is evident, with early targets including the Asian citrus psyllid (Baltzegar et al. 2018). In parallel, efforts are also being made to investigate CRISPR-/Cas-based gene drives for weed control. However, challenges such as self-fertilization and prolific seed production in many species remain significant hurdles (Neve 2018). Therefore, it is suggested that future gene drive systems will likely emphasize species specificity, geographical confinement, and reversibility (Marshall and Akbari 2018), which align with SynGen's emphasis on precision and biosafety.

Beyond pest suppression, synthetic apomixis systems have demonstrated a novel gene drive-like approach in plants. For example, multiplex CRISPR-/Cas9-mediated editing of four genes in rice, including *OsPLDα2* and three meiosis and fertilization-related genes (*OsOSD1*, *OsPAIR1*, and *OsREC8*), which results in the development of the *Fix4* system (Hu et al. 2025). This system enables clonal seed production with high fertility and heritable diploid progeny that allow the fixation of elite hybrid genotypes without the need for repeated hybridization cycles. As a game-changing application of SynGen, *Fix4* corresponds to a novel form of gene drive that locks an elite hybrid genotype and offers a scalable strategy for hybrid propagation in agriculture and a potential paradigm shift in crop breeding (Hu et al. 2025).

Synthetic ecology, a complementary approach, involves designing unique microbial consortia (e.g., nitrogen-fixing endophytes or biocontrol agents) using SynBio principles to enhance plant health and tolerance. These engineered communities can be programmed for beneficial functions such as nutrient acquisition, stress mitigation, or pathogen suppression, which serve as ecological dimensions for genomic innovation (De Souza et al. 2020; Delgado-Baquerizo et al. 2025). Recent advances also suggest the feasibility of developing microbe-based delivery platforms that enable in situ editing of host genomes or gene expression, including the application of synthetic symbionts equipped with CRISPR/Cas machinery. For example, engineered endophytes have been used for targeted genome modification *in planta*, which represents a breakthrough in non-transgenic genome manipulation approaches (Verma et al. 2023; Bard et al. 2024). This opens up possibilities for next-generation plant–microbe partnerships where beneficial microbes support plant health and serve as dynamic vectors for precise trait improvement, as reviewed by Raza et al. (2025a).

Recent developments suggest that recoding plastid genomes, owing to their compact size and independent inheritance, offer a favorable path for implanting novel resistance traits at the synthetic genome level. The plastome, targeted by many plant viruses, could be synthetically re-engineered (e.g., via synplastomes) to express defense pathways or even be recoded for viral resistance by altering stop codons, rendering incoming viral RNA untranslatable (Ma and Isaacs 2016; Ostrov et al. 2016; Piatek et al. 2018; Zhang et al. 2020). This approach introduces the possibility of a universal, preemptive virus resistance system in crops, which is a significant leap from the current one-pathogen-one-resistance-gene paradigm (Lajoie et al. 2013; Ma and Isaacs 2016).

In principle, gene drives and synthetic ecology extend the SynGen structure from editing single loci to reshaping entire ecological and evolutionary interactions (Pixley et al. 2019). These frontier tools could highlight the next generation of climate-smart, pest-resilient, and self-regulating agricultural systems.

7.4.5 Synthetic Gene Circuits: Towards Programmable Regulatory Networks in Plants

Another emerging frontier of SynGen is the construction of reversible and programmable "synthetic gene circuits," which offer precise, orthogonal control over gene expression *in planta* (Andres et al. 2019; Brophy et al. 2022; Nayak et al. 2025). A recent breakthrough study developed and characterized a CRISPRi-based logic gate platform that enables modular, reversible, and cross-species gene repression using a comprehensive toolkit of engineered synthetic promoters (Khan et al. 2024). This platform was validated in Arabidopsis and extended to other species, including moss (*Physcomitrium patens*), wheat, and rapeseed protoplasts, which demonstrated its broad applicability. This system supports the construction of multi-layered logic gates (e.g., NOT, NOR, AND, OR, and NIMPLY), which enable complex Boolean operations and programmable spatiotemporal control of gene expression. The reversible nature of these circuits further allows plants to mount transient, dynamic responses to internal or environmental stimuli, further expanding the toolkit for engineering conditional and tissue-specific responses (Khan et al. 2024).

The study compares logic gate performance against perfect gate states by introducing a quantitative benchmarking basis. This strong evaluation system significantly strengthens the DBTL cycle in plant SynBio, which provides a reliable method for iterating and improving gene circuit designs. The use of CRISPRi for effective repression of both strong and moderate promoters also ensures broad applicability for diverse regulatory needs (Khan et al. 2024). This work highlights a significant advancement in plant SynGen by delivering a compact, modular, and orthogonal circuit platform that enables sophisticated, programmable control of gene regulatory pathways. It lays the groundwork for precision engineering of crop regulatory networks, which can significantly contribute to future efforts in developing environmentally responsive and stress-adaptive plants.

7.5 Challenges and Bottlenecks Associated with SynGen Applications

Despite its transformative potential, SynGen faces several challenges before it can be broadly applied to agriculture (Fig. 7.3). Technically, synthesizing and assembling long plant DNA sequences remains error-prone, as chemical DNA synthesis frequently introduces base substitutions, insertions, and deletions that accumulate over kilobase scales (Masaki et al. 2022). Unlike microbial genomes, plant genomes are more complex and challenging to assemble and integrate, and compatibility issues (such as codon usage bias or epigenetic modifications) can compromise host viability (Zhang et al. 2020; Masaki et al. 2022). Scaling these innovations to practical agriculture introduces further complexity. Many synthetic constructs that function well under lab conditions fail under varying field environments, where climate, soil, and pathogen pressures interact erratically. In this context, field-level validation across generations is critical to ensure trait stability and efficacy (Wurtzel et al. 2019; Zhang et al. 2020, 2024; Du et al. 2024). Moreover, while gene synthesis costs fall, constructing entire chromosomes or metabolic pathways in crops remains expensive and logistically intensive. Reducing synthesis errors and lowering costs remain key engineering goals. Finally, regulatory and biosafety concerns present significant bottlenecks (Zhang et al. 2020). While many DNA synthesis providers

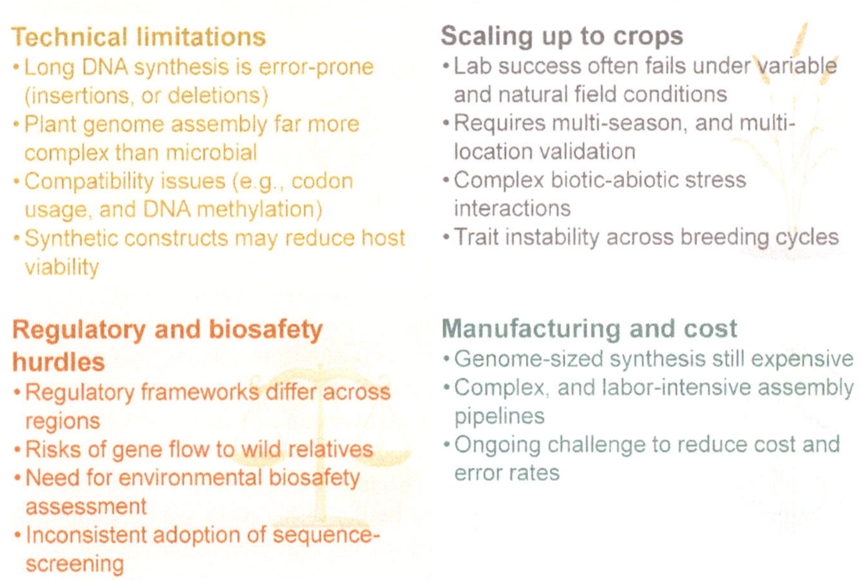

Fig. 7.3 Key challenges for applying synthetic genomics (SynGen) in agriculture and future crop improvement programs

use sequence screening to prevent misuse of potentially hazardous genes (Hoffmann et al. 2023), oversight bases for synthetic crops remain fragmented globally. On the other hand, environmental release raises valid concerns about horizontal gene flow or unintended ecological effects (Gómez-Tatay and Hernández-Andreu 2019; Hoffmann et al. 2023). Insecurity around regulatory approvals and public perception continues to constrain the deployment of SynGen technologies in agriculture (Fig. 7.3).

7.6 Key Implications of Redesigned Products

SynGen introduces significant ethical, legal, and societal implications that must be addressed in parallel with its technological progress (Fig. 7.4). Ethically, the creation of synthetic life forms prompts concerns over the perceived boundary between natural and engineered organisms, particularly in relation to unexpected ecological impacts or dual-use risks such as pathogen reconstruction (Zhang et al. 2020; Hoffmann et al. 2023). Legally, intellectual property rights around synthetic genes and genomes remain complex, which further raises issues of restricted access, biopiracy, and the need for equitable benefit-sharing when indigenous biodiversity is involved. Socially, public trust is critical. Despite cautious support for high-benefit applications like healthcare, agricultural uses of SynBio face more profound

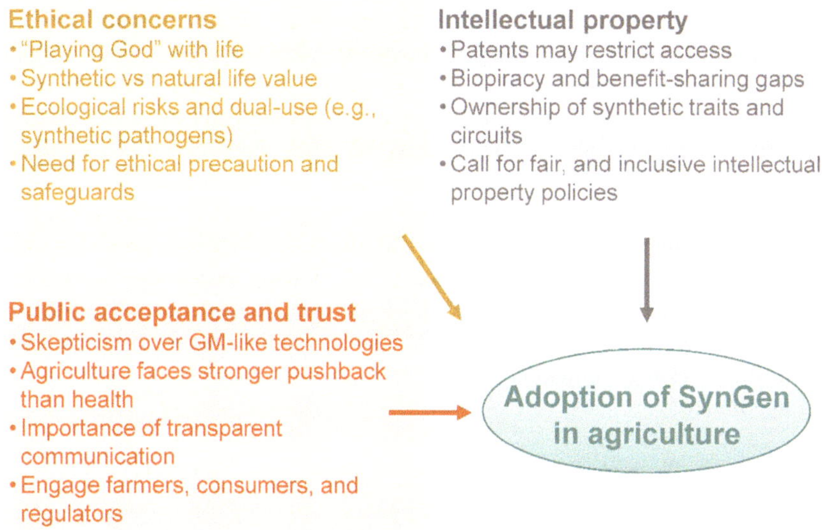

Fig. 7.4 Major factors associated with the rapid adoption of synthetic genomics (SynGen) in agriculture and beyond

skepticism due to cultural and historical concerns. Transparent risk communication, inclusive stakeholder engagement, and ethical provision will be essential to ensure responsible innovation and broader societal acceptance (Rook et al. 2025). Without deliberate efforts to address these concerns, the adoption of SynGen in agriculture may be hindered, regardless of its technical potential.

7.7 Conclusion and Future Recommendations

SynGen has emerged as a game-changing convergence of biology, engineering, and computational sciences that offers unique power to redesign life for specific, purposeful outcomes. At its core, SynGen relies on rational genome design and the iterative DBTL cycle, now increasingly enhanced by AI, big data, and omics. These foundations have enabled rapid progress toward reprogramming plants with synthetic chromosomes, metabolic circuits, and minimal genome chassis, ultimately geared toward designing climate-smart, high-yield, and nutritionally enhanced crops. Its ongoing research applications set the stage for a new era in crop improvement that extends beyond conventional breeding and transgenic approaches. Despite its promise, SynGen faces significant hurdles associated with SynGen applications and the implications of redesigned productions (Figs. 7.3 and 7.4), which demand more effort and collaborations.

Soon, several key frontiers will continue to emerge, and we recommend that more research efforts be made to harness their full potential within the context of crop SynGen. For instance, Xenobiology and synthetic XNA systems could provide genetic "firewalls," preventing gene flow to natural populations (Schmidt et al. 2018). Minimal genomes and plant chassis offer new testbeds for understanding life and optimizing agronomic efficiency (Xu et al. 2023). Enzymatic DNA synthesis or "DNA printers" may soon automate long-sequence construction that enables faster iteration cycles. On the policy side, adaptive governance mechanisms, harmonized DNA screening, transparent approval routes, and biosafety protocols must keep pace with innovation (Hoffmann et al. 2023). Finally, public–private partnerships and open-source education platforms will be essential to democratize SynGen and train the next-generation bioengineers. If developed responsibly, SynGen holds the potential to change agriculture and help secure sustainable food systems in a changing climate.

Acknowledgment We thank all the authors whose relevant work has been cited in this chapter.

References

Ahmar S, Gill RA, Jung K-H, Faheem A, Qasim MU, Mubeen M, Zhou W (2020) Conventional and molecular techniques from simple breeding to speed breeding in crop plants: recent advances and future outlook. Int J Mol Sci 21:2590

Allen RS, Tilbrook K, Warden AC, Campbell PC, Rolland V, Singh SP, Wood CC (2017) Expression of 16 nitrogenase proteins within the plant mitochondrial matrix. Front Plant Sci 8:287

An D, Kim H, Ju S, Go YS, Kim HU, Suh MC (2017) Expression of *Camelina* WRINKLED1 isoforms rescue the seed phenotype of the *Arabidopsis wri1* mutant and increase the triacylglycerol content in tobacco leaves. Front Plant Sci 8:34

Andres J, Blomeier T, Zurbriggen MD (2019) Synthetic switches and regulatory circuits in plants. Plant Physiol 179:862–884

Armanhi JSL, de Souza RSC, Biazotti BB, Yassitepe JECT, Arruda P (2021) Modulating drought stress response of maize by a synthetic bacterial community. Front Microbiol 12:747541

Baltzegar J, Cavin Barnes J, Elsensohn JE, Gutzmann N, Jones MS, King S, Sudweeks J (2018) Anticipating complexity in the deployment of gene drive insects in agriculture. J Responsib Innov 5:S81–S97

Bard NW, Cronk QC, Davies TJ (2024) Fungal endophytes can modulate plant invasion. Biol Rev 99:1652–1671

Barrangou R, Doudna JA (2016) Applications of CRISPR technologies in research and beyond. Nat Biotechnol 34:933–941

Becker MM, Graham RL, Donaldson EF, Rockx B, Sims AC, Sheahan T, Pickles RJ, Corti D, Johnston RE, Baric RS (2008) Synthetic recombinant bat SARS-like coronavirus is infectious in cultured cells and in mice. Proc Natl Acad Sci 105:19944–19949

Bhogireddy S, Mangrauthia SK, Kumar R, Pandey AK, Singh S, Jain A, Budak H, Varshney RK, Kudapa H (2021) Regulatory non-coding RNAs: a new frontier in regulation of plant biology. Funct Integr Genomics 21:313–330

Breuer M, Earnest TM, Merryman C, Wise KS, Sun L, Lynott MR, Hutchison CA, Smith HO, Lapek JD, Gonzalez DJ (2019) Essential metabolism for a minimal cell. elife 8:e36842

Brophy JA, Magallon KJ, Duan L, Zhong V, Ramachandran P, Kniazev K, Dinneny JR (2022) Synthetic genetic circuits as a means of reprogramming plant roots. Science 377:747–751
Burt A, Crisanti A (2018) Gene drive: evolved and synthetic. ACS Chem Biol 13:343–346
Byrne FJ, Devonshire AL (1996) Biochemical evidence of haplodiploidy in the whitefly Bemisia tabaci. Biochem Genet 34:93–107
Cameron DE, Bashor CJ, Collins JJ (2014) A brief history of synthetic biology. Nat Rev Microbiol 12:381–390
Cao Y, Ma J, Han S, Hou M, Wei X, Zhang X, Zhang ZJ, Sun S, Ku L, Tang J (2023) Single-cell RNA sequencing profiles reveal cell type-specific transcriptional regulation networks conditioning fungal invasion in maize roots. Plant Biotechnol J 21:1839–1859
Carbonell P, Jervis AJ, Robinson CJ, Yan C, Dunstan M, Swainston N, Vinaixa M, Hollywood KA, Currin A, Rattray NJ (2018) An automated Design-Build-Test-Learn pipeline for enhanced microbial production of fine chemicals. Commun Biol 1:66
Casini A, Christodoulou G, Freemont PS, Baldwin GS, Ellis T, MacDonald JT (2014) R2oDNA designer: computational design of biologically neutral synthetic DNA sequences. ACS Synth Biol 3:525–528
Cavanagh AP, South PF, Bernacchi CJ, Ort DR (2022) Alternative pathway to photorespiration protects growth and productivity at elevated temperatures in a model crop. Plant Biotechnol J 20:711–721
Cello J, Paul AV, Wimmer E (2002) Chemical synthesis of poliovirus cDNA: generation of infectious virus in the absence of natural template. Science 297:1016–1018
Chen L-G, Lan T, Zhang S, Zhao M, Luo G, Gao Y, Zhang Y, Du Q, Lu H, Li B (2024) A designer synthetic chromosome fragment functions in moss. Nat Plants 10:228–239
Chen X, Yao R, Hua X, Du K, Liu B, Yuan Y, Wang P, Yan Q, Dong L, Groen SC (2025) Identification of maize genes that condition early systemic infection of sugarcane mosaic virus through single-cell transcriptomics. Plant Commun 6(5):101297
Datta K, Baisakh N, Oliva N, Torrizo L, Abrigo E, Tan J, Rai M, Rehana S, Al-Babili S, Beyer P (2003) Bioengineered '*golden*' indica rice cultivars with β-carotene metabolism in the endosperm with hygromycin and mannose selection systems. Plant Biotechnol J 1:81–90
De Souza RSC, Armanhi JSL, Arruda P (2020) From microbiome to traits: designing synthetic microbial communities for improved crop resiliency. Front Plant Sci 11:1179
Delgado-Baquerizo M, Singh BK, Liu YR, Sáez-Sandino T, Coleine C, Muñoz-Rojas M, Bastida F, Trivedi P (2025) Integrating ecological and evolutionary frameworks for SynCom success. New Phytol 246:1922–1933
Depuydt T, De Rybel B, Vandepoele K (2023) Charting plant gene functions in the multi-omics and single-cell era. Trends Plant Sci 28:283–296
Diretto G, Al-Babili S, Tavazza R, Papacchioli V, Beyer P, Giuliano G (2007) Metabolic engineering of potato carotenoid content through tuber-specific overexpression of a bacterial mini-pathway. PLoS One 2:e350
Du F, Dai J, Jiao Y (2024) Insights into a functional synthetic plant genome. New Phytol 244:46–50
Du Z, Zhang B, Weng H, Gao L (2025) Single-cell RNA sequencing reveals the developmental landscape of wheat roots. Plant Cell Environ 48:3431–3447
Durán P, Vailleau F, Roux F (2025) Building microbial synthetic communities: get inspired by the design of synthetic plant communities. New Phytol 246:402–405
Ermakova M, Danila FR, Furbank RT, von Caemmerer S (2020) On the road to C4 rice: advances and perspectives. Plant J 101:940–950
Feng X, Jia L, Cai Y, Guan H, Zheng D, Zhang W, Xiong H, Zhou H, Wen Y, Hu Y (2022) ABA-inducible *DEEPER ROOTING 1* improves adaptation of maize to water deficiency. Plant Biotechnol J 20:2077–2088
Fitzpatrick TB, Dalvit I, Chang FH, Wang K, Fudge JB, Chang SH, Maillot B, Gruissem W (2024) Vitamin B_1 enhancement in the endosperm of rice through thiamine sequestration. Plant Biotechnol J 22:2330

Flores-Duarte NJ, Navarro-Torre S, Mateos-Naranjo E, Redondo-Gómez S, Pajuelo E, Rodríguez-Llorente ID (2023) Nodule synthetic bacterial community as legume biofertilizer under abiotic stress in estuarine soils. Plants 12:2083

Fredens J, Wang K, De La Torre D, Funke LF, Robertson WE, Christova Y, Chia T, Schmied WH, Dunkelmann DL, Beránek V (2019) Total synthesis of *Escherichia coli* with a recoded genome. Nature 569:514–518

Gibson DG, Benders GA, Andrews-Pfannkoch C, Denisova EA, Baden-Tillson H, Zaveri J, Stockwell TB, Brownley A, Thomas DW, Algire MA (2008) Complete chemical synthesis, assembly, and cloning of a *Mycoplasma genitalium* genome. Science 319:1215–1220

Gibson DG, Young L, Chuang R-Y, Venter JC, Hutchison CA III, Smith HO (2009) Enzymatic assembly of DNA molecules up to several hundred kilobases. Nat Methods 6:343–345

Gomez MA, Lin ZD, Moll T, Chauhan RD, Hayden L, Renninger K, Beyene G, Taylor NJ, Carrington JC, Staskawicz BJ (2019) Simultaneous CRISPR/Cas9-mediated editing of cassava *eIF4E* isoforms *nCBP-1* and *nCBP-2* reduces cassava brown streak disease symptom severity and incidence. Plant Biotechnol J 17:421–434

Gómez-Tatay L, Hernández-Andreu JM (2019) Biosafety and biosecurity in synthetic biology: a review. Crit Rev Environ Sci Technol 49:1587–1621

Goold HD, Wright P, Hailstones D (2018) Emerging opportunities for synthetic biology in agriculture. Genes (Basel) 9:341

Goshisht MK (2024) Machine learning and deep learning in synthetic biology: key architectures, applications, and challenges. ACS Omega 9:9921–9945

Guo D, Zhu Q, Yang D, Xie Z, Dong K, Zhang W, Chen G, Bi X, Wu Y, Li Y (2024) DeepSeek-coder: when the large language model meets programming--the rise of code intelligence. arXiv preprint arXiv:240114196

Guo X, Zhao A, Han J, Yuping L, Chen X, Cheng Z, Hou L, Lv L (2025) Single-cell transcriptome reveals the cellular response to PEG-induced stress in wheat leaves. J Agric Food Chem 73(18): 10764–10777

Gurdo N, Volke DC, McCloskey D, Nikel PI (2023) Automating the design-build-test-learn cycle towards next-generation bacterial cell factories. New Biotechnol 74:1–15

Han L, Usher S, Sandgrind S, Hassall K, Sayanova O, Michaelson LV, Haslam RP, Napier JA (2020) High level accumulation of EPA and DHA in field-grown transgenic camelina–a multi-territory evaluation of TAG accumulation and heterogeneity. Plant Biotechnol J 18:2280–2291

Hillson N, Caddick M, Cai Y, Carrasco JA, Chang MW, Curach NC, Bell DJ, Le Feuvre R, Friedman DC, Fu X (2019) Building a global alliance of biofoundries. Nat Commun 10:2040

Hoffmann SA, Diggans J, Densmore D, Dai J, Knight T, Leproust E, Boeke JD, Wheeler N, Cai Y (2023) Safety by design: biosafety and biosecurity in the age of synthetic genomics. iScience 26: 106165

Hu F, Liu C, Jin X, Sun T, Hong L, Rao Y, Qian Q, Wang K (2025) *OsPLDα2*-dependent synthetic apomixis enables normal seed setting in hybrid rice via genome editing. Sci Bull 70(23): 3957–3959

Huang K, Qu Y, Cousins H, Johnson WA, Yin D, Shah M, Zhou D, Altman R, Wang M, Cong L (2024) Crispr-gpt: an llm agent for automated design of gene-editing experiments. arXiv preprint arXiv:240418021

Hutchison CA III, Chuang R-Y, Noskov VN, Assad-Garcia N, Deerinck TJ, Ellisman MH, Gill J, Kannan K, Karas BJ, Ma L (2016) Design and synthesis of a minimal bacterial genome. Science 351:aad6253

Jacobson AL, Nault BA, Vargo EL, Kennedy GG (2016) Restricted gene flow among lineages of *Thrips tabaci* supports genetic divergence among cryptic species groups. PLoS One 11: e0163882

Jairin J, Kobayashi T, Yamagata Y, Sanada-Morimura S, Mori K, Tashiro K, Kuhara S, Kuwazaki S, Urio M, Suetsugu Y (2013) A simple sequence repeat-and single-nucleotide polymorphism-based genetic linkage map of the brown planthopper, *Nilaparvata lugens*. DNA Res 20:17–30

Jakhanwal S, Cress BF, Maguin P, Lobba MJ, Marraffini LA, Doudna JA (2021) A CRISPR-Cas9–integrase complex generates precise DNA fragments for genome integration. Nucleic Acids Res 49:3546–3556

James JS, Dai J, Chew WL, Cai Y (2024) The design and engineering of synthetic genomes. Nat Rev Genet 26:298–319

Joshi R, Sahoo KK, Singh AK, Anwar K, Pundir P, Gautam RK, Krishnamurthy S, Sopory S, Pareek A, Singla-Pareek SL (2020) Enhancing trehalose biosynthesis improves yield potential in marker-free transgenic rice under drought, saline, and sodic conditions. J Exp Bot 71:653–668

Kaur S, Egidi E, Qiu Z, Macdonald CA, Verma JP, Trivedi P, Wang J, Liu H, Singh BK (2022) Synthetic community improves crop performance and alters rhizosphere microbial communities. J Sustain Agric Environ 1:118–131

Kempinski C, Chappell J (2019) Engineering triterpene metabolism in the oilseed of *Arabidopsis thaliana*. Plant Biotechnol J 17:386–396

Khan MA, Herring G, Zhu JY, Oliva M, Fourie E, Johnston B, Zhang Z, Potter J, Pineda L, Pflueger J (2024) CRISPRi-based circuits to control gene expression in plants. Nat Biotechnol 43:416–430

Konig H, Frank D, Heil R, Coenen C (2013) Synthetic genomics and synthetic biology applications between hopes and concerns. Curr Genomics 14:11–24

Kosuri S, Church GM (2014) Large-scale de novo DNA synthesis: technologies and applications. Nat Methods 11:499–507

Kuppu S, Mishra N, Hu R, Sun L, Zhu X, Shen G, Blumwald E, Payton P, Zhang H (2013) Water-deficit inducible expression of a cytokinin biosynthetic gene *IPT* improves drought tolerance in cotton. PLoS One 8:e64190

Kurai T, Wakayama M, Abiko T, Yanagisawa S, Aoki N, Ohsugi R (2011) Introduction of the *ZmDof1* gene into rice enhances carbon and nitrogen assimilation under low-nitrogen conditions. Plant Biotechnol J 9:826–837

Lajoie MJ, Rovner AJ, Goodman DB, Aerni H-R, Haimovich AD, Kuznetsov G, Mercer JA, Wang HH, Carr PA, Mosberg JA (2013) Genomically recoded organisms expand biological functions. Science 342:357–360

Li R, Li R, Li X, Fu D, Zhu B, Tian H, Luo Y, Zhu H (2018) Multiplexed CRISPR/Cas9-mediated metabolic engineering of γ-aminobutyric acid levels in *Solanum lycopersicum*. Plant Biotechnol J 16:415–427

Liao X, Ma H, Tang YJ (2022) Artificial intelligence: a solution to involution of design–build–test–learn cycle. Curr Opin Biotechnol 75:102712

Lim SD, Lee S, Choi W-G, Yim WC, Cushman JC (2019) Laying the foundation for crassulacean acid metabolism (CAM) biodesign: expression of the C_4 metabolism cycle genes of CAM in *Arabidopsis*. Front Plant Sci 10:101

Liu W, Stewart CN (2015) Plant synthetic biology. Trends Plant Sci 20:309–317

Liu Q, Liang Z, Feng D, Jiang S, Wang Y, Du Z, Li R, Hu G, Zhang P, Ma Y (2021) Transcriptional landscape of rice roots at the single-cell resolution. Mol Plant 14:384–394

Liu A, Feng B, Xue B, Wang B, Wu B, Lu C, Zhao C, Deng C, Zhang C, Ruan C (2024) Deepseek-v3 technical report. arXiv preprint arXiv:241219437

Ma NJ, Isaacs FJ (2016) Genomic recoding broadly obstructs the propagation of horizontally transferred genetic elements. Cell Syst 3:199–207

Marand AP, Chen Z, Gallavotti A, Schmitz RJ (2021) A cis-regulatory atlas in maize at single-cell resolution. Cell 184:3041–3055. e3021

Marshall JM, Akbari OS (2018) Can CRISPR-based gene drive be confined in the wild? A question for molecular and population biology. ACS Chem Biol 13:424–430

Masaki Y, Onishi Y, Seio K (2022) Quantification of synthetic errors during chemical synthesis of DNA and its suppression by non-canonical nucleosides. Sci Rep 12:12095

Medema MH, Van Raaphorst R, Takano E, Breitling R (2012) Computational tools for the synthetic design of biochemical pathways. Nat Rev Microbiol 10:191–202

Naqvi S, Zhu C, Farre G, Ramessar K, Bassie L, Breitenbach J, Perez Conesa D, Ros G, Sandmann G, Capell T (2009) Transgenic multivitamin corn through biofortification of endosperm with three vitamins representing three distinct metabolic pathways. Proc Natl Acad Sci 106:7762–7767

Nayak N, Mehrotra S, Karamchandani AN, Santelia D, Mehrotra R (2025) Recent advances in designing synthetic plant regulatory modules. Front Plant Sci 16:1567659

Neve P (2018) Gene drive systems: do they have a place in agricultural weed management? Pest Manag Sci 74:2671–2679

Noyce RS, Lederman S, Evans DH (2018) Construction of an infectious horsepox virus vaccine from chemically synthesized DNA fragments. PLoS One 13:e0188453

Ortiz-Ramírez C, Guillotin B, Xu X, Rahni R, Zhang S, Yan Z, Coqueiro Dias Araujo P, Demesa-Arevalo E, Lee L, Van Eck J (2021) Ground tissue circuitry regulates organ complexity in maize and *Setaria*. Science 374:1247–1252

Ostrov N, Landon M, Guell M, Kuznetsov G, Teramoto J, Cervantes N, Zhou M, Singh K, Napolitano MG, Moosburner M (2016) Design, synthesis, and testing toward a 57-codon genome. Science 353:819–822

Paine JA, Shipton CA, Chaggar S, Howells RM, Kennedy MJ, Vernon G, Wright SY, Hinchliffe E, Adams JL, Silverstone AL (2005) Improving the nutritional value of Golden Rice through increased pro-vitamin A content. Nat Biotechnol 23:482–487

Park S, Mani V, Kim JA, Lee SI, Lee K (2022) Combinatorial transient gene expression strategies to enhance terpenoid production in plants. Front Plant Sci 13:1034893

Piatek AA, Lenaghan SC, Stewart CN Jr (2018) Advanced editing of the nuclear and plastid genomes in plants. Plant Sci 273:42–49

Pixley KV, Falck-Zepeda JB, Giller KE, Glenna LL, Gould F, Mallory-Smith CA, Stelly DM, Stewart CN Jr (2019) Genome editing, gene drives, and synthetic biology: will they contribute to disease-resistant crops, and who will benefit? Annu Rev Phytopathol 57:165–188

Radivojević T, Costello Z, Workman K, Garcia Martin H (2020) A machine learning automated recommendation tool for synthetic biology. Nat Commun 11:4879

Rai K, Wang Y, O'Connell RW, Patel AB, Bashor CJ (2024) Using machine learning to enhance and accelerate synthetic biology. Curr Opin Biomed Eng 31:100553

Raza A, Khare T, Zhang X, Rahman MM, Hussain M, Gill SS, Chen ZH, Zhou M, Hu Z, Varshney RK (2025a) Novel strategies for designing climate-smart crops to ensure sustainable agriculture and future food security. J Sustain Agric Environ 4:e70048

Raza A, Li Y, Prakash CS, Hu Z (2025b) Panomics to manage combined abiotic stresses in plants. Trends Plant Sci 30(10):1079–1084

Raza A, Zaman QU, Shabala S, Tester M, Munns R, Hu Z, Varshney RK (2025c) Genomics-assisted breeding for designing salinity-smart future crops. Plant Biotechnol J 23(8):3119–3151

Rialle S, Felicori L, Dias-Lopes C, Pérès S, El Atia S, Thierry AR, Amar P, Molina F (2010) BioNetCAD: design, simulation and experimental validation of synthetic biochemical networks. Bioinformatics 26:2298–2304

Richardson SM, Mitchell LA, Stracquadanio G, Yang K, Dymond JS, DiCarlo JE, Lee D, Huang CLV, Chandrasegaran S, Cai Y (2017) Design of a synthetic yeast genome. Science 355:1040–1044

Roell M-S, Zurbriggen MD (2020) The impact of synthetic biology for future agriculture and nutrition. Curr Opin Biotechnol 61:102–109

Rogers C, Oldroyd GE (2014) Synthetic biology approaches to engineering the nitrogen symbiosis in cereals. J Exp Bot 65:1939–1946

Rook O, Zwart H, Dogterom M (2025) Public attitudes to potential synthetic cells applications: pragmatic support and ethical acceptance. PLoS One 20:e0319337

Sargent D, Conaty WC, Tissue DT, Sharwood RE (2022) Synthetic biology and opportunities within agricultural crops. J Sustain Agric Environ 1:89–107

Schmidt M, Pei L, Budisa N (2018) Xenobiology: state-of-the-art, ethics, and philosophy of new-to-nature organisms. In: Zhao H, Zeng AP (eds) Synthetic biology–metabolic engineering. Advances in Biochemical Engineering/Biotechnology, vol 162. Springer, Cham. pp 301–315

Schwille P (2011) Bottom-up synthetic biology: engineering in a tinkerer's world. Science 333: 1252–1254

Shao Z, Zhao H, Zhao H (2009) DNA assembler, an *in vivo* genetic method for rapid construction of biochemical pathways. Nucleic Acids Res 37:e16–e16

Sharma S, Partap A, Balaguer MAdL, Malvar S, Chandra R (2022) DeepG2P: fusing multi-modal data to improve crop production. arXiv preprint arXiv:221105986

South PF, Cavanagh AP, Liu HW, Ort DR (2019) Synthetic glycolate metabolism pathways stimulate crop growth and productivity in the field. Science 363:eaat9077

Stra A, Almarwaey LO, Alagoz Y, Moreno JC, Al-Babili S (2023) Carotenoid metabolism: new insights and synthetic approaches. Front Plant Sci 13:1072061

Su X, Xu J, Rhodes D, Shen Y, Song W, Katz B, Tomich J, Wang W (2016) Identification and quantification of anthocyanins in transgenic purple tomato. Food Chem 202:184–188

Sun T, Zhu Q, Wei Z, Owens LA, Fish T, Kim H, Thannhauser TW, Cahoon EB, Li L (2021) Multi-strategy engineering greatly enhances provitamin A carotenoid accumulation and stability in Arabidopsis seeds. aBIOTECH 2:191–214

Van Dis EA, Bollen J, Zuidema W, Van Rooij R, Bockting CL (2023) ChatGPT: five priorities for research. Nature 614:224–226

van Lent P, Schmitz J, Abeel T (2023) Simulated design–build–test–learn cycles for consistent comparison of machine learning methods in metabolic engineering. ACS Synth Biol 12:2588–2599

Vanhercke T, Divi UK, El Tahchy A, Liu Q, Mitchell M, Taylor MC, Eastmond PJ, Bryant F, Mechanicos A, Blundell C (2017) Step changes in leaf oil accumulation via iterative metabolic engineering. Metab Eng 39:237–246

Varshney RK, Bohra A, Yu J, Graner A, Zhang Q, Sorrells ME (2021) Designing future crops: genomics-assisted breeding comes of age. Trends Plant Sci 26:631–649

Verma V, Batta A, Singh HB, Srivastava A, Garg SK, Singh VP, Arora PK (2023) Bioengineering of fungal endophytes through the CRISPR/Cas9 system. Front Microbiol 14:1146650

Wang W, Zhang X, Zhang Y, Zhang Z, Yang C, Cao W, Liang Y, Zhou Q, Hu Q, Zhang Y (2025) Single-cell and spatial transcriptomics reveals a stereoscopic response of rice leaf cells to Magnaporthe oryzae infection. Adv Sci 12(19):e2416846

Wurtzel ET, Vickers CE, Hanson AD, Millar AH, Cooper M, Voss-Fels KP, Nikel PI, Erb TJ (2019) Revolutionizing agriculture with synthetic biology. Nat Plants 5:1207–1210

Xu X, Crow M, Rice BR, Li F, Harris B, Liu L, Demesa-Arevalo E, Lu Z, Wang L, Fox N (2021) Single-cell RNA sequencing of developing maize ears facilitates functional analysis and trait candidate gene discovery. Dev Cell 56:557–568. e556

Xu X, Meier F, Blount BA, Pretorius IS, Ellis T, Paulsen IT, Williams TC (2023) Trimming the genomic fat: minimising and re-functionalising genomes using synthetic biology. Nat Commun 14:1984

Yang S, Hu G, Wang J, Song J (2024) CRISPR/Cas-based gene editing tools for large DNA fragment integration. ACS Synth Biol 14:57–71

Ye X, Al-Babili S, Kloti A, Zhang J, Lucca P, Beyer P, Potrykus I (2000) Engineering the provitamin A (β-carotene) biosynthetic pathway into (carotenoid-free) rice endosperm. Science 287:303–305

Yin R, Xia K, Xu X (2023) Spatial transcriptomics drives a new era in plant research. Plant J 116: 1571–1581

Yu W, Zhang S, Zhao S, Chen L-g, Cao J, Ye H, Yan J, Zhao Q, Mo B, Wang Y (2024) Designing a synthetic moss genome using GenoDesigner. Nat Plants 10:848–856

Yurchenko O, Shockey JM, Gidda SK, Silver MI, Chapman KD, Mullen RT, Dyer JM (2017) Engineering the production of conjugated fatty acids in *Arabidopsis thaliana* leaves. Plant Biotechnol J 15:1010–1023

Zeng S, Mao Z, Ren Y, Wang D, Xu D, Joshi T (2021) G2PDeep: a web-based deep-learning framework for quantitative phenotype prediction and discovery of genomic markers. Nucleic Acids Res 49:W228–W236

Zhang W, Mitchell LA, Bader JS, Boeke JD (2020) Synthetic genomes. Annu Rev Biochem 89:77–101

Zhang J, Chen Y, Fu L, Guo E, Wang B, Dai L, Si T (2021a) Accelerating strain engineering in biofuel research via build and test automation of synthetic biology. Curr Opin Biotechnol 67: 88–98

Zhang T-Q, Chen Y, Liu Y, Lin W-H, Wang J-W (2021b) Single-cell transcriptome atlas and chromatin accessibility landscape reveal differentiation trajectories in the rice root. Nat Commun 12:2053

Zhang L, He C, Lai Y, Wang Y, Kang L, Liu A, Lan C, Su H, Gao Y, Li Z (2023) Asymmetric gene expression and cell-type-specific regulatory networks in the root of bread wheat revealed by single-cell multiomics analysis. Genome Biol 24:65

Zhang D, Xu F, Wang F, Le L, Pu L (2024) Synthetic biology and artificial intelligence in crop improvement. Plant Commun 6:101220

Zhang R, Wang Y, Yang W, Wen J, Liu W, Zhi S, Li G, Chai N, Huang J, Xie Y (2025) PlantGPT: an arabidopsis-based intelligent agent that answers questions about plant functional genomics. Adv Sci 12(30):e03926

Zhong L, Geng L, Xiang Y, Guang X, Cao L, Shi J, Li W, Wang J, He W, Huang L (2025) Comparative spatial transcriptomics reveals root dryland adaptation mechanism in rice and HMGB1 as a key regulator. Mol Plant 18:797–819

Zhu W, Han R, Shang X, Zhou T, Liang C, Qin X, Chen H, Feng Z, Zhang H, Fan X (2024) The CropGPT project: call for a global, coordinated effort in precision design breeding driven by AI using biological big data. Mol Plant 17:215–218

Zhu J, Jia Q, Tang Q-Y, Osman G, Gu M-Y, Wang N, Zhang Z-D (2025) Application of synthetic microbial communities of *Kalidium schrenkianum* in enhancing wheat salt stress tolerance. Int J Mol Sci 26:860

Zong J, Wang L, Zhu L, Bian L, Zhang B, Chen X, Huang G, Zhang X, Fan J, Cao L (2022) A rice single cell transcriptomic atlas defines the developmental trajectories of rice floret and inflorescence meristems. New Phytol 234:494–512

Part III
Advanced Applications in Crop Improvement

Chapter 8
Big Data, Machine Learning, and Artificial Intelligence for Crop Improvement

Vasileia Spyridaki, Cassandria Geraldine Tay Fernandez, Vanika Garg, and Rajeev K. Varshney

Abstract Recent technological advances have produced vast and complex data sets for crop improvement and plant breeding, termed big data. The challenge arising from this plethora of information is to extract meaningful information to be applied for developing next-generation crops and advancing current agricultural practices for crop production. Artificial intelligence (AI) and machine learning (ML), a subfield of AI, offer a set of computational approaches that aim to transform the information from large-scale datasets into breeding decisions by bringing different data types together and identifying hidden patterns. Here, we present an overview of big data, ML and AI in the context of crop improvement. We discuss what big data is, the challenges arising from large-scale datasets, and the big data types encountered in crop improvement research. We introduce the basics of AI and ML, focusing on examples drawn from agriculture. We highlight the advantages and scopes for applying ML and AI in crop improvement and gain insight into how big data, ML, and AI are integrated into research presently. We show how combining big data with AI and ML can revolutionise decision-making processes in crop improvement.

Keywords Complex datasets · Smart agriculture · Artificial intelligence · Machine learning · Deep learning · Plant breeding · Big data integration

Learning Objectives

- Defining big data, types of big data encountered in crop improvement research, challenges arising from large-scale datasets, and how big data can enable informed agricultural decisions.
- Types of artificial intelligence (AI) and machine learning (ML), and how they can be applied in the context of plant agricultural sciences and crop improvement.

V. Spyridaki · C. G. Tay Fernandez · V. Garg · R. K. Varshney (✉)
WA State Agricultural Biotechnology Centre, Centre for Crop and Food Innovation, Food Futures Institute, Murdoch University, Murdoch, WA, Australia
e-mail: Rajeev.Varshney@murdoch.edu.au

V. Garg et al. (eds.), *DNA of Sustainability*, Sustainability Sciences in Asia and Africa,
https://doi.org/10.1007/978-981-95-5109-5_8

- Why is there a rise in integrating big data, ML and AI in crop improvement research?
- Goals of big data, ML, and AI integration in crop improvement applications.

8.1 Introduction

Farming and agriculture have relied on data analytics to optimise production for centuries. Whether it be measuring water quality, fertiliser rates, or market demands, processing and understanding data have been crucial in agricultural production. Over the last decade, data from a variety of sources have become more readily available on a scale previously unseen. Petabytes of both quantitative and qualitative data can be gathered in days and be used to improve agricultural production, reduce environmental impact, develop next-generation crops, and inform management decisions.

Tools such as AI and ML are being used to analyse these large-scale datasets and identify hidden patterns within the data. This information has been used to enable farmers and businesses to make informed decisions to optimise management decisions and enhance productivity, transforming the information into actionable outputs.

This paper aims to present an overview of big data, ML and AI in the context of agriculture. Specifically, we discuss what big data is and its potential applications, as well as the challenges that come with using big data for crop improvement. We also highlight the scope for applying ML and AI in crop improvement and show how combining big data with AI and ML can revolutionise the decision-making processes in crop improvement.

8.2 Big Data

8.2.1 What Is Big Data and Why Do We Use It?

'Big Data' refers to large-scale datasets that are too large to fit into local memory and are often highly complex, requiring specialised pipelines and programmes for processing (Cox and Ellsworth 1997). As the name suggests, it encompasses a substantial body of information defined by several key characteristics, including the size of the dataset to be analysed and processed (volume), the speed at which the data is generated or updated (velocity), the types of structured and unstructured data produced (variety), and extent the data is changing (variability; McAfee and Brynjolfsson 2012; Xu et al. 2022).

With the sheer volume and variability of data, big data can be used to distinguish the signal from the noise; the meaningful information trying to be detected from random, unwanted variation. Big data analysis incorporates a large swath of data from a variety of sources, including both structured and unstructured data. As the name suggests, structured data are organised with defined formats and lengths and can often be arranged as Excel spreadsheets and databases. Most high-throughput data are structured, including genotype and phenotype data for germplasm populations, expression data, and mass spectrometry data (Xu et al. 2022). In contrast, unstructured data are datasets that cannot be as easily categorised, such as published literature, audio, and visual media. By integrating both types of data, big data analyses allow for nonlinear relationships, causal effects, and other trends that would not be visible in smaller or narrower datasets are able to be identified on both the micro and macro scale (Xu et al. 2022).

The main advantage of big data lies in its ability to transform raw information into actionable insights (Khan et al. 2024b). Big data analysis has successfully been used across many domains for economic development, industrial progress, and scientific innovation (Kamilaris et al. 2017; Wang and Wang 2020; Hidayat and Khabibi 2020; Pal et al. 2020). Big data have also been used to understand complex agricultural ecosystems, including farming systems.

Studying environmental data can lead farmers and companies to change and improve their agricultural practices, increasing the productivity of the crops. Genomics big data can also be used for crop improvement, facilitating phenotype and genotype predictions for genomic selection.

8.2.2 *The Big Data Era in Plant Sciences*

8.2.2.1 What Types of Data Do We Encounter in Crop Improvement Research?

The 'big data era' refers to how large-scale data collection and analysis are being integrated into all facets of research and industry. In plant sciences, big data is now being used to enhance our understanding of plant biology, improve agricultural practices, and improve crops.

The cost of sequencing has rapidly decreased over the last two decades thanks to advancements in high-throughput sequencing technologies. As a result, there has been an influx of DNA and RNA sequencing data, making large-scale biological datasets such as genomics and transcriptomics data more readily available for a plethora of species, including ones with complex, polyploid, and large genomes (Pazhamala et al. 2021). This has led to fully annotated, high-quality genome assemblies in several crop plants such as wheat (Bayer et al. 2022a), rice (Wang et al. 2018), tomato (Su et al. 2021), sugarcane (Healey et al. 2024), pea (Yang et al. 2022), peanut (Wang et al. 2025), and maize (Jiao et al. 2012; Chen et al. 2023a). Alongside individuals being sequenced more precisely, new sequencing techniques have led to population-wide analysis of data. Single-nucleotide polymorphism (SNP) calling can be performed over a population of large, genetically diverse breeding populations, recording thousands of genetic markers and millions of SNPs (Chen et al. 2023b). SNPs play a vital role in shaping the genetic diversity of crops, and understanding these variants is important for breeding, conservation, and developing new crops with improved characteristics. Similarly, genetic markers can be used to understand the genetic properties as well as networks that contribute to developing superior crop varieties (Varshney et al. 2019).

With climate change reducing overall crop output and a demand for food, induced by a growing population (van Dijk et al. 2021), there is a need to maximise crop yield and make crops more resistant to both biotic and abiotic stress factors. Omics experiments studying the genome, transcriptome, proteome, metabolome, etc., provide huge swathes of information and provide insights into molecular and biochemical processes. By integrating information garnered by omics technologies, complex plant regulatory networks can be identified and dissected to infer and describe causal links between different components of the system, providing an understanding of system-wide responses (systems biology) during growth, development, and stresses (Urano et al. 2010; Hong et al. 2019; Pazhamala et al. 2021). Currently, big data have contributed to the research of several major crops, including wheat, rice, soybean, chickpea, and maize (Table 8.1).

Systems biology can be used to dissect complex agronomic traits, predict the outcome of mutations or modifications, and predict phenotypes in different environmental conditions (Lavarenne et al. 2018). Systems biology and other big data analytics can also be used to identify high-yielding, resistant crops, aiding in genomic selection, and genomic-assisted breeding (Varshney et al. 2005, 2021a). It can be used to model and analyse multigenic traits such as plant architecture, nitrogen use and water use efficiency, and abiotic and biotic stress tolerance. Producing and interpreting omics big data are crucial for harnessing and maximising genetic gains (Kim et al. 2020).

However, the main issue with big data in biology is that there is a lot of data and it is difficult to search through, aggregate, and visualise large datasets (Wolfert et al. 2017). There are no standard protocols shared among the global community, especially for collecting multi-omic data (Macklin 2019). There are also difficulties in managing the increasing complexity of biological data, transferring, scaling, and analysing the data to derive meaningful analyses from it (Pal et al. 2020). The rapid increase in data demands efficient infrastructure for big data storage and analysis. NCBI (http://www.ncbi.nlm.nih.gov/) and EMBL (www.ebi.ac.uk/) are two well-known publicly accessible biological repositories available online that have risen to meet the challenge of big data in biology. These databases accept data submissions from different laboratories worldwide and store massive amounts of sequencing, assembly, and taxonomical data. NCBI also stores unstructured data in the form of academic literature. Cloud storage and cloud-based services have also risen in popularity to try and accommodate the growing data and computational demands of big data, such as Amazon Web Services, OmicsBox (OmicsBox 2019), and Galaxy (Jalili et al. 2019). Above all else, full utilisation of this data requires a strong capacity to handle multidimensional big data.

Table 8.1 Examples of big data types and their contributions to crop improvement

Data type	Dataset	Contributions to crop improvement	References
Phenotypic, genotypic, weather, and field management data	Database	Understanding and managing maize performance	McFarland et al. (2020)
Genotypic and phenotypic data for 3k accessions	Database	Identification of superior haplotypes for key chickpea traits and genomic prediction	Varshney et al. (2021b)
Nested association mapping of genes controlling soybean yield and other key traits	Database	Yield potential improvement of soybean varieties	Xavier et al. (2016)
Synthetic dataset of canopy reflectance	Database	Estimation of wheat traits	Chen et al. (2022)
Aerial image database	Database	Research in computer vision for agriculture	Chiu et al. (2020)
Wheat pangenome	Pangenome	Facilitation of wheat genomics-assisted breeding	Bayer et al. (2022a)
Chickpea super-pangenome	Pangenome	Identification of loci for agronomic traits of chickpea and facilitation of chickpea breeding and improvement	Khan et al. (2024a)
Chili pangenome	Pangenome	Genomic diversity and population structure of wild and domesticated peppers: a model for evolutionary distribution	Tripodi et al. (2021)
Soybean pangenome	Pangenome	Identification of gene loss during domestication and breeding	Bayer et al. (2022b)
Rice genome assembly	Reference genome	Reference genome for future analysis	Wang et al. (2018)
Sugarcane genome assembly	Reference genome	Accurate SNP calling and genomics-assisted improvement of sugarcane; previous studies were using a reference genome from other crop species	Healey et al. (2024)

8.2.2.2 Agricultural Uses of Big Data

Big data have primarily been used to improve agricultural practices to enhance crop productivity. Remote sensing and drone usage have made it easier for farmers to gather analytics on their crops, capturing a wide spread of environmental data, and allowing them to monitor soil health or track weather patterns. This information can be used for a variety of purposes: in soil science, big data extracted from the WA Department of Agriculture and Food (AGRIC) soil database was used to establish meaningful relationships between various soil profiles at different locations (Armstrong et al. 2007). In another study, big data were used to refine the geospatial targeting of new drought-tolerant maize varieties in southern Africa and showed how analytical tools enhance targeting and uptake of new agricultural technologies for boosting food security in developing countries (Tesfaye et al. 2016).

Along with drones, satellite data and Global Positioning System (GPS) devices are becoming standard for gathering new data, even in poor or remote countries, increasing the availability, frequency, and quantity of big data (Carletto 2021). By studying this data, researchers can identify trends and gather insights into existing issues like operational inefficiencies. This has been implemented in rural India, where the Progressive Rural Integrated Digital Enterprise (PRIDE™) business model has been used to help farmers get access to scientific agriculture practices and climate data on their mobile phones, helping them make informed farming decisions for farm management (Sawant et al. 2016). Data-driven decision-making has allowed farmers to improve their productivity by increasing efficiency and optimising every aspect of farming operations (such as weeding, disease detection, and harvesting), which has also minimised environmental impact, thereby improving sustainability.

The challenges and limitations of big data analytics are similar in agriculture as they are in research. While unstructured data are easier to collect and allows for a greater scope of what can be measured, this sort of data is complex and often subject to selectivity and coverage bias (Carletto et al. 2015). Additionally, due to the complexity of the agricultural process, agricultural data are often of poor quality. The data often extends over long periods of time with significant seasonal and annual variation, and despite tracking data being easier, the data often lacks a proper record or measurement errors (Carletto et al. 2015). Data also needs to be aggregated into a single matrix after extensive preprocessing and normalisation (Pazhamala et al. 2021). Fortunately, several platforms to integrate multidimensional omics datasets are now available, such as mixOmics, OnPLS modelling, Integromics, and sparse Multi-Block Partial Least Squares (Misra et al. 2018; Pinu et al. 2019).

Big data solutions can be prohibitive for small-scale farmers who do not have the resources to invest in and maintain these technologies. However, as technologies advance and improved tools increase productivity, governments and sectors have been investing in big data technology, slowly making this more available for smaller farms (Carletto et al. 2015; Kilic et al. 2021). Programmes are being developed to help make big data more accessible, such as the LandPKS project, which has been providing smartphone applications for measuring soil quality for farmers to interpret data on their land, such as soil type, soil health, and vegetation cover, for over a decade (Herrick et al. 2013).

Big data facilitates actionable insights, but to get there, complex datasets must first be processed, normalised, interpreted, and then translated. This often requires different technologies and software platforms to be integrated across a large variety of data sources to make sense. Big data can transform raw information into actionable insights that drive smarter farming decisions. There are many examples of big data improving crops (Table 8.1). Further development of big data technologies has allowed for accurate crop yield prediction and early disease detection, and technologies such as artificial intelligence and machine learning can push this even further.

8.3 The Concepts of Machine Learning and Artificial Intelligence

8.3.1 Artificial Intelligence

Artificial intelligence (AI) is a branch of computer science aiming to develop systems or algorithms able to perform tasks that humans are better at the moment (Rich 1983). Nowadays, such tasks include interpreting visual signals from the world, understanding and responding to human language, making decisions, learning, and predicting future events based on previous experiences (Fig. 8.1).

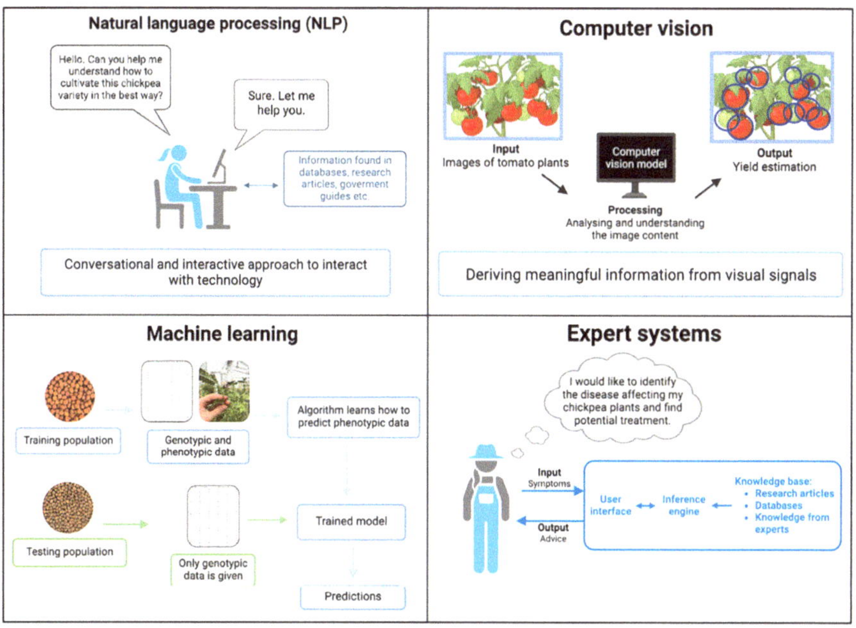

Fig. 8.1 Major branches of artificial intelligence (AI). (**a**) Natural language processing models aim at enabling machines to recognise, understand, and respond to human language. (**b**) Computer vision aims to develop machines that derive meaningful information from visual signals, such as videos and images. Computer vision tasks include image classification, object recognition, and object tracking, among others. (**c**) Machine learning aims at enabling computers to learn without being explicitly programmed. (**d**) Expert systems aim to solve decision-making problems typically performed by human experts, such as disease diagnosis

8.3.2 *Machine Learning*

Machine learning (ML) is a subfield of AI aiming to develop computer algorithms that improve themselves by learning from the provided dataset and ultimately making decisions. The models can perform regression or classification of new (unseen) data points based on previous experience gained by training in the provided dataset. Examples of ML models include random forest (RF), support vector machines (SVMs), k-nearest neighbours (kNN), artificial neural networks (ANN), and deep learning (DL). Recently, DL has been highlighted by the research community due to its flexibility in learning nonlinear patterns in vast, unstructured datasets resulting from its multilayered structure (deep neural networks) and its ability to handle different data types (Montesinos-López et al. 2021a; Danilevicz et al. 2022; Murphy et al. 2024). Commonly used DL topologies are multilayer perceptrons (MLPs), convolutional neural networks (CNNs), and recurrent neural networks (RNNs), among others.

The learning process can further be distinguished into supervised, unsupervised, and reinforcement learning (Fig. 8.2). Supervised learning is the most popular

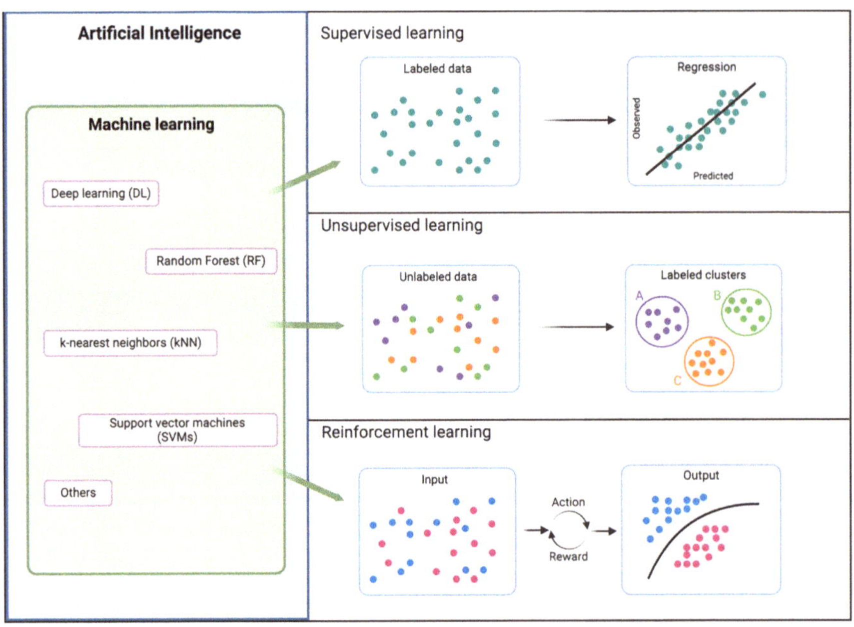

Fig. 8.2 Artificial intelligence, machine learning, and types of learning. Machine learning (ML) is a subdivision of artificial intelligence (AI). Among ML models, examples include deep learning, random forest, and support vector machines, among others. The learning process can be characterised as (a) supervised, in which the algorithm is being trained using labelled data aiming to predict the output and recognise patterns; (b) unsupervised, the algorithm is given unlabelled data and discovers patterns without guidance; and (c) reinforcement learning, in which the algorithm is trained following a trial-and-error approach

training method and involves training the ML model using labelled data. Examples of supervised learning include regression and classification. During the supervised learning process, the algorithm optimises and fits its parameters to perform well in assigning the input information to the given output. Specifically, an objective function is used to measure the error between the desired (known) output and the predictions. During training, the machine modifies its internal parameters (weights) to minimise the objective function; thus, the error (Hastie et al. 2009; LeCun et al. 2015). After the training, the model can predict the output of unseen input points. An example in this category is genomic prediction (GP), aiming to predict phenotypic traits, such as grain yield or time for flowering, of an individual crop based on its marker profile, usually composed of SNPs (Crossa et al. 2017).

On the other hand, unsupervised learning seeks to identify hidden patterns and similarities and cluster the data being trained using an unlabelled dataset. Unlike supervised learning, unsupervised learning has no direct way to measure the success of predictions (Hastie et al. 2009). An example includes principal component analysis (PCA), which is used to discover unknown relationships and visualise individual variability in population genetics studies.

Lastly, in reinforcement learning, the algorithm is trained using a trial-and-error approach in which the performance of the algorithm is increased by giving a positive reward to learn the pattern of interest. The goal of the training process is to maximise a reward signal. In reinforcement learning, the algorithm must discover which actions taken will lead to the most reward by trying them out, unlike supervised learning, in which the algorithm is provided with the true output values in the training set (Sutton and Barto 2018). One way to train an algorithm for image recognition tasks can be achieved by adopting a reinforcement learning approach.

8.4 Making Sense of Big Data for Crop Improvement

8.4.1 Advantages of AI and ML for Big Data

Plant breeding programmes are producing abundant data points of genomic, phenotypic, field management, soil, and other multi-omic datasets. The challenge arises as to how to use these vast datasets in a smart way to achieve higher genetic gains. There are multiple reasons why AI and ML approaches can be proven advantageous in comparison to conventional ones.

Conventional statistical approaches cannot easily handle big data due to their large-scale, diverse sources, and unstructured nature. In modern plant breeding, processing and combining of large and varied biological data sets are crucial to facilitate efficiency and enable decision-making processes. Unlike traditional approaches, AI and ML offer modern tools to enable the analysis of such complex datasets by extracting meaningful information and even identifying hidden and unexpected patterns and associations between data. The strict hypotheses characterising traditional statistical methods often burden the exploration of the datasets by introducing biases (Pérez-Enciso and Zingaretti 2019; Montesinos-López et al. 2021a, b). Moreover, ML and DL can integrate data types that are not suited to classical tabular formats, such as images resulting from high-throughput phenotyping in phenomics (Danilevicz et al. 2022). In the case of genomic prediction, nonlinear models such as ML can capture nonadditive gene interactions, such as dominance and epistasis, which have been proven to result in more accurate predictions of complex traits (Dwivedi et al. 2024). In addition, ML and DL approaches can account for the genotype-to-environment ($G \times E$) interactions, which is crucial given the environmental dependency characterising plant complex traits (Holland 2007), such as drought tolerance or yield.

8.4.2 Applications of Big Data and AI and ML Integration

In the context of plant breeding, AI and ML are systematising the information that is already available in the form of large-scale datasets by bringing different data types together and transforming them into breeding decisions (Fig. 8.3). Additionally, AI

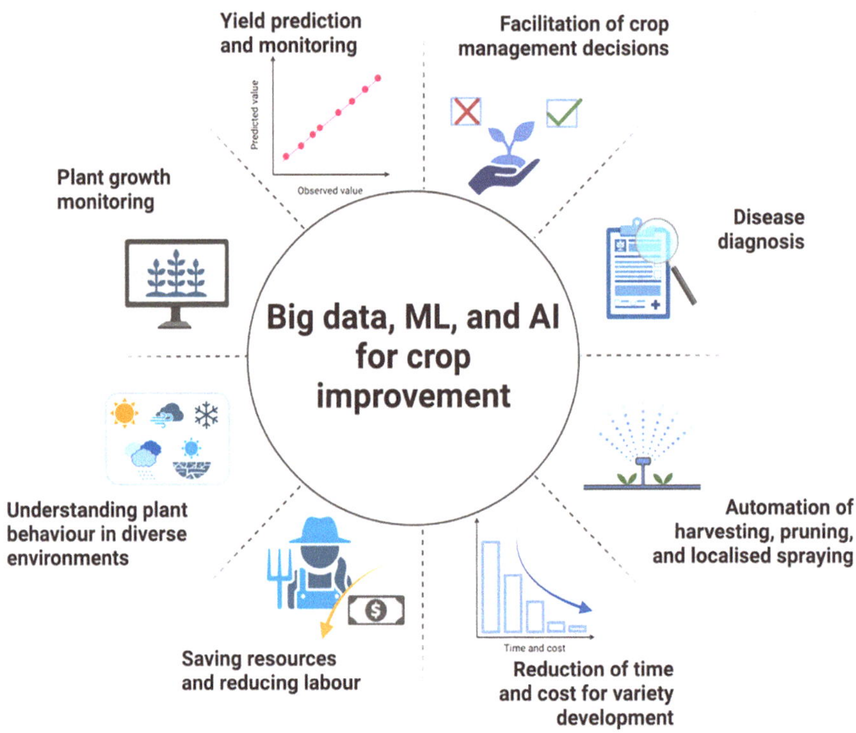

Fig. 8.3 Goals of big data, machine learning (ML), and artificial intelligence (AI) integration in crop improvement research

and ML are facilitating farm monitoring, thus saving time in data gathering and processing. Systems based on AI and ML can also aid the understanding of plant behaviour in diverse environments. The environment plays a vital role in plant development, often dominating the genetic effects in complex traits, such as yield or drought tolerance (Khan et al. 2022).

The application of ML and AI in crop improvement aims to leverage existing big data and transform information into agricultural decisions that will boost productivity and sustainability and reduce environmental impact.

In phenomics research, the use of AI and ML is being explored for the conversion of sensory data into phenotypic information, the development of models for phenotypic prediction, and database management for the storage and sharing of valuable information to be utilised in crop improvement (Khan et al. 2022). The ultimate goal of integrating AI and ML in phenomics is the automation of agricultural processes such as monitoring, spraying, localising, and harvesting, thus saving time and resources towards large-scale production or variety development (Afonso et al. 2020).

Various researchers investigate the integration of AI and ML for object detection, counting, and classification of crops in agricultural programmes to replace the currently followed labour-intensive practices (Table 8.2). Afonso et al. (2020) investigated tomato detection and counting in images taken in production greenhouses using the Mask-RCNN model (He et al. 2017), a DL algorithm of a CNN-based architecture. The promising results of the study indicate that using a set of simple cameras in a challenging experimental setup has the potential to advance fruit harvesting and yield estimation practices. Wu et al. (2019) integrated image processing and DL for counting the grain number per panicle of rice to replace manual counting and overcome the challenges of other image-based counting systems, which have difficulty in separating overlapped grains. The DL method showed accuracy >99% but was more time-consuming than the conventional approaches tested. Bauer et al. (2019) developed an open-source software, Airsurf-L, which combines computer vision and DL for the facilitation of large-scale aerial phenotyping and yield-related traits of lettuce. The method showed 98% accuracy for scoring and categorising; thus, acting as a valuable tool for quantifying the yield for harvest and facilitating crop management decision-making.

Accurate disease detection plays a crucial role in enhancing agricultural yield and avoiding losses. Current disease detection methods are resource-intensive, prone to errors, and require expert observation. Phenomics data combined with computer vision and ML have been investigated to systematise rust and senescence prediction in maize (DeSalvio et al. 2022); two important biotic, abiotic, and physiological factors affecting maize growth. The researchers used unoccupied aerial system (UAS) based high-throughput phenotyping for the collection of high-resolution images (RGB camera). Eight models, including ML, were compared for their prediction accuracy, indicating successful early indication of senescence and rust. In the sub-Saharan Africa region, maize lethal necrosis (MLN) and maize streak virus (MSV) are two diseases significantly affecting maize production. O'Halloran et al. (2024) investigated the integration of DL in the disease detection and

Table 8.2 Examples of studies that combine artificial intelligence (AI) and machine learning (ML) with big data for crop improvement

Study	Crop	Big data	AI and ML methods	References
Tomato detection and counting in the glasshouse	Tomato	123 images corresponding to 1612 fruits in total	Computer vision using the deep learning (DL) algorithm Mask-RCNN	Afonso et al. (2020)
Rice identification and grain number per panicle counting	*Japonica* and *Indica* rice	540 panicles, corresponding to 1337 images for DL model training	Faster RCNN (Ren et al. 2017), ResNet101 convolutional network (He et al. 2016)	Wu et al. (2019)
Counting and classification of lettuce in the field	*Lactuca sativa*	100,000 labelled lettuce signals for DL model training	Computer vision, DL (convolutional neural network; CNN), machine learning (ML)	Bauer et al. (2019)
Prediction of disease traits	Maize	36 vegetation indices from 2-year field trials	Ridge, lasso, elastic net, random forest, support vector machine with radial and linear kernels, partial least squares, and k-nearest neighbours	DeSalvio et al. (2022)
Disease detection (maize lethal necrosis; MLN, and maize streak virus; MSV)	Maize	15,344 images comprising MSV, MLN, and healthy maize leaves	CNN, EfficientNet V2 B0, and B1, LeNet-5, VGG-16, and ResNet50	O'Halloran et al. (2024)
Monitoring of wheat biophysical traits	Wheat	25,201 samples	DL (multilayer perceptron; MLP)	Chen et al. (2022)
Yield prediction	Maize	Field management, weather, and soil data from 96,137 genotypes in 41 locations across 6 years	Multimodal DL	Kick et al. (2023)
Prediction of wheat cultivar performance, fruit soluble solid content (SCC) for genomic selection	Wheat	Single nucleotide polymorphism (SNP) and phenotypic data for 599 and 2000 samples	DL (three CNN layers)	Wang et al. (2023)
	Maize	SNP and phenotypic data for 6210 samples. Also, SNP, gene expression, and phenotypic data for 1404 samples.		

(continued)

Table 8.2 (continued)

Study	Crop	Big data	AI and ML methods	References
	Tomato	SNP, gene expression, structural variation, and insertion–deletion (InDel) data for 332 samples.		
Prediction of yield and yield-related traits	Maize	SNP information (158,281 SNPs) for 309 maize lines evaluated in three environments	DL (feedforward neural network)	Montesinos-López et al. (2018)
	Wheat	SNP information (12,083 SNPs) for 250 wheat lines evaluated in three environments		
	Wheat	SNP information (39,758 SNPs) for 2374 wheat lines evaluated in heat and drought environments		

classification process. Six different DL architectures were tested, and a dataset comprised of 15,344 maize images categorised as infected by MLN or MSV and healthy was used for model training.

Crop monitoring of biophysical variables and plant growth can facilitate the optimisation of agricultural practices, maintenance of sustainability in crop production, and estimation of crop yield. Chen et al. (2022) integrated a crop growth model with a radiative transfer model to construct a diverse synthetic training dataset of canopy reflectance measurements. The synthetic dataset was used to train an MLP-based DL model for multi-trait prediction of biophysical variables; the DL model outperformed the compared models, including RF.

There are many factors affecting soil–water content and soil–moisture content, including the rate of mineralisation in the soil (Ma et al. 2022), the water-holding capacity of the soil, pH balance, nutrient content and soil texture (Dragone et al. 2020). AI and ML algorithms have been used to study the different sets of data available and have been used to remove noise from soil–water content and electrical conductivity data for soil parameters (Ratshiedana et al. 2023). AI has also been used to predict soil texture and to measure soil–water content because it is able to reduce the time, cost, and labour involved with processing different datasets. Gao et al. (2022) trained deep neural networks over 2 years using nine variables in a neural network to manage water and soil-state with high performance and statistical correctness.

Climate change and unpredictable weather patterns, such as irregular rainfall distribution, can further exacerbate the challenges faced by farmers, including water scarcity and poor soil fertility. To circumvent this, AI can be used to amalgamate data from satellites, ground sensors, and weather stations to forecast weather in real time, creating more accurate weather forecasts. This advanced modelling can be used to predict the best times for sowing and harvesting or for watering and fertilising the plants, overall optimising agricultural productivity. Pierre et al. (2023) designed a weather-based smart irrigation and fertigation system using Internet of Things in conjunction with AI, specifically with an adaptive-neuro fuzzy inference system-based rain prediction model (Bushara and Abraham 2015). The developed prototype showed that half of the water and fertilisers can be saved when the system is used. Pu-Yun et al. (2025) also designed a hybrid convolutional autoencoder and LSTM-based transformer network (CAE–LSTMT) to downcast simulation data into precise, hourly forecasts using 55,538 temperature and relative humidity records. The system predicted forecasts 5–20% more accurately than traditional LSTM models and was extremely accurate in predicting vapour pressure deficits and heat indexes. Knowing high temperatures are coming will allow farmers to protect crops and livestock against heat stress. Accurate forecasting allows farmers to make informed decisions in real time, reducing climate-related risks.

Managing weeds is important for optimising crop yields. ML algorithms have been coupled with imaging processing technologies to allow for real-time identification and mapping of target weeds, and several systems have been developed to automatically weed large fields. ML algorithms, such as CNN, DCNN, SVM, ANNs, RF classifier and KNN have been used to predict and map where weeds will arise (Muchhadiya et al. 2024). For example, Schmidt-Lebuhn et al. (2024) trained a weed identification model on 7161 images and were able to identify over 400 species of weeds, which can now able to be used on mobile devices. By integrating AI and ML algorithms into farm machinery, herbicides can also be applied to specific zones instead of over-spraying fields. This will significantly be reducing herbicide volumes and lower both cost and environmental impact.

Genomic selection (GS) is the application of genomic prediction (GP) models to select candidate individuals based on the estimation of their genomic estimated breeding values (GEBVs) or phenotypic values (Alemu et al. 2024). GP models make predictions based on genome-wide information, mainly capturing the SNP profile of the individuals (Crossa et al. 2017). There is a pressing need to develop GP methods with increased predictive ability to achieve food security goals. Attention is currently being paid to using ML and DL methods for capturing complex nonadditive gene effects characterising complex traits such as yield or drought tolerance (Montesinos-López et al. 2021a) and being flexible in incorporating diverse datasets. In addition, the integration of large-scale multi-omics datasets is believed to bridge the gap between genotypic variation and phenotypic outcomes, thus leading to more accurate results (Crossa et al. 2021).

Research efforts currently focus on the integration of AI and ML in genomic prediction schemes for increasing genetic gains. Aiming to capture the link between genotypic variation and phenotype, researchers include diverse information in the prediction process. Kick et al. (2023) developed a multimodal DL model that integrated different multi-omics big data; genotype, weather, soil, and field management data for 96,137 maize genotypes evaluated in 41 locations across 6 years were used for the model training. Wang et al. (2023) proposed a DL-based method for genomic prediction by integrating multi-omics datasets, such as SNPs, gene expression, structural variation, and insertions–deletions information for the prediction of tomato, maize, and wheat traits. DNNGP performed better than previously developed DL methods and conventional approaches. Montesinos-López et al. (2018) investigated how correlated traits and information from multiple environments can enhance phenotype predictions when combined with DL models. The use of multi-trait and multi-environment DL models resulted in similar prediction accuracies with the compared Bayesian multi-trait and multi-environment model (BMTME; Montesinos-López et al. 2016), although the DL model was advantageous due to the fact that it required fewer computational resources.

8.4.3 Challenges of AI and Big Data

AI and big data have the potential to transform the agricultural sphere by optimising productivity and fostering sustainability. However, deriving meaning from big datasets remains a challenge. While the data's scale can be used to correlate relationships between variables, the data's density makes it difficult to find these relationships in the first place. With more data, there are more chances for more false positives to arise, especially when looking for correlations. This is especially true in predictive models, where bias can lead to false positive results.

High-throughput biological data is accumulating at an unprecedented rate, and technologies are needed to manage these large datasets. These technologies need to be able to query and integrate both structured and unstructured data. Additionally, analysing and visualising big datasets is also difficult due to how memory-intensive big data is. While having a lot of data is good for research, it can also be challenging to store the data efficiently. As the amount of big data increases, so does the need for more space and more efficient compression techniques that can facilitate storage and retrieval of huge data, all without the loss of information (Durand and Ladner 2004). There is a need for a substantial digital infrastructure that can accommodate ML, AI and all the data it needs to process.

AI has multiple uses in agriculture, but risks widening the digital divide between large and small-scale farmers due to how expensive the technology is. Small and medium-sized farms often struggle to implement the sensors, drones or machinery required to make use of AI and ML algorithms with the same ease as larger commercial farms can. Additionally, farms that are marginalised and have poor Internet connections are less likely to benefit from AI, as the cost of data remains prohibitive, widening the gap between subsistence farmers and commercial farmers (Mehrabi et al. 2021). AI risks establishing a 'big data divide' between data-rich and data-poor individuals, creating and reinforcing a system where less data and less means of gathering it would put smaller farms at a huge disadvantage (Rotz et al. 2019). Additionally, farmers need to be educated in how to use these advanced technologies, and some may be apprehensive about changing long-standing methods, especially if existing equipment or workflows need to be changed.

Privacy and security are two of the biggest concerns surrounding AI. If AI is to be adopted into common agricultural practices, strict cybersecurity measures and data management practices are needed to protect sensitive information of farm operations and yields. It is essential to ensure the privacy and security of farmers' data and proprietary rights, even if the data is chosen to be shared. Additionally, it needs to be clear who controls the data generated between AI systems and who gets to access data to protect intellectual properties in methods of farming.

On the flipside, if AI is adopted by many farms, it poses the risk of replacing human workers in the industry. Agricultural robots and automation enabled by AI threaten the agency of farmers and the jobs of people performing manual labour, at the very least downsizing the industry. In the short term, large impacts are unlikely given the existing limitations of AI; the ethical use of AI and automation in farming must be considered. Big data and AI may lead to a loss of farmer autonomy, as increased usage of machines may subjugate them to both commercial and institutional demands (Fleming et al. 2018).

While there are many benefits AI can provide in the agricultural industry, there are many hurdles it faces before being adopted on a widespread scale. Big data requires a lot of time, resources, and effort to implement, and can be cost-prohibitive to smaller farms. AI is also prone to biases when not properly implemented, and there is a need for substantial digital infrastructure. Equally, there is the issue of data privacy and the risk of widening the gap between large and small-scale farmers.

New infrastructure and computing resources are slowly being developed to further crop development. However, laws and regulations must be implemented so there is equitable access to technology and to ensure AI can be sustainably and ethically used for diverse farming needs, not just for large corporations.

8.5 Future Prospects

The influx of data has been facilitating rapid developments in technology, much of which is able to be implemented in crop development. For instance, recent advancements in blockchain technologies have been used to modernise agriculture. Blockchain is a transparent, electronic system that allows for electronic record keeping, validation, and verification. It has been used as a tool to trace the journey of agricultural products without the presence of an intermediary organisation. Additionally, it can track transactions among anonymous parties, allowing for transparency amongst stakeholders and facilitating reliability (Krithika 2022). There have been discussions on how AI algorithms can analyse blockchain data to provide consumers with real-time information on the condition of food products (Mukherjee et al. 2021). These systems could also potentially help reduce food waste, improve food safety, and increase efficiency in the food supply chain by storing data collected by sensors and then later using said data for optimising the greenhouse environment (Jamil et al. 2022).

With more software tools being produced to facilitate biological big data and AI technology, AI-based breeding pipelines are growing in popularity. Aside from data collection, AI can be used to unlock gene diversity within genebanks and potentially be used to bridge the genotype–phenotype gap to facilitate crop breeding (Zhang et al. 2025). AI holds promise for refining crop traits by predicting the potential effects of gene variants on plant phenotypes, overall increasing the efficiency for breeding programmes (Farooq et al. 2024). Work is also going into improving the precision of gene editing systems using AI. For gene editing technologies such as CRISPR-Cas9, AI models such as DeepCRISPR (Chuai et al. 2018) have been used to design the optimal guide RNAs for a specified target sequence. These tools consider factors like genomic context, Cas protein type, desired mutation type, and off-target sites to design the optimal guide RNAs for a specified target sequence. However, more research needs to be done to refine AI's integration with gene editing, as it is costly and there is still a need for a more comprehensive CRISPR database (Dixit et al. 2023).

Ongoing research and development will make these technologies more accessible and effective. However, these tools will never be put fully to use if they do not have the support and protection from the government. For AI to be effectively implemented in the agriculture sector, the government needs to implement regulations to ensure that companies fairly and ethically use AI-powered solutions. Additionally, steps need to be taken so farmers can access data and make informed decisions. The private and public sectors need to educate and provide the resources and training to farmers to use AI-powered solutions so AI can facilitate data-driven decision-making and optimise resource allocation.

8.6 Summary

In this chapter, we provided an overview of how ML and AI can be utilised to interpret the continuously generated large-scale, complex datasets, termed big data, with the aim of crop improvement. Big data can be encountered in different forms, from genome assemblies, whole genome re-sequencing data, pangenomes, and super-pangenomes to phenotypic records obtained from multi-environment field trials, field management information, and high-resolution images of crops, to name a few. Despite their benefits, big data is hard to store, manage, and process, and they are not very useful if they remain unprocessed. ML and AI integration into crop improvement aims to facilitate the handling of big data and transform diverse or hidden information into valuable decisions through yield prediction, disease diagnosis, plant growth monitoring, and automation of localised spaying. Also, their use can reduce the cost and time for variety development. Leveraging big data, ML and AI can facilitate addressing the current agricultural challenges arising from population growth and climate change and achieving greater sustainability and profitability in crop production.

References

Afonso M, Fonteijn H, Fiorentin FS et al (2020) Tomato fruit detection and counting in greenhouses using deep learning. Front Plant Sci 11:571299. https://doi.org/10.3389/fpls.2020.571299

Alemu A, Åstrand J, Montesinos-López OA et al (2024) Genomic selection in plant breeding: key factors shaping two decades of progress. Mol Plant 17(4):552–578. https://doi.org/10.1016/j.molp.2024.03.007

Armstrong LJ, Diepeveen D, Maddern R (2007) The application of data mining techniques to characterize agricultural soil profiles. AusDM 785–100

Bauer A, Bostrom AG, Ball J et al (2019) Combining computer vision and deep learning to enable ultra-scale aerial phenotyping and precision agriculture: a case study of lettuce production. Hortic Res 6. https://doi.org/10.1038/s41438-019-0151-5

Bayer PE, Petereit J, Durant É et al (2022a) Wheat panache: a pangenome graph database representing presence–absence variation across sixteen bread wheat genomes. Plant Genome 15(3):e20221. https://doi.org/10.1002/tpg2.20221

Bayer PE, Valliyodan B, Hu H et al (2022b) Sequencing the USDA core soybean collection reveals gene loss during domestication and breeding. Plant Genome 15(1):e20109. https://doi.org/10.1002/tpg2.20109. Epub 2021 Jun 24. PMID: 34169673.

Bushara NO, Abraham A (2015) Using adaptive neuro-fuzzy inference system (ANFIS) to improve the long-term rainfall forecasting. J Netw Innov Comput 3:146–158. [Online]. Available: www.mirlabs.net/jnic/index.html

Carletto C (2021) Better data, higher impact: improving agricultural data systems for societal change. Eur Rev Agric Econ 48(4):719–740. https://doi.org/10.1093/erae/jbab030

Carletto C, Gourlay S, Winters P (2015) From guesstimates to GPStimates: land area measurement and implications for agricultural analysis. J Afr Econ 24(5):593–628. https://doi.org/10.1093/jae/ejv011

Chen Q, Zheng B, Chen T et al (2022) Integrating a crop growth model and radiative transfer model to improve estimation of crop traits based on deep learning. J Exp Bot 73(19):6558–6574. https://doi.org/10.1093/jxb/erac291

Chen J, Wang Z, Tan K et al (2023a) A complete telomere-to-telomere assembly of the maize genome. Nat Genet 55:1221–1231. https://doi.org/10.1038/s41588-023-01419-6

Chen N-C, Kolesnikov A, Goel S et al (2023b) Improving variant calling using population data and deep learning. BMC Bioinformatics 24(1):197. https://doi.org/10.1186/s12859-023-05294-0

Chiu MT, Xu X, Wei Y et al (2020) Agriculture-vision: a large aerial image database for agricultural pattern analysis. In: IEEE/CVF conference on computer vision and pattern recognition (CVPR), pp 2828–2838

Chuai G, Ma H, Yan J et al (2018) DeepCRISPR: optimized CRISPR guide RNA design by deep learning. Genome Biol 19:80–88. https://doi.org/10.1186/s13059-018-1459-4

Cox M, Ellsworth D (1997) Managing big data for scientific visualization

Crossa J, Pérez-Rodríguez P, Cuevas J et al (2017) Genomic selection in plant breeding: methods, models, and perspectives. Trends Plant Sci 22(11):961–975. https://doi.org/10.1016/j.tplants.2017.08.011

Crossa J, Fritsche-Neto R, Montesinos-Lopez OA et al (2021) The modern plant breeding triangle: optimizing the use of genomics, phenomics, and enviromics data. Front Plant Sci 12:651480. https://doi.org/10.3389/fpls.2021.651480

Danilevicz MF, Gill M, Anderson R et al (2022) Plant genotype to phenotype prediction using machine learning. Front Genet 13:822173. https://doi.org/10.3389/fgene.2022.822173

DeSalvio AJ, Adak A, Murray SC et al (2022) Phenomic data-facilitated rust and senescence prediction in maize using machine learning algorithms. Sci Rep 12(1):7571. https://doi.org/10.1038/s41598-022-11591-0

Dixit S, Kumar A, Srinivasan K et al (2023) Advancing genome editing with artificial intelligence: opportunities, challenges, and future directions. Front Bioeng Biotechnol 11(1). https://doi.org/10.3389/fbioe.2023.1335901

Dragone G, Kerssemakers AAJ, Driessen JLSP, Yamakawa CK, Brumano LP, Mussatto SI (2020) Innovation and strategic orientations for the development of advanced biorefineries. Bioresour Technol 302:122847

Durand N, Ladner RE (2004) Grammar-based compression of DNA sequences

Dwivedi SL, Heslop-Harrison P, Amas J et al (2024) Epistasis and pleiotropy-induced variation for plant breeding. Plant Biotechnol J:1–20. https://doi.org/10.1111/pbi.14405

Farooq MA, Gao S, Hassan MA et al (2024) Artificial intelligence in plant breeding. Trends Genet 40(10):891–908

Fleming A, Jakku E, Lim-Camacho L et al (2018) Is big data for big farming or for everyone? Perceptions in the Australian grains industry. Agron Sustain Dev 38(3):24–34. https://doi.org/10.1007/s13593-018-0501-y

Gao L et al (2022) A deep neural network based SMAP soil moisture product. Remote Sens Environ 277:113059

Hastie T, Tibshirani R, Friedman J (2009) The elements of statistical learning: data mining, inference, and prediction, 2nd edn. Springer, New York

He K, Zhang X, Ren S et al (2016) Deep residual learning for image recognition. In: IEEE conference on computer vision and pattern recognition (CVPR), 27–30 June 2016, pp 770–778. https://doi.org/10.1109/CVPR.2016.90

He K, Gkioxari G, Dollár P et al (2017) Mask R-CNN. In: IEEE international conference on computer vision (ICCV), Venice, Italy, 22–29 Oct 2017, pp 2980–2988. https://doi.org/10.1109/ICCV.2017.322

Healey AL, Garsmeur O, Lovell JT et al (2024) The complex polyploid genome architecture of sugarcane. Nature 628(8009):804–810. https://doi.org/10.1038/s41586-024-07231-4

Herrick J, Sala O, Karl J (2013) Land degradation and climate change: a sin of omission? Front Ecol Environ 11. https://doi.org/10.2307/23470470

Hidayat R, Khabibi A (2020) The big data era: what are the implications for the world of accounting? Majalah Ilmiah Bijak 17:248–254. https://doi.org/10.31334/bijak.v17i2.781

Holland JB (2007) Genetic architecture of complex traits in plants. Curr Opin Plant Biol 10(2): 156–161. https://doi.org/10.1016/j.pbi.2007.01.003

Hong W-J, Kim Y-J, Chandran AKN et al (2019) Infrastructures of systems biology that facilitate functional genomic study in rice. Rice 12(1):15. https://doi.org/10.1186/s12284-019-0276-z

Jalili V, Afgan E, Taylor J et al (2019) Cloud bursting galaxy: federated identity and access management. Bioinformatics 36(1):1–9. https://doi.org/10.1093/bioinformatics/btz472

Jamil F, Ibrahim M, Ullah I, Kim S, Kahng HK, Kim DH (2022) Optimal smart contract for autonomous greenhouse environment based on IoT blockchain network in agriculture. Comput Electron Agric 192:106573

Jiao Y, Zhao H, Ren L et al (2012) Genome-wide genetic changes during modern breeding of maize. Nat Genet 44(7):812–815. https://doi.org/10.1038/ng.2312

Kamilaris A, Kartakoullis A, Prenafeta-Boldú FX (2017) A review on the practice of big data analysis in agriculture. Comput Electron Agric 143:23–37. https://doi.org/10.1016/j.compag.2017.09.037

Khan MHU, Wang S, Wang J et al (2022) Applications of artificial intelligence in climate-resilient smart-crop breeding. Int J Mol Sci 23(19):11156. https://doi.org/10.3390/ijms231911156

Khan AW, Garg V, Sun S et al (2024a) *Cicer* super-pangenome provides insights into species evolution and agronomic trait loci for crop improvement in chickpea. Nat Genet 56(6): 1225–1234. https://doi.org/10.1038/s41588-024-01760-4

Khan R, Usman M, Moinuddin M (2024b) From raw data to actionable insights: navigating the world of data analytics. IJAETI 1(4):142–166

Kick DR, Wallace JG, Schnable JC et al (2023) Yield prediction through integration of genetic, environment, and management data through deep learning. G3 (Bethesda) 13(4). https://doi.org/10.1093/g3journal/jkad006

Kilic T, Moylan H, Ilukor J et al (2021) Root for the tubers: extended-harvest crop production and productivity measurement in surveys. Food Policy 102:102033. https://doi.org/10.1016/j.foodpol.2021.102033

Kim KD, Kang Y, Kim C (2020) Application of genomic big data in plant breeding: past, present, and future. Plants (Basel) 9(11). https://doi.org/10.3390/plants9111454
Krithika LB (2022) Survey on the applications of blockchain in agriculture. Agriculture 12(9):1333. https://doi.org/10.3390/agriculture12091333
Lavarenne J, Guyomarc'h S, Sallaud C et al (2018) The spring of systems biology-driven breeding. Trends Plant Sci 23(8):706–720. https://doi.org/10.1016/j.tplants.2018.04.005
LeCun Y, Bengio Y, Hinton G (2015) Deep learning. Nature 521(7553):436–444. https://doi.org/10.1038/nature14539
Ma C et al (2022) The role and mechanism of commercial macroalgae for soil conditioner and nutrient uptake catalyzer. Plant Growth Regul 97(3):455–476
Macklin P (2019) Key challenges facing data-driven multicellular systems biology. Gigascience 8(10). https://doi.org/10.1093/gigascience/giz127
McAfee A, Brynjolfsson E (2012) Big data: the management revolution. Harv Bus Rev 90(10):60-6, 68, 128
McFarland BA, AlKhalifah N, Bohn M et al (2020) Maize genomes to fields (G2F): 2014-2017 field seasons: genotype, phenotype, climatic, soil, and inbred ear image datasets. BMC Res Notes 13(1):71. https://doi.org/10.1186/s13104-020-4922-8
Mehrabi Z, McDowell MJ, Ricciardi V et al (2021) The global divide in data-driven farming. Nat Sustain 4:154–160. https://doi.org/10.1038/s41893-020-00631-0
Misra BB, Langefeld C, Olivier M et al (2018) Integrated omics: tools, advances, and future approaches. J Mol Endocrinol 62(1):R21–R45. https://doi.org/10.1530/JME-18-0055
Montesinos-López OA, Montesinos-López A, Crossa J et al (2016) A genomic Bayesian multi-trait and multi-environment model. G3 (Bethesda) 6(9):2725–2744. https://doi.org/10.1534/g3.116.032359
Montesinos-López OA, Montesinos-López A, Crossa J et al (2018) Multi-trait, multi-environment deep learning modeling for genomic-enabled prediction of plant traits. G3 (Bethesda) 8(12):3829–3840. https://doi.org/10.1534/g3.118.200728
Montesinos-López OA, Montesinos-López A, Hernandez-Suarez CM et al (2021a) Deep-learning power and perspectives for genomic selection. Plant Genome 14(3):e20122
Montesinos-López OA, Montesinos-López A, Pérez-Rodríguez P et al (2021b) A review of deep learning applications for genomic selection. BMC Genomics 22(19). https://doi.org/10.1186/s12864-020-07319-x
Muchhadiya RM, Kumawat PD, Gohil BS et al (2024) Precision weed management using artificial intelligence tools and techniques for sustainable agriculture. Agric Food e-Newsletter 6:1
Mukherjee U, Dutta S, Bandyopadhyay SK (2021) Assembling blockchain and IoT for smart food-supply chain. Asian J Adv Agric Res 16:49–58
Murphy KM, Ludwig E, Gutierrez J et al (2024) Deep learning in image-based plant phenotyping. Annu Rev Plant Biol 75:771–795. https://doi.org/10.1146/annurev-arplant-070523-042828
O'Halloran T, Obaido G, Otegbade B et al (2024) A deep learning approach for maize lethal necrosis and maize streak virus disease detection. Mach Learn Appl 16:100556. https://doi.org/10.1016/j.mlwa.2024.100556
OmicsBox—Bioinformatics made easy. BioBam Bioinformatics March 3, 2019. https://www.biobam.com/omicsbox
Pal S, Mondal S, Das G et al (2020) Big data in biology: the hope and present-day challenges in it. Gene Rep 21:100869. https://doi.org/10.1016/j.genrep.2020.100869
Pazhamala LT, Kudapa H, Weckwerth W et al (2021) Systems biology for crop improvement. Plant Genome 14(2):e20098. https://doi.org/10.1002/tpg2.20098
Pérez-Enciso M, Zingaretti LM (2019) A guide on deep learning for complex trait genomic prediction. Genes 10(7):553. https://doi.org/10.3390/genes10070553
Pierre N, Viviane I, Lambert UJ et al (2023) AI based real-time weather condition prediction with optimized agricultural resources. Eur J Technol 7(2):36–49
Pinu FR, Beale DJ, Paten AM et al (2019) Systems biology and multi-omics integration: viewpoints from the metabolomics research community. Meta 9(4):76. https://doi.org/10.3390/metabo9040076

Pu-Yun K, Yun-Ting W, Yu-Wen C, Meng-Hsin I, Ming-Hwi Y, Li-Chiu C, Fi-John C (2025) AI-driven weather downscaling for smart agriculture using autoencoders and transformers Comput Electr Agric 232:110129. https://doi.org/10.1016/j.compag.2025.110129
Ratshiedana PE, Abd Elbasit MAM, Adam E, Chirima JG et al (2023) Determination of soil electrical conductivity and moisture on different soil layers using electromagnetic techniques in irrigated arid environments in South Africa. Water. https://doi.org/10.3390/w15101911
Ren S, He K, Girshick R et al (2017) Faster R-CNN: towards real-time object detection with region proposal networks. IEEE Trans Pattern Anal Mach Intell 39(6):1137–1149. https://doi.org/10.1109/TPAMI.2016.2577031
Rich E (1983) Artificial intelligence. McGraw-Hill, New York
Rotz S, Gravely E, Mosby I et al (2019) Automated pastures and the digital divide: how agricultural technologies are shaping labour and rural communities. J Rural Stud 68:112–122. https://doi.org/10.1016/j.jrurstud.2019.01.023
Sawant M, Urkude R, Jawale S (2016) Organized data and information for efficacious agriculture using PRIDE™ model. Int Food Agribusiness Manag Rev 19(A):1–16. https://doi.org/10.22004/ag.econ.240701
Schmidt-Lebuhn AN, Bell M, Eckelmann C et al (2024) WeedScan, a weed reporting system for Australia using an image classification model for identification. Invasive Plant Sci Manag 17(3): 219–227. https://doi.org/10.1017/inp.2024.19
Su X, Wang B, Geng X et al (2021) A high-continuity and annotated tomato reference genome. BMC Genomics 22:898. https://doi.org/10.1186/s12864-021-08212-x
Sutton R, Barto A (2018) Reinforcement learning: an introduction, 2nd edn. The MIT Press, Cambridge
Tesfaye K, Sonder K, Caims J et al (2016) Targeting drought-tolerant maize varieties in southern Africa: a geospatial crop modeling approach using big data. Int Food Agribusiness Manag Rev 19(A):75–92
Tripodi P, Rabanus-Wallace M, Barchi L et al (2021) Global range expansion history of pepper (Capsicum spp.) revealed by over 10,000 genebank accessions. Proc Natl Acad Sci USA 118(34):e2104315118. https://doi.org/10.1073/pnas.2104315118
Urano K, Kurihara Y, Seki M et al (2010) 'Omics' analyses of regulatory networks in plant abiotic stress responses. Curr Opin Plant Biol 13(2):132–138. https://doi.org/10.1016/j.pbi.2009.12.006
van Dijk M, Morley T, Rau ML et al (2021) A meta-analysis of projected global food demand and population at risk of hunger for the period 2010–2050. Nat Food 2(7):494–501. https://doi.org/10.1038/s43016-021-00322-9
Varshney RK, Graner A, Sorrells ME (2005) Genomics-assisted breeding for crop improvement. Trends Plant Sci 10(12):621–630. https://doi.org/10.1016/j.tplants.2005.10.004
Varshney RK, Pandey MK, Bohra A et al (2019) Toward the sequence-based breeding in legumes in the post-genome sequencing era. Theor Appl Genet 132(3):797–816. https://doi.org/10.1007/s00122-018-3252-x
Varshney RK, Bohra A, Yu J et al (2021a) Designing future crops: genomics-assisted breeding comes of age. Trends Plant Sci 26(6):631–649. https://doi.org/10.1016/j.tplants.2021.03.010
Varshney RK, Roorkiwal M, Sun S et al (2021b) A chickpea genetic variation map based on the sequencing of 3,366 genomes. Nature 599(7886):622–627. https://doi.org/10.1038/s41586-021-04066-1
Wang WYC, Wang Y (2020) Analytics in the era of big data: the digital transformations and value creation in industrial marketing. Ind Mark Manag 86:12–15. https://doi.org/10.1016/j.indmarman.2020.01.005
Wang W, Mauleon R, Hu Z et al (2018) Genomic variation in 3,010 diverse accessions of Asian cultivated rice. Nature 557(7703):43–49. https://doi.org/10.1038/s41586-018-0063-9
Wang K, Abid MA, Rasheed A et al (2023) DNNGP, a deep neural network-based method for genomic prediction using multi-omics data in plants. Mol Plant 16(1):279–293. https://doi.org/10.1016/j.molp.2022.11.004

Wang X, Sun Z, Qi F et al (2025) A telomere-to-telomere genome assembly of the cultivated peanut. Mol Plant 18(1):5–8. https://doi.org/10.1016/j.molp.2024.12.001

Wolfert S, Ge L, Verdouw C et al (2017) Big data in smart farming—a review. Agric Syst 153:69–80. https://doi.org/10.1016/j.agsy.2017.01.023

Wu W, Liu T, Zhou P et al (2019) Image analysis-based recognition and quantification of grain number per panicle in rice. Plant Methods 15(1):122. https://doi.org/10.1186/s13007-019-0510-0

Xavier A, Muir WM, Rainey KM (2016) Assessing predictive properties of genome-wide selection in soybeans. G3 (Bethesda) 6(8):2611–2616. https://doi.org/10.1534/g3.116.032268

Xu Y, Zhang X, Li H et al (2022) Smart breeding driven by big data, artificial intelligence, and integrated genomic-enviromic prediction. Mol Plant 15(11):1664–1695. https://doi.org/10.1016/j.molp.2022.09.001

Yang T, Liu R, Luo Y et al (2022) Improved pea reference genome and pan-genome highlight genomic features and evolutionary characteristics. Nat Genet 54:1553–1563. https://doi.org/10.1038/s41588-022-01172-2

Zhang Y, Chen L, Huang R (2025) The application of artificial intelligence in crop gene research: current status and prospects. Adv Resour Res 5:720–735

Chapter 9
Genomic Selection for Crop Improvement in the Post-NGS Era

Vitor Seiti Sagae, Shatabdi Deb Proma, Sehijpreet Kaur, Ana Carolina Campana Nascimento, Moysés Nascimento, and Diego Jarquin

Abstract The advancements in DNA sequencing technologies brought new perspectives to the crop improvement process right after the first human genome was read. Next-generation sequencing (NGS) techniques are characterised by a high-throughput and highly accurate process to obtain RNA and DNA sequencing data. The post-NGS era in plant breeding is marked by the employment of high-density molecular markers spread throughout the crop genome. Genomic Selection (GS) is a tool that predicts the genetic merit of individuals using all available information derived from molecular markers, and it was proposed before the NGS era. This prediction tool leverages the linkage disequilibrium between markers and quantitative-trait-locus (QTL), and it is being applied mainly for highly quantitative traits using single or multi-trait approaches. The power of GS in crop improvement is highlighted in this chapter by discussing the potential of this tool in impacting the breeder's equation components, such as selection intensity (i), selection accuracy (r), genetic variance (σ), and the selection time length (L), aiming to increase the response to selection or genetic gains. In addition, GS can help breeders to improve their understanding of the genotype-by-environment (G × E) interactions observed in multi-environment trials for optimising the allocation of resources. At the end of the chapter, we discuss the integration of information obtained through NGS techniques with other data types (layers), enabling the so-called multi-omics approach to increase the efficiency of prediction models, enhancing the applicability of this tool in breeding strategies.

Keywords Genomic selection · Quantitative traits · Prediction models · Genetic gains · Multi-omics integration

V. S. Sagae · A. C. C. Nascimento · M. Nascimento
Agronomy Department, University of Florida, Gainesville, FL, USA

Laboratory of Intelligence Computational and Statistical Learning (LICAE), Department of Statistics, Federal University of Viçosa, Viçosa, MG, Brazil

S. D. Proma · S. Kaur · D. Jarquin (✉)
Agronomy Department, University of Florida, Gainesville, FL, USA
e-mail: jhernandezjarqui@ufl.edu

V. Garg et al. (eds.), *DNA of Sustainability*, Sustainability Sciences in Asia and Africa,
https://doi.org/10.1007/978-981-95-5109-5_9

9.1 Introduction

Plant breeding drives humankind as one alternative to fill the gap between food production and demand in response to the global population growth, through the development of new varieties with higher yield and adaptability (Bhat et al. 2016; He and Li 2020). Also, the crop improvement process has the potential to respond to climate change, enabling the release of crops for new and challenging environmental conditions (Wang et al. 2017). In the past, crop improvement was primarily based on phenotypic evaluations to identify superior individuals (Parveen et al. 2023). However, the observed genetic gains obtained through phenotypic evaluations are not enough to fulfill the required production expected for the next few years for several of the most important crops, such as rice, wheat, maize, and soybeans (Ray et al. 2013; Cobb et al. 2019). Mainly because the traditional breeding strategies based on phenotypic information are lengthy (time-consuming) and costly. Therefore, new techniques should be incorporated into the crop improvement process to accelerate variety development, while maintaining genetic diversity and reducing operational costs (Wang et al. 2018; Cobb et al. 2019; He and Li 2020).

The conventional cross improvement process is based on evaluating morphological traits controlled by a few or many genes, also known as phenotypic selection (PS) (Jaganathan et al. 2020). Nevertheless, phenotypic selection has some intrinsic limitations mainly for complex traits, such as a high environmental effect and consequently low heritability in the presence of pronounced genotype-by-environment (G × E) interactions, and the need to visualise or measure trait expression, which can take several years in perennial species (Bhat et al. 2016; Jonas and De Koning 2016). In fact, most economic traits in crops are controlled by many genes with small effect (Xu et al. 2017), and the subsequent evaluations throughout the years turn crop improvement into a long process even for annual species (Tuberosa 2012).

Molecular markers are a powerful tool to help crop improvement in both crucial points of plant breeding: creating genetic diversity and selecting for traits of interest (Yang et al. 2015; Xu et al. 2017). This tool has been developed for more than 40 years of study and is increasingly being incorporated in the crop improvement process (Grover and Sharma 2016). In the 1980s, at the beginning of the so-called pre-NGS era, the most used sequencing protocol to obtain molecular markers was based on the Sanger sequencing proposed by Fredrick Sanger in 1977, known as the chain termination-based sequencing method (Sanger et al. 1977; Liu et al. 2012). This method uses retroviral RT enzymes to incorporate chain-terminating dideoxynucleotides into molecules, being a simple approach but still highly accurate and considerably at low cost to obtain RNA and DNA sequences (Guo and Pyle 2023).

Several areas of biology and medical sciences were drastically changed after the discoveries enabled by Sanger sequencing (Goodwin et al. 2016). The following decade was important due to the surging of other types of molecular markers such as the restriction fragment length polymorphism (RFLP), amplified fragment length polymorphism (AFLP) and the microsatellites (SSR) (Jaganathan et al. 2020). These findings were developed into new tools to be employed in crop improvement, being used primarily for indirect selection or for trait introgression, associating markers with phenotypes (Collard and Mackill 2008). The marker-assisted selection (MAS) was widely used to detect plants with desirable alleles, initiating what is known as molecular breeding (Wang et al. 2018). However, MAS is limited to traits controlled by one or a few major genes, being not suitable for most complex traits controlled by many genes with small effects (Hasan et al. 2021).

The human genome project started in 1998 and finished in 2001. It brought significant advancements in sequencing techniques, where new genotyping technologies were developed in a few years, aiming to increase the speed and accuracy while reducing costs for obtaining sequencing data (Hu et al. 2021). The development of the next-generation sequencing (NGS) was based on a high-throughput and highly parallelisation process, consequently generating more data at an affordable cost (Liu et al. 2012). The use of NGS enabled the obtention of high-density genomic data through some approaches, such as whole-genome sequencing (WGS) and reduced representation sequencing, as the genotype-by-sequencing (GBS) (Poland and Rife 2012). These approaches enabled the identification of minor genetic variations widely spread throughout the genome as the single-nucleotide polymorphisms (SNPs), and larger structural changes as indels (insertion–deletion variations), being especially useful for crop improvement even for species that lack a reference genome (Satam et al. 2023). The table presents a few of the new sequencing platforms developed in the post-NGS era (Hu et al. 2021) (Table 9.1).

Table 9.1 Examples of sequencing platforms in the post-NGS era

Technology	Company	System platforms	Maximum read length
Short-read	Illumina	iSeq	2 × 150 bp
		Miniseq	2 × 300 bp
		MiSeq	2 × 150 bp
		NextSeq550	2 × 150 bp
		NovaSeq 6000	2 × 250 bp
		MiSeqDx	2 × 300 bp
		NextSeq550 Dx	2 × 150 bp
	ThermoFisher	GeneStudio S5	600 bp
		Genexus	400 bp
		Ion PGM-Dx	200 bp
Long-read	Pacific Biosciences	Sequel	300 kb
		Sequel II	
		Sequel IIe	
	Oxford Nanopore	Flongle	>4 Mb
		MiniON	
		GridION	
		PromethION	

Source: Adapted from Hu et al. (2021)

Next-generation sequencing techniques made it possible to obtain high-density marker panels faster and with relatively less cost (Hu et al. 2021). Hence, in this post-NGS era, genomic selection GS surged as a proposed method to overcome the drawbacks from MAS, which, instead of selecting based on marker-phenotype associations, all available dense marker information is used to predict the genetic merit of even unphenotyped individuals (Bhat et al. 2016). This methodology brought to plant breeding new perspectives to increase the understanding of complex traits, holding the potential to impact the crop improvement process through several approaches that will be discussed below in this chapter. In a scenario with increasing food demand and more often and intense climate changes, GS can be a crucial tool to overcome these challenges.

9.2 Statistical Models Applied to Genomic Selection (GS)

GS models utilise high-density panels of molecular markers, predominantly SNPs, to predict the genetic merit of individuals. These models establish a functional relationship between genotypes and phenotypes in a reference population to estimate the genetic merit of new individuals with higher accuracy, especially for complex and difficult-to-measure traits. This technique is particularly useful for traits with low heritability (i.e. those where the environment has a large influence on phenotypic expression) as well as for traits that are expensive to evaluate (e.g. baking time) or that manifest late in the organism's life cycle such as yield (Bonnett et al. 2022) and rust resistance (Saavedra et al. 2023). One of the main advantages of GS lies in the ability to identify genetically superior individuals at early stages of the selection process, accelerating genetic gain and reducing costs (Meuwissen et al. 2001).

The initial linear predictor adopted to compute the genomic breeding values is described as:

$$y_i = \mu + \sum_{j=1}^{p} x_{ij}\beta_j + e_i,$$

where μ is an intercept, x_{ij} is the genotype of the ith individual at the jth marker ($j = 1, 2, \ldots, p$), β_j is the corresponding marker effect, and e_i is the error term.

The GS models often face the challenge of having more parameters to estimate than available records (large-p and small-n problem) for model fitting, making it infeasible to obtain a solution for the p marker effects. To address this, it is crucial to employ statistical methods that incorporate shrinkage estimation techniques for regularisation, either within a frequentist or a Bayesian framework (de los Campos et al. 2013).

Shrinkage estimation techniques, also denoted as penalised methods, aim to find an optimal trade-off between estimating variance and bias. Here, the mean squared error (MSE) is used as a proxy to measure this balance between the variance and the bias. In the GS context, where in general p (markers) $>>>> n$ (records), the use of standard estimation methods as ordinary least squares (OLS) and maximum likelihood (ML), increases the variance estimates as the number of markers considered in the modelling (p) increases. In practice, the estimation process becomes unstable. To address this problem, shrinkage or penalised methods force (or reduce) the estimates towards zero to obtain a lower variance, but in return increase the bias (Montgomery et al. 2020). In other words, this approach promotes regularisation, reducing the effective number of parameters and increasing the stability of estimates.

Under the frequentist approach, the penalised methods consider as estimates of regression coefficients the solution to the following optimisation problem:

$$\left(\widehat{\mu}, \widehat{\beta}\right) = \operatorname{argmin}_{\left(\widehat{\mu}, \widehat{\beta}\right)} \left\{ \sum_{i=1}^{n} \left(y_i - \mu - \sum_{j=1}^{p} x_{ij}\beta_j \right)^2 + \lambda J\left(\widehat{\beta}\right) \right\},$$

where $J\left(\widehat{\beta}\right)$ is some penalty function, λ is the parameter that controls the strength of regularisation. When $\lambda = 0$ the optimisation problem reduces to the OLS approach.

Penalised estimation methods differ according to the implemented penalty function, resulting in different levels of coefficient shrinkage. Hoerl and Kennard (1970) introduced ridge regression (RR), which uses an L2 norm penalty function $J\left(\widehat{\beta}\right) = \sum_{j=1}^{p} \beta_j^2$ to shrink estimates towards zero, but never exactly zero. Tibshirani (1996) proposed the least absolute shrinkage and selection operator (LASSO), denoted as L1 norm, and it is defined as $J\left(\widehat{\beta}\right) = \sum_{j=1}^{p} |\beta_j|$. Unlike RR, LASSO performs both shrinkage of estimates and variable selection simultaneously. Nevertheless, the LASSO solution is constrained to have a maximum amount of p nonzero coefficients, with p representing the number of predictor variables (Park and Casella 2008).

Bayesian inference provides an alternative approach to the frequentist shrinkage estimation, and some of these models correspond to the Bayesian version of the penalised models. The Bayesian framework considers the unknown model parameters as random variables. Hence, any prior knowledge about these parameters can be incorporated into the analysis through probability distributions. This characteristic allows this method a higher thorough understanding of the underlying structure of the evaluated trait.

To perform the inference, prior information is combined with the observed data (likelihood). This combination is done through Bayes' theorem, which can be simplified in a Bayesian framework as follows: Posterior $\propto$ Likelihood $\times$ Prior (Gamerman and Lopes 2006).

The use of different prior distributions can lead to different methods. Bayesian ridge regression assumes a normal prior distribution with equal variance for all marker effects, $\beta_j \mid \sigma_\beta^2 \sim N\left(0, \sigma_\beta^2\right)$. This assumption results in a uniform shrinkage of marker effects, consistent with an infinitesimal model where the trait is explained by many QTLs with small effects. Bayes A and B methods allow for adaptive modelling of the genetic architecture. In Bayes A, it is achieved by using a normal prior distribution with varying variances for marker effects, $\beta_j \mid \sigma^2_{\beta_j} \sim N\left(0, \sigma^2_{\beta_j}\right)$. Unlike Bayes A, Bayes B employs a normal mixture distribution as a prior, $\beta_j \mid \sigma^2_{\beta_j} \sim (1-\pi)N(0, \sigma^2_{\beta_j}) + \pi N(0, \sigma^2_{\beta_j})$, allowing for variable selection. In this prior distribution, π represents the probability that a marker is drawn from the second component of the mixture, $N\left(0, \sigma^2_{\beta_j}\right)$.

To reduce the impact of hyperparameters on the prior distributions, Habier et al. (2011) introduced Bayes Cπ and Dπ. In these methods, the hyperparameter π is treated as a parameter itself, with its own prior distribution. Specifically, in Bayes Cπ, the hyperparameter π is assigned a uniform prior distribution, while in Bayes Dπ, a gamma distribution is used as the prior. For all methods, the inverted chi-square distribution is considered as the prior distribution for the variance components. More details can be found in Gianola et al. (2009, 2013). Another method, the Bayesian LASSO proposed by Park and Casella (2008), considers a double exponential as prior distribution of the marker effects, effectively shrinking small effects towards zero (Fig. 9.1).

One of the most widely used methods in genomic selection (GS) is the genomic best linear unbiased prediction (G-BLUP) (Habier et al. 2007; VanRaden 2008; Zhang et al. 2021; Nascimento et al. 2024). GBLUP can be considered an alternative parameterisation of the Bayesian ridge regression (BRR), where the linear combination between p markers and their corresponding effects $\sum_{j=1}^{p} x_{ij}\beta_j$ is replaced by a genomic effect $g_i = \sum_{j=1}^{p} x_{ij}\beta_j$. Stacking all the genomic effects g_i into a single vector $\mathbf{g} = \{g_i\}$ we have $\mathbf{g} = \mathbf{X\beta}$. Considering the properties of the multivariate

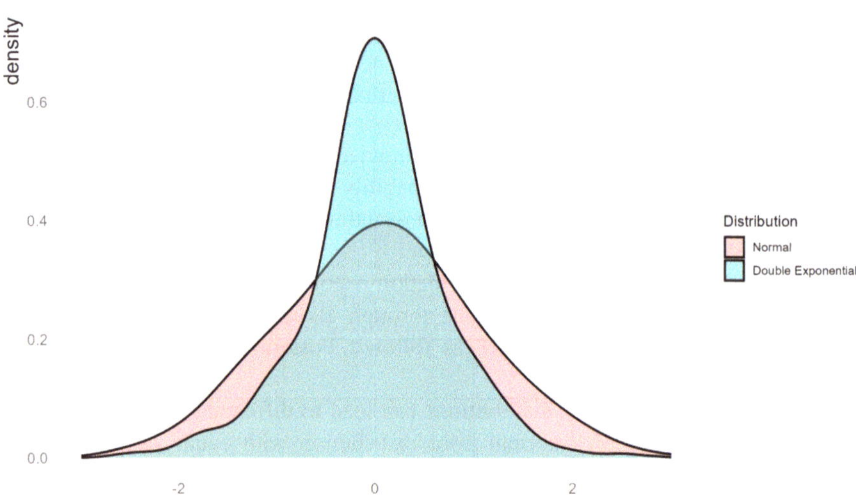

Fig. 9.1 Representation of the prior distribution assumed for marker effects for two different genomic selection models: Bayesian ridge regression (normal distribution) vs. Bayesian least absolute shrinkage and selection operator (double exponential distribution)

normal distribution, $\mathbf{g} \sim N(\mathbf{0}, \mathbf{G}\sigma_a^2)$, where $\mathbf{G} = \boldsymbol{XX'}k$ for some k (de los Campos et al. 2013) representing the corresponding variance component. Here, **G** is denoted as the kinship matrix and its entries describe genomic similarities between pairs of genotypes. The G-BLUP model can be written as:

$$\mathbf{Y} = \mathbf{Xb} + \mathbf{Zg} + \mathbf{e},$$

where **Y** is the vector of observations; **X** and **Z** are the incidence matrices connecting the phenotypes to the fixed effects in **b** and the random genetic effects in **g**. The probability distributions from the random effects are defined as:

$$\mathbf{g} \sim N(\mathbf{0}, \mathbf{G}\sigma_g^2) \text{ and } \mathbf{e} \sim N(\mathbf{0}, \sigma_e^2 \mathbf{I}_n).$$

Under the frequentist approach, the estimation of random effects (**g**) is conducted using Best Linear Unbiased Prediction (BLUP), a method that provides optimal linear unbiased predictions. Fixed effects (**b**) are estimated using generalised least squares, which accounts for the correlation structure among observations. Variance components, which quantify the variability associated with random effects, are estimated using Restricted Maximum Likelihood (REML), a method that provides robust and efficient estimates (Mrode and Thompson 2005). The system of equations resulting from these models is nonlinear. Consequently, iterative numerical methods, such as the Newton–Raphson algorithm, must be employed to approximate the solution to the system of equations (Mrode and Thompson 2005).

Another paradigm implemented to conduct predictions considers the use of nonparametric models, such as artificial neural networks (ANNs) (González-Camacho et al. 2012; Crossa et al. 2017; de Sousa et al. 2022; Ficht et al. 2023), decision trees (DTs), and ensemble methods like Bagging (Ba), Random Forest (RF), and Boosting (Bo) (González-Recio and Forni 2011; Ogutu et al. 2011; Barreto et al. 2024; Kaur et al. 2024).

Nonparametric approaches make no assumptions about the underlying model structure. This flexibility allows for the capture of complex relationships, such as nonadditive effects and G × E interactions, which do not require prior knowledge or distributional assumptions about the phenotypic values (Sousa et al. 2021). These methods are inspired by human behaviour, such as in artificial intelligence (e.g. ANNs), or leverage computational power through machine learning approaches (e.g. Bagging, Random Forest, and Boosting).

In addition, the semi-parametric reproducing Kernel Hilbert spaces (RKHS) modelling is another option (de los Campos et al. 2009; Jiang and Reif 2015; Chen et al. 2023).

The semi-parametric RKHS modelling uses kernel functions to define the similarity or relationship between records. This allows the data to be mapped into a high-dimensional feature space where similarities and nonlinear relationships between records are more easily captured (Schölkopf and Smola 2002; Jiang and Reif 2015).

In any of the previous approaches, the availability of abundant genomic information derived from NGS is critical to establishing accurate relationships between phenotyped and un-phenotyped individuals. Hence, the success of selecting accurately superior individuals depends in part on the genomic information implemented to build these relationships.

9.3 Multi-Trait Genomic Prediction (MTGP)

A comprehensive understanding and consideration of interrelationships among traits is crucial in plant breeding programmes, as superior genotypes should exhibit optimal attributes for multiple traits (Bernardo 2020). Incorporating genetically correlated traits during modelling enables the exploration of pleiotropic effects or linkage disequilibrium (LD) between genes, thereby sometimes improving model accuracy (Jia and Jannink 2012). Hence, improving multiple traits of interest (e.g. disease resistance, plant height, yield performance, protein content, oil content, etc.) could result in improved overall fitness of the individuals in the long term.

Multi-trait genomic prediction (MTGP) enables the joint exploitation of information derived from multiple traits. The conventional linear MTGP model for t traits is described as:

$$\begin{bmatrix} \mathbf{X}_1 & \cdots & \mathbf{0} \\ \vdots & \ddots & \vdots \\ \mathbf{0} & \cdots & \mathbf{X}_T \end{bmatrix} \begin{bmatrix} \mathbf{b}_1 \\ \vdots \\ \mathbf{b}_T \end{bmatrix} + \begin{bmatrix} \mathbf{Z}_1 & \cdots & \mathbf{0} \\ \vdots & \ddots & \vdots \\ \mathbf{0} & \cdots & \mathbf{Z}_T \end{bmatrix} \begin{bmatrix} \mathbf{g}_1 \\ \vdots \\ \mathbf{g}_T \end{bmatrix} + \begin{bmatrix} \mathbf{e}_1 \\ \vdots \\ \mathbf{e}_T \end{bmatrix},$$

where $\mathbf{y}_t$ is the vector of observations for the tth trait (t=1,2,...,T); $\mathbf{X}_t$ and $\mathbf{Z}_t$ are the incidence matrices relating the observations in $\mathbf{y}_t$ to the fixed effects in $\mathbf{b}_t$ and the random genetic effects in $\mathbf{g}_t$. The probability distributions from the random effects are defined as:

$$\begin{bmatrix} \mathbf{g}_1 \\ \vdots \\ \mathbf{g}_T \end{bmatrix} \sim \mathbf{N}(\mathbf{0}, \Sigma_{\mathrm{g}} \otimes \mathbf{G}) \text{ and } \begin{bmatrix} \mathbf{e}_1 \\ \vdots \\ \mathbf{e}_T \end{bmatrix} \sim \mathbf{N}(\mathbf{0}, \Sigma_{\mathrm{e}} \otimes \mathbf{I}_T) \text{ where } \Sigma_{\mathrm{a}} = \begin{bmatrix} \sigma_{\mathrm{g}_{1,1}} & \cdots & \sigma_{\mathrm{g}_{1,T}} \\ \vdots & \ddots & \vdots \\ \sigma_{\mathrm{g}_{T,1}} & \cdots & \sigma_{\mathrm{g}_{T,T}} \end{bmatrix}$$

$$\text{and } \Sigma_{\mathrm{e}} = \begin{bmatrix} \sigma_{\mathrm{e}_{1,1}} & \cdots & \sigma_{\mathrm{e}_{1,T}} \\ \vdots & \ddots & \vdots \\ \sigma_{\mathrm{e}_{T,1}} & \cdots & \sigma_{\mathrm{e}_{T,T}} \end{bmatrix}.$$

The application of MTGP models is of interest for enhancing the accuracy of low-heritability or difficult-to-measure traits when analysed simultaneously with traits exhibiting strong correlations and high heritability estimates (Jia and Jannink 2012). Gaire et al. (2022) illustrated this by demonstrating increased accuracy for deoxynivalenol content (DON) in wheat when FDK is evaluated on selection candidates. These authors considered four correlated traits (incidence—INC, severity—SEV, Fusarium damaged kernels—FDK, and deoxynivalenol content—DON) for modelling. Although all four traits were considered for analysis, the primary focus of the study was on FDK and DON. These traits are of particular interest due to their significant impact on wheat quality (FDK) and the challenges associated with their measurement (DON) (Gaire et al. 2022). Previous studies have demonstrated substantial increases in prediction accuracy, ranging from 20% to 100%, by using MTGP relative to single-trait (ST) models (Sun et al. 2017; Arojju et al. 2020).

The estimation of MTGP models within a frequentist framework requires the solution of nonlinear equations, which often relies on iterative numerical methods. Therefore, the inclusion of numerous traits can bring some challenges during the estimation process, as iterative algorithms need to reach convergence to a solution (Muvunyi et al. 2022). Alternatively, the Bayesian inference framework offers the flexibility to incorporate prior information about the parameters in the form of probability distributions. This can be particularly beneficial in situations where sample sizes are insufficient. It is worth noting that Markov chain Monte Carlo (MCMC) methods, employed to generate samples from the marginal posterior distribution, must converge following some criterion statistics, as Geweke (1992) and Raftery and Lewis (1992), to ensure accurate inference (Gelman and Rubin 1992).

The decision to employ an MTGP model is influenced by the extent of the genetic correlations between traits. In cases where the traits present strong and significant correlations, the MTGP model could provide more accurate and precise estimates of the genetic parameters compared to the ST models. However, it is important to consider the sample size and the number of evaluated traits, as these two factors can affect the convergence of iterative algorithms used to estimate the model parameters (Lado et al. 2018). Therefore, a balance between increasing the number of traits to capture complex genetic relationships and maintaining sufficient statistical information is important to ensure reliable parameter estimation. In practice, if the sample size is too small or the number of traits is too large, the model could not converge, resulting in unreliable estimations. The key point is that while more traits can improve the model's accuracy, enough data is essential to support the additional complexity. In practice, the use of MTGP models can mitigate the negative impact of small sample sizes on prediction accuracy, optimising phenotyping resource allocation (Lado et al. 2018; Arojju et al. 2020; Muvunyi et al. 2022). In any case, the MTGP models will outperform ST models in a parametric framework, when partial information across traits (e.g. recording two out of five traits) is observed for a specific genotype (Persa et al. 2020).

9.4 GS in Crop Improvement

Crop improvement is essential for addressing global food security and climate change adaptation. The traditional methods of breeding have made significant strides, yet they are limited in terms of speed and efficiency. Genomic selection emerged as a transformative approach that leverages advances in genome-wide marker data to enhance the selection of superior genotypes, allowing breeders to make precise predictions of an individual's potential. This approach allows more accurate and efficient identification of desirable traits within breeding populations. The impact of genomic selection on crop improvement can be understood through the lens of the breeder's equation (Falconer and Mackay 2009), which is expressed as:

$$R = \frac{i \times r \times \sigma_g}{L}$$

where R represents the response to selection, and it is obtained as the product of the selection intensity i, the selection accuracy r, and the σ_g genetic variance in the population divided by the length of the selection cycle L. Each component of the equation is discussed in detail with emphasis on the role of GS in maximising genetic gains. GS enhances genetic progress by optimising all the components of response to selection, including selection intensity, accuracy, genetic variance, and cycle length, allowing breeders to achieve faster and more efficient improvements in crop traits.

9.4.1 Response to Selection (R)

Response to selection represents a change in the mean phenotype of a population as a result of the selection of individuals with superior values for traits of interest, reflecting the rate of genetic gain achieved over a breeding cycle. For any breeding programme, the goal is to increase R as it will lead to improved crop varieties over successive generations.

Several studies have shown the advantages of using GS to accelerate genetic gains across diverse crop species. For example, Beyene et al. (2015) reported an increase in genetic gain for grain yield in eight bi-parental maize populations using GS compared to the pedigree selection scheme. Furthermore, Zhang et al. (2015) demonstrated that low-density SNPs (~200 markers) provided good prediction accuracy for simple traits with moderate-to-high heritability in biparental maize populations, while genotyping-by-sequencing (GBS, ~2000 markers) performed better for complex traits and simple traits under stress conditions with low-to-moderate heritability. Xu et al. (2020) found that high-density markers improve prediction accuracy in genomic selection, with the required marker density varying by species and population types.

Similarly, genotyping-by-sequencing identified a large number of SNPs in elite soybean lines, achieving a prediction accuracy of 0.64 for grain yield, highlighting the potential of genomic selection in soybean breeding (Jarquin et al. 2014a). Bernardo and Yu (2007) showed that using a biparental maize population, ridge-regression BLUP achieved 18% higher genetic gain for highly heritable traits ($h^2 = 0.8$) and 43% for low heritability traits ($h^2 = 0.2$), compared to marker-assisted selection.

9.4.2 *Selection Intensity* (i)

Selection intensity quantifies the proportion of individuals selected for breeding relative to the overall population. A higher selection intensity indicates that a smaller proportion of individuals with the most desirable traits are selected, whereas lower selection intensity involves selecting a larger proportion of the population for breeding.

In a high selection intensity strategy, selection pressure is increased, focusing on top-performing candidates, leading to greater response to selection, ultimately higher genetic gain (Silva et al. 2021). With GS, breeders estimate the genomic estimated breeding values (GEBVs) using genome-wide marker data. This allows for identification of individuals with high genetic merit early in the selection process before full phenotypic expression, thus allowing breeders to select top-performing individuals before extensive field testing (Bassi et al. 2016; Voss-Fels et al. 2019).

Along with increasing selection intensity, resource efficiency is also improved by reducing the reliance on time-consuming and expensive phenotypic evaluations. Moreover, the use of genome-wide marker data allows precise identification of individuals with favourable alleles, resulting in higher selection differential. Silva et al. (2021) showed that higher selection intensity results in greater genetic gains, especially when combined with advanced prediction models like Bayes A, Bayes B, and GBLUP in soybean breeding. Rutkoski et al. (2017) also reported that high selection intensity on wheat breeding programmes, when combined with GS, leads to faster genetic gains for complex traits.

9.4.3 *Selection Accuracy* (r)

Selection accuracy measures the correlation between an individual's true breeding values and the estimated breeding value. It is one of the key factors in the breeder's equation. Higher accuracy means that breeders can more reliably select the best candidates, which will lead to greater improvements in desired traits, enabling faster genetic gain. GS uses genomic markers spread across the genome, thus capturing more genetic variation. Also, advanced models used in GS, such as GBLUP and machine learning approaches, account for complex genetic architecture.

Meuwissen et al. (2001) were the first who introduce an approach to integrate dense molecular markers into predictive models by regressing phenotypes on all available markers using a straightforward yet effective linear model. Previously, Bernardo (1994) introduced this concept when the availability of genomic data was limited to very few features using RLFP as a proxy to predict maize single crosses. Using data generated with NGS, de los Campos et al. (2013) also reported that incorporating a large number of genome-wide markers significantly increased prediction accuracy compared to using fewer markers or focusing solely on QTLs, highlighting the advantage of dense marker coverage in genomic selection in increasing the prediction accuracy.

9.4.4 Genetic Variance (σ_g)

Genetic variance represents the variability of breeding values for a trait. Higher genetic variance means that the breeders can identify transgressive individuals with advantageous genetic profiles to select, contributing to higher response to selection. Employing GS, rare beneficial alleles can also be identified, helping to retain genetic diversity. Moreover, crosses can be designed to maximise the genetic variance, sustaining the genetic diversity. GS can also monitor and minimise inbreeding by maintaining diversity, while helping to maximise the genetic potential of breeding populations, leading to greater and more consistent genetic gains over time.

Jannink et al. (2010) reported that by applying GS in crop breeding, breeders were able to better balance selection for high-performing traits while maintaining overall genetic diversity. Additionally, Hickey et al. (2014) showed that using GS-based strategies, those using optimal cross-selection retained higher levels of genetic variance in livestock populations, a principle that is applicable to plant breeding as well. Crossa et al. (2017) also reported that GS in pre-breeding programmes could enhance genetic variance by accelerating the incorporation of diverse genes from gene bank accessions into elite breeding lines.

9.4.5 Selection Cycle Length (L)

Selection cycle length refers to the time required to complete one round of selection, recombination, and evaluation in a breeding programme (Rutkoski et al. 2022). GS allows breeders to conduct early selections of outperforming genotypes based on GEBVs before reaching maturity or even before planting these in the field. This approach is particularly advantageous for perennial crops, which often require several years to reach maturity and manifest traits of interest, such as height or yield (e.g. *Miscanthus*, *Elaeis*). This early selection reduces the time required for field trials and multi-year evaluations. Consequently, GS holds the potential to reduce the cycle lengths by half of the time or more, accelerating the delivery of improved crop varieties.

Heffner et al. (2010) reported that the annual gain achieved using GS was up to threefold for the high-intensity maize breeding programme and up to twofold for the low-intensity winter breeding programme compared to MAS. These gains were primarily due to shortening the breeding cycle length. Kumar et al. (2012) found that using GS in perennial crops like apple trees reduced the breeding cycle length by enabling earlier selection of seedlings with desirable traits, accelerating the overall process of varietal development.

9.5 Genomic Selection in Genetic Diversity Management

The crop improvement process begins by generating genetic variability through crossing parents, and then selecting promisor progenies according to a breeding strategy. However, this process per se, only explores the genetic variability already existent within the germplasm by combining favourable alleles from different loci (Bernardo 2014). As a result, the recurrent selection process to develop elite genotypes causes a narrowing in genetic diversity, limiting the efficiency in creating new improved breeding populations (Böhm et al. 2017) in the long term.

Genetic diversity played a crucial role in crop improvement, being a source of favourable alleles for traits of interest (Wang et al. 2017). In addition to developing economically important traits, currently new challenges are emerging, increasing the importance of preserving a wide genetic base to maintain potential alleles to be used against biotic and abiotic stresses, and to adapt crops to new environmental conditions (Mickelbart et al. 2015). Developing work on maintaining available potential genetic variability and finding ways to use it efficiently in crop improvement becomes crucial for addressing these future new challenges (Wang et al. 2017; He and Li 2020; Varshney et al. 2021).

Introducing exotic germplasm into breeding programmes through landraces or wild relative species is one of the most common ways to increase the genetic variability in a breeding germplasm (Muleta et al. 2017; Allier et al. 2020). Efforts in collecting and conserving exotic germplasm are needed in places known as genebanks, which focus on maintaining them available for future utilisation. All over the world, there are more than 1750 genebanks working on preserving this crucial genetic diversity; some important examples are presented in Table 9.2 (FAO

Table 9.2 List of the current largest genebanks (11) around the world by the number of stored accessions

Genebank	Country	Number of accessions
National Plant Germplasm System (NPGS)	USA	>617,000
National Crop Genebank of China (NCGC)	China	~392,000
ICAR-National Bureau of Plant Genetic Resources (ICAR-NBPGR)	India	>392,000
N.I. Vavilov All Russian Institute of Plant Genetic Resources (VIR)	Russia	>340,000
National Institute of Agrobiological Sciences (NIAS)	Japan	~240,000
International Maize and Wheat Improvement Centre (CIMMYT)	México	>180,000
Federal Ex situ Gene Bank at IPK (Leibniz Institute of Plant Genetics and Crop Plant Research)	Germany	>151,000
International Centre for Agricultural Research in the Dry Areas (ICARDA)	Lebanon	~145,000
International Rice Research Institute	Philippines	~128,000
Brazilian Agricultural Research Corporation (EMBRAPA)	Brazil	~102,000
International Centre for Tropical Agriculture (CIAT)	Colombia	>67,000

2010). Also, it is well worth highlighting the genebank Svalbard Global Seed Vault, located in Norway, which maintains duplicates of more than 1.25 million samples from several genebanks and institutions distributed globally, serving as an insurance for the world's food supply through preservation of crop diversity. In 2024, the World Food Prize was awarded to Cary Fowler and Geoffrey Hawtin, scientists who had crucial roles in developing the seed vault, being a proof of the importance of safeguarding potential alleles to overcome future challenges.

Due to the large number of samples stored in collections it might turn difficult to characterise and evaluate in fields all accessions properly, limiting the available information to incorporate exotic germplasm into the crop improvement process (He and Li 2020). Indeed, the post-NGS technologies could have a critical role enabling a better understanding about germplasm collections safeguarded in genebanks through cost-effective and efficient large scale genotyping processes.

The introgression of favourable alleles from exotic sources is already a common process implemented in crop improvement through backcross techniques using molecular markers (Weckwerth et al. 2020). Some cases using the *Fhb-genes* in wheat (Brar et al. 2019), *Sub1* in rice (Bailey-Serres et al. 2010), *Rph-genes* in barley (Martin et al. 2020), *ahFAD2* in peanuts (Nawade et al. 2019), and *crtRB1* in maize (Liu et al. 2015) are examples of the success in the use of exotic germplasm. Nevertheless, usually there is a high difference between the performance of exotic donors and elite genotypes for economically important complex traits as yield, limiting the obtained genetic gains (Bernardo 2009). For instance, some strategies were developed to create bridging populations to avoid direct crosses between exotic and elite genotypes. Instead, the best individuals in bridging populations are used in crosses to obtain the breeding populations. For example, in Allier et al. (2020), genomic selection was used to predict the best combinations to obtain bridging populations, combining both alleles of interest from exotic germplasm and good performance to be introduced later into elite populations.

Although many genebanks lack phenotypic information about their accessions, GS can be integrated with the characterisation of the available genetic diversity using NGS, enabling a large-scale screening to mine exotic germplasm to be incorporated into the crop improvement process (He and Li 2020). Also, many studies suggest that increasing the variability in reference populations to train GS models can improve the accuracy of prediction, benefiting crop improvement in both ways by increasing diversity and the selection accuracy (Muleta et al. 2017; Li et al. 2022; Parveen et al. 2023). Developing elite genotypes using the genomic estimated breeding values can accelerate the narrowing of genetic diversity (Jannink 2010). However, employing GS to incorporate new individuals into the germplasm pool or manage inbreeding rates can overcome some of the drawbacks caused by the increasing selection intensity and trends in selecting related individuals, maintaining for a longer term the genetic gains (Lorenz et al. 2011; De Beukelaer et al. 2017; Allier et al. 2020). Hence, characterising the genebanks with high precision using NGS opens a venue to exploit the genetic diversity to continue genetic gain rates for longer periods of time and offers the possibility to breed for future environmental conditions.

9.6 GS to Capture G × E Interactions Effects in Multi-Environment Trials (MET)

Breeding programmes aim to develop genotypes showing an optimum performance across multiple environments, ensuring a stable behaviour compared to other genotypes. The change in the response patterns is associated with the rank change of one group of genotypes in varying environmental conditions (Crossa et al. 2011). When a significant change in the response patterns is present, this well-known phenomenon is regarded as the genotype-by-environment (G × E) interaction or GEI. The study of the GEI is a key feature for understanding broader adaptation of genotypes under different environmental influences. GEI helps decide whether a particular environment represents all trials or if more trials are required to capture environmental variability. Van Eeuwijk et al. (2016) showed how the genotypes react to the environmental stimuli, while GEI occurs when their norms are not parallel but intersect, converge, or cross over. The most complex form of GEI occurs when the genotypes have crossovers or nonlinear relationships with environments. Environments having high genetic correlations show small crossover interactions, while significant crossover happens from low correlation. Hence, GEI is a key factor in assessing the performance of genotypes for targeted breeding trials, helping breeders to conduct trials in a more informed manner. For this, it is necessary to conduct multi-environment trials (METs).

Here, the METs capture GEI by testing multiple genotypes in diverse environments, providing information to the breeders about the number of environments they should consider for crop evaluation. In METs, the overall performance of the genotypes is assessed across environments. This helps select cultivars that perform consistently well in diverse environmental conditions or genotypes significantly outperforming others showing local adaptation. However, conducting METs across many environments is resource-intensive, costly, and time-demanding. Also, large and complex MET datasets are difficult to analyse to demonstrate accurate results (Olivoto and Lúcio 2020).

GS, since developed, has been used to consider GEI in METs to avoid replication of genotypes across trials as well as predict the performance of unobserved genotypes (Jarquin et al. 2020; Ankamah-Yeboah et al. 2020; Wilson et al. 2024). Previous studies have described different ways and advantages of incorporating GEI in the GP models (Burgueño et al. 2012; Jarquin et al. 2014b; Lopez-Cruz et al. 2015). A common approach to implementing GEI in the model is using an unstructured variance–covariance (VCOV) matrix. It includes a particular genetic variance for each environment in the diagonal and pairwise covariances between environments in the off-diagonal (Dalsente Krause et al. 2020). However, fitting the models using an unstructured variance–covariance matrix with a large number of parameters experiences computational issues (De Oliveira et al. 2020). Compared to the earlier approaches of GBLUP models with only main effects, when both genomic markers and GEI are included in the GP models, high prediction accuracies were obtained (Jarquin et al. 2014b; Acosta-Pech et al. 2017). In addition, accounting for G × E interactions could also be beneficial for the optimisation of the allocation of resources in METs. The initial idea of GS was to screen genotypes at earlies stages of the breeding pipeline where one environment is only observed. However, the inclusion of G × E interaction in METs holds the potential to make up the unobserved combinations of genotypes-in-environments in late stages of the breeding pipeline when evaluating genotypes' stability.

Jarquin et al. (2020) presented a GP-based sparse testing method to reduce the cost associated with METs by selecting a subset of genotypes to be evaluated across environments at a fixed cost. It increases selection intensity which indirectly improves the genetic gain. Results based on the sparse testing designs implemented in multiple crop species have shown that it is possible to decrease the phenotyping cost up to 90% for a target set of genotypes and environments. Alternatively, it is possible to increase the testing capacity by ninefold for a given budget (Persa et al. 2023). For large breeding programmes, it is infeasible to observe all genotypes in all environments due to budget restrictions, seed, land, and water availability. Hence, the availability of genomic information derived from NGS could help to reduce environmental footprint by optimising the allocation of resources while returning accurate results. However, sparse testing analysis requires balanced datasets for making predictions and therefore, yet to be improved for unbalanced datasets in METs.

9.7 Multi-Trait Multi-Environment Genomic Prediction Models MTME

In general, when the METs are established, it is a common practice to collect phenotypic information on other traits besides the main trait of interest. Hence, MET datasets have been used from a multivariate context that can borrow information across all traits in all environments in a multi-trait multi-environment (MTME) GP model (Gill et al. 2021). Montesinos-López et al. (2016) proposed an implementation under the Bayesian approach to analyse multiple traits and multiple environments for genomic prediction. This approach allows for the analysis of data from multiple traits measured across multiple environments in a unified framework. This framework is particularly useful for understanding G × E interaction and improving the predictions of new genotypes in untested environments, aiming to guide breeding programmes in identifying promising genotypes and selecting appropriate parents for future crosses, as demonstrated in studies on soybean by Volpato et al. (2019) and wheat by Gill et al. (2021).

9.8 Integration of Genomic Selection with Other OMICs

The advancements related to technologies in other fields have enabled GP models to be integrated with different types of data, such as those derived from weather sensors collecting environmental data, high-throughput phenotyping platforms via remote and proximal sensing, and features generated by crop-growth models (CGM), using both linear and nonlinear kernels to capture GEI (de los Campos et al. 2020; Costa-Neto et al. 2021; Adak et al. 2024). The continuous increase of the amount of available information we can use to characterise trait performance increases our expectation for improving predictive ability (Jarquin et al. 2024). Hence, all the available information could be leveraged to enhance the prediction accuracy of the GP models for METs, allowing the selection of genotypes at early stages of the breeding programmes.

9.9 Enviromics

Phenotypic variation from GEI turns enviromics crucial to be incorporated into GP models across environments. Jarquin et al. (2014b) proposed a reaction norm model to include multidimensional environmental covariates (ECs) by computing a kinship covariance structure between ECs to account for the contrast between genomic markers and ECs. Through simulation, de los Campos et al. (2020) studied the GEI by using historical weather records and evaluated the performance of cultivars under different weather conditions while considering uncertainty regarding the model parameters. New tools as the EnvRtype R-package (R Core Team 2024), allow for gathering worldwide macro-environmental data and take advantage of the GEI estimation using different kernel methods (Costa-Neto et al. 2021). In situations where genotypes have already been observed and there is a high environmental correlation, incorporating macro-environmental data might not improve model accuracy. However, it can be informative when there is no available information regarding the target environments besides the performance of genotypes. Hence, by leveraging information from large-scale envirotyping data and historical data in the GP models, it is possible to predict new genotypes in unobserved environments and design more sophisticated METs.

9.10 Phenomics

Traditional phenomics relies on studying all plants over a range of environments. In contrast, modern phenomics can evaluate plant characteristics using various spectral cameras at different points of time during their growth stages (Crossa et al. 2021). Unlike GP, modern phenotyping or high-throughput phenotyping (HTP) requires planting the genotypes in fields to measure these. However, associated traits with the main trait of interest can be measured at the initial stage of the breeding programmes without waiting for the plants' maturity stage. HTP makes phenotyping faster and nondestructive in controlled environments and fields as well (Jangra et al. 2021). It involves high-resolution vision-based sensors (hyperspectral imaging and NIR spectral imaging), UAV flights, remote sensing, etc., providing biomarkers that can be used as secondary traits (Jin et al. 2021). For example, biomass and chlorophyll contents are considered important biomarkers for studying nitrogen response in wheat (Banerjee et al. 2020). Vegetation indices and plant heights are considered biomarkers for yield (Adak et al. 2024).

Nowadays, HTP datasets are incorporated into GP models that serve as secondary traits to predict complex traits (Adak et al. 2023; Kaushal et al. 2024; Adak et al. 2024). Models including multi-kernels for integrating genomic and phenomics information have shown better performance compared to single-trait GBLUP (Galán et al. 2020; Adak et al. 2023). However, a major drawback of HTP in MET is aligning the phenotyping schedules for experiments conducted over multiple years and environments. For example, field-based UAV flights provide biomarkers that do not match the flight dates (Adak et al. 2024). The varying flight dates can difficult the understanding of how complex traits change in response to environmental factors. Although temporal phenotyping across different time points could be a solution for understanding complex GEI, precise phenotyping besides aligning schedules would be necessary for accurate predictive modelling.

9.11 Crop Growth Models (CGM)

CGM is centred on understanding the interplay between physiological factors of daily crop growth and development to predict genotype-by-environment-by-management interactions (G × E × M) (Cooper et al. 2009; Washburn et al. 2020). It uses biophysical functions that reflect the influence of environmental covariates on the interaction of crop growth and development (Voss-Fels et al. 2019). CGM needs large sets of multi-environment phenotypic data for each genotype to estimate the genotype-specific parameters (GSPs) to capture G × E × M (Washburn et al. 2020). Recently, CGM has been fitted with GP models (CGM-WGP) where the effects of markers and genotype-specific-parameters can be combined to estimate the shared information between individuals through genomic data (Technow et al. 2015; Jighly et al. 2023). Messina et al. (2018) explained how to expand the CGM–GP method to calibrate models with multi-environmental data using a Bayesian approach. Jighly et al. (2023) showed the advantage of using CGM–WGP to evaluate the prediction accuracy of the unobserved genotypes in wheat. These studies demonstrate the potential of combining CGM with GP, which can make a significant contribution to capturing GEI considering the recent climatic conditions.

The information collected from enviromics, phenomics, or CGM helps build data-driven breeding pipelines for METs to capture GEI other than conventional phenotyping. The information to use in the pipeline depends on their availability, structure, and objectives of the programmes. In the future, optimising GP models with these sources would enhance breeding decisions, reduce phenotyping time, and cost in METs, aiming to develop climate resilient crops. In any of the mentioned scenarios, the availability of reliable NGS information plays the main role in driving variability, helping to ensure sustainable genetic gains in the long term.

References

Acosta-Pech R, Crossa J, de los Campos G et al (2017) Genomic models with genotype × environment interaction for predicting hybrid performance: an application in maize hybrids. Theor Appl Genet 130:1431–1440. https://doi.org/10.1007/s00122-017-2898-0

Adak A, Kang M, Anderson SL et al (2023) Phenomic data-driven biological prediction of maize through field-based high-throughput phenotyping integration with genomic data. J Exp Bot 74: 5307–5326. https://doi.org/10.1093/jxb/erad216

Adak A, DeSalvio AJ, Arik MA, Murray SC (2024) Field-based high-throughput phenotyping enhances phenomic and genomic predictions for grain yield and plant height across years in maize. G3: genes, genomes. Genetics 14:jkae092. https://doi.org/10.1093/g3journal/jkae092

Allier A, Teyssèdre S, Lehermeier C et al (2020) Optimized breeding strategies to harness genetic resources with different performance levels. BMC Genomics 21:349. https://doi.org/10.1186/s12864-020-6756-0

Ankamah-Yeboah T, Janss LL, Jensen JD et al (2020) Genomic selection using pedigree and marker-by-environment interaction for barley seed quality traits from two commercial breeding programs. Front Plant Sci 11:539. https://doi.org/10.3389/fpls.2020.00539

Arojju SK, Cao M, Trolove M et al (2020) Multi-trait genomic prediction improves predictive ability for dry matter yield and water-soluble carbohydrates in perennial ryegrass. Front Plant Sci 11:1197. https://doi.org/10.3389/fpls.2020.01197

Bailey-Serres J, Fukao T, Ronald P et al (2010) Submergence tolerant Rice: SUB1's journey from landrace to modern cultivar. Rice 3:138–147. https://doi.org/10.1007/s12284-010-9048-5

Banerjee BP, Joshi S, Thoday-Kennedy E et al (2020) High-throughput phenotyping using digital and hyperspectral imaging-derived biomarkers for genotypic nitrogen response. J Exp Bot 71: 4604–4615. https://doi.org/10.1093/jxb/eraa143

Barreto CAV, Das Graças Dias KO, De Sousa IC et al (2024) Genomic prediction in multi-environment trials in maize using statistical and machine learning methods. Sci Rep 14:1062. https://doi.org/10.1038/s41598-024-51792-3

Bassi FM, Bentley AR, Charmet G et al (2016) Breeding schemes for the implementation of genomic selection in wheat (Triticum spp.). Plant Sci 242:23–36. https://doi.org/10.1016/j.plantsci.2015.08.021

Bernardo R (1994) Prediction of maize single-cross performance using RFLPs and information from related hybrids. Crop Sci 34:20–25. https://doi.org/10.2135/cropsci1994.0011183X003400010003x

Bernardo R (2009) Genomewide selection for rapid introgression of exotic germplasm in maize. Crop Sci 49:419–425. https://doi.org/10.2135/cropsci2008.08.0452

Bernardo R (2014) Genomewide selection of parental inbreds: classes of loci and virtual biparental populations. Crop Sci 54:2586–2595. https://doi.org/10.2135/cropsci2014.01.0088

Bernardo R (2020) Breeding for quantitative traits in plants, 3rd edn. Stemma Press, Woodbury, Minnesota

Bernardo R, Yu J (2007) Prospects for genomewide selection for quantitative traits in maize. Crop Sci 47:1082–1090. https://doi.org/10.2135/cropsci2006.11.0690

Beyene Y, Semagn K, Mugo S et al (2015) Genetic gains in grain yield through genomic selection in eight bi-parental maize populations under drought stress. Crop Sci 55:154–163. https://doi.org/10.2135/cropsci2014.07.0460

Bhat JA, Ali S, Salgotra RK et al (2016) Genomic selection in the era of next generation sequencing for complex traits in plant breeding. Front Genet 7. https://doi.org/10.3389/fgene.2016.00221

Böhm J, Schipprack W, Utz HF, Melchinger AE (2017) Tapping the genetic diversity of landraces in allogamous crops with doubled haploid lines: a case study from European flint maize. Theor Appl Genet 130:861–873. https://doi.org/10.1007/s00122-017-2856-x

Bonnett D, Li Y, Crossa J et al (2022) Response to early generation genomic selection for yield in wheat. Front Plant Sci 12:718611. https://doi.org/10.3389/fpls.2021.718611

Brar GS, Brûlé-Babel AL, Ruan Y et al (2019) Genetic factors affecting Fusarium head blight resistance improvement from introgression of exotic Sumai 3 alleles (including Fhb1, Fhb2, and Fhb5) in hard red spring wheat. BMC Plant Biol 19:179. https://doi.org/10.1186/s12870-019-1782-2

Burgueño J, de los Campos G, Weigel K, Crossa J (2012) Genomic prediction of breeding values when modeling genotype × environment interaction using pedigree and dense molecular markers. Crop Sci 52:707–719. https://doi.org/10.2135/cropsci2011.06.0299

Chen S-P, Tung C-W, Wang P-H, Liao C-T (2023) A statistical package for evaluation of hybrid performance in plant breeding via genomic selection. Sci Rep 13:12204. https://doi.org/10.1038/s41598-023-39434-6

Cobb JN, Juma RU, Biswas PS et al (2019) Enhancing the rate of genetic gain in public-sector plant breeding programs: lessons from the breeder's equation. Theor Appl Genet 132:627–645. https://doi.org/10.1007/s00122-019-03317-0

Collard BCY, Mackill DJ (2008) Marker-assisted selection: an approach for precision plant breeding in the twenty-first century. Philos Trans R Soc B 363:557–572. https://doi.org/10.1098/rstb.2007.2170

Cooper M, Van Eeuwijk FA, Hammer GL et al (2009) Modeling QTL for complex traits: detection and context for plant breeding. Curr Opin Plant Biol 12:231–240. https://doi.org/10.1016/j.pbi.2009.01.006

Costa-Neto G, Galli G, Carvalho HF et al (2021) *EnvRtype*: a software to interplay enviromics and quantitative genomics in agriculture. G3 Genes|Genomes|Genetics 11:jkab040. https://doi.org/10.1093/g3journal/jkab040

Crossa J, Perez-Elizalde S, Jarquin D et al (2011) Bayesian estimation of the additive main effects and multiplicative interaction model. Crop Sci 51:1458–1469. https://doi.org/10.2135/cropsci2010.06.0343

Crossa J, Pérez-Rodríguez P, Cuevas J et al (2017) Genomic selection in plant breeding: methods, models, and perspectives. Trends Plant Sci 22:961–975. https://doi.org/10.1016/j.tplants.2017.08.011

Crossa J, Fritsche-Neto R, Montesinos-Lopez OA et al (2021) The modern plant breeding triangle: optimizing the use of genomics, phenomics, and enviromics data. Front Plant Sci 12:651480. https://doi.org/10.3389/fpls.2021.651480

Dalsente Krause M, Dias KODG, Pedroso Rigal Dos Santos J et al (2020) Boosting predictive ability of tropical maize hybrids via genotype-by-environment interaction under multivariate GBLUP models. Crop Sci 60:3049–3065. https://doi.org/10.1002/csc2.20253

De Beukelaer H, Badke Y, Fack V, De Meyer G (2017) Moving beyond managing realized genomic relationship in long-term genomic selection. Genetics 206:1127–1138. https://doi.org/10.1534/genetics.116.194449

de los Campos G, Gianola D, Rosa GJM (2009a) Reproducing kernel Hilbert spaces regression: a general framework for genetic evaluation. J Anim Sci 87:1883–1887. https://doi.org/10.2527/jas.2008-1259

de los Campos G, Hickey JM, Pong-Wong R et al (2013) Whole-genome regression and prediction methods applied to plant and animal breeding. Genetics 193:327–345. https://doi.org/10.1534/genetics.112.143313

de los Campos G, Pérez-Rodríguez P, Bogard M et al (2020) A data-driven simulation platform to predict cultivars' performances under uncertain weather conditions. Nat Commun 11:4876. https://doi.org/10.1038/s41467-020-18480-y

De Oliveira AA, Resende MFR, Ferrão LFV et al (2020) Genomic prediction applied to multiple traits and environments in second season maize hybrids. Heredity 125:60–72. https://doi.org/10.1038/s41437-020-0321-0

de Sousa IC, Nascimento M, De Castro Sant'anna I et al (2022) Marker effects and heritability estimates using additive-dominance genomic architectures via artificial neural networks in Coffea canephora. PLoS One 17:e0262055. https://doi.org/10.1371/journal.pone.0262055

Falconer DS, Mackay T (2009) Introduction to quantitative genetics, 4th edn, [16. print.]. Pearson, Prentice Hall, Harlow
FAO (ed) (2010) The second report on the state of the world's plant genetic resources for food and agriculture. FAO, Rome
Ficht A, Konkin DJ, Cram D et al (2023) Genomic selection for agronomic traits in a winter wheat breeding program. Theor Appl Genet 136:38. https://doi.org/10.1007/s00122-023-04294-1
Gaire R, De Arruda MP, Mohammadi M et al (2022) Multi-trait genomic selection can increase selection accuracy for deoxynivalenol accumulation resulting from fusarium head blight in wheat. Plant Genome 15:e20188. https://doi.org/10.1002/tpg2.20188
Galán RJ, Bernal-Vasquez A-M, Jebsen C et al (2020) Integration of genotypic, hyperspectral, and phenotypic data to improve biomass yield prediction in hybrid rye. Theor Appl Genet 133: 3001–3015. https://doi.org/10.1007/s00122-020-03651-8
Gamerman D, Lopes HF (2006) Markov Chain Monte Carlo: stochastic simulation for Bayesian inference, 2nd edn. Chapman and Hall/CRC
Gelman A, Rubin DB (1992) Inference from iterative simulation using multiple sequences. Stat Sci 7. https://doi.org/10.1214/ss/1177011136
Geweke J (1992) Evaluating the accuracy of sampling-based approaches to the calculation of posterior moments. In: Bernardo JM, Berger JO, Dawid P, Smith AFM (eds) Bayesian statistics 4. Oxford University PressOxford, pp 169–194
Gianola D (2013) Priors in whole-genome regression: the Bayesian alphabet returns. Genetics 194: 573–596. https://doi.org/10.1534/genetics.113.151753
Gianola D, de los Campos G, Hill WG et al (2009) Additive genetic variability and the Bayesian alphabet. Genetics 183:347–363. https://doi.org/10.1534/genetics.109.103952
Gill HS, Halder J, Zhang J et al (2021) Multi-trait multi-environment genomic prediction of agronomic traits in advanced breeding lines of winter wheat. Front Plant Sci 12:709545. https://doi.org/10.3389/fpls.2021.709545
González-Camacho JM, de los Campos G, Pérez P et al (2012) Genome-enabled prediction of genetic values using radial basis function neural networks. Theor Appl Genet 125:759–771. https://doi.org/10.1007/s00122-012-1868-9
González-Recio O, Forni S (2011) Genome-wide prediction of discrete traits using bayesian regressions and machine learning. Genet Sel Evol 43:7. https://doi.org/10.1186/1297-9686-43-7
Goodwin S, McPherson JD, McCombie WR (2016) Coming of age: ten years of next-generation sequencing technologies. Nat Rev Genet 17:333–351. https://doi.org/10.1038/nrg.2016.49
Grover A, Sharma PC (2016) Development and use of molecular markers: past and present. Crit Rev Biotechnol 36:290–302. https://doi.org/10.3109/07388551.2014.959891
Guo L-T, Pyle AM (2023) RT-based Sanger sequencing of RNAs containing complex RNA repetitive elements. In: Methods in enzymology. Elsevier, pp 17–27
Habier D, Fernando RL, Dekkers JCM (2007) The impact of genetic relationship information on genome-assisted breeding values. Genetics 177:2389–2397. https://doi.org/10.1534/genetics.107.081190
Habier D, Fernando RL, Kizilkaya K, Garrick DJ (2011) Extension of the bayesian alphabet for genomic selection. BMC Bioinformatics 12:186. https://doi.org/10.1186/1471-2105-12-186
Hasan N, Choudhary S, Naaz N et al (2021) Recent advancements in molecular marker-assisted selection and applications in plant breeding programmes. J Genet Eng Biotechnol 19:128. https://doi.org/10.1186/s43141-021-00231-1
He T, Li C (2020) Harness the power of genomic selection and the potential of germplasm in crop breeding for global food security in the era with rapid climate change. Crop J 8:688–700. https://doi.org/10.1016/j.cj.2020.04.005
Heffner EL, Lorenz AJ, Jannink J, Sorrells ME (2010) Plant breeding with genomic selection: gain per unit time and cost. Crop Sci 50:1681–1690. https://doi.org/10.2135/cropsci2009.11.0662
Hickey JM, Dreisigacker S, Crossa J et al (2014) Evaluation of genomic selection training population designs and genotyping strategies in plant breeding programs using simulation. Crop Sci 54:1476–1488. https://doi.org/10.2135/cropsci2013.03.0195

Hoerl AE, Kennard RW (1970) Ridge regression: biased estimation for nonorthogonal problems. Technometrics 12(1):55–67

Hu T, Chitnis N, Monos D, Dinh A (2021) Next-generation sequencing technologies: an overview. Hum Immunol 82:801–811. https://doi.org/10.1016/j.humimm.2021.02.012

Jaganathan D, Bohra A, Thudi M, Varshney RK (2020) Fine mapping and gene cloning in the post-NGS era: advances and prospects. Theor Appl Genet 133:1791–1810. https://doi.org/10.1007/s00122-020-03560-w

Jangra S, Chaudhary V, Yadav RC, Yadav NR (2021) High-throughput phenotyping: a platform to accelerate crop improvement. Phenomics 1:31–53. https://doi.org/10.1007/s43657-020-00007-6

Jannink J-L (2010) Dynamics of long-term genomic selection. Genet Sel Evol 42:35. https://doi.org/10.1186/1297-9686-42-35

Jannink J-L, Lorenz AJ, Iwata H (2010) Genomic selection in plant breeding: from theory to practice. Brief Funct Genomics 9:166–177. https://doi.org/10.1093/bfgp/elq001

Jarquin D, Kocak K, Posadas L et al (2014a) Genotyping by sequencing for genomic prediction in a soybean breeding population. BMC Genomics 15:740. https://doi.org/10.1186/1471-2164-15-740

Jarquin D, Crossa J, Lacaze X et al (2014b) A reaction norm model for genomic selection using high-dimensional genomic and environmental data. Theor Appl Genet 127:595–607. https://doi.org/10.1007/s00122-013-2243-1

Jarquin D, Howard R, Crossa J et al (2020) Genomic prediction enhanced sparse testing for multi-environment trials. G3 Genes|Genomes|Genetics 10:2725–2739. https://doi.org/10.1534/g3.120.401349

Jarquin D, Roy A, Clarke B, Ghosal S (2024) Combining phenotypic and genomic data to improve prediction of binary traits. J Appl Stat 51:1497–1523. https://doi.org/10.1080/02664763.2023.2208773

Jia Y, Jannink J-L (2012) Multiple-trait genomic selection methods increase genetic value prediction accuracy. Genetics 192:1513–1522. https://doi.org/10.1534/genetics.112.144246

Jiang Y, Reif JC (2015) Modeling epistasis in genomic selection. Genetics 201:759–768. https://doi.org/10.1534/genetics.115.177907

Jighly A, Weeks A, Christy B et al (2023) Integrating biophysical crop growth models and whole genome prediction for their mutual benefit: a case study in wheat phenology. J Exp Bot 74:4415–4426. https://doi.org/10.1093/jxb/erad162

Jin X, Zarco-Tejada PJ, Schmidhalter U et al (2021) High-throughput estimation of crop traits: a review of ground and aerial phenotyping platforms. IEEE Geosci Remote Sens Mag 9:200–231. https://doi.org/10.1109/MGRS.2020.2998816

Jonas E, De Koning DJ (2016) Goals and hurdles for a successful implementation of genomic selection in breeding programme for selected annual and perennial crops. Biotechnol Genet Eng Rev 32:18–42. https://doi.org/10.1080/02648725.2016.1177377

Kaur S, Kakani VG, Carver B et al (2024) Hyperspectral imaging combined with machine learning for high-throughput phenotyping in winter wheat. Plant Phenome J 7:e20111. https://doi.org/10.1002/ppj2.20111

Kaushal S, Gill HS, Billah MM et al (2024) Enhancing the potential of phenomic and genomic prediction in winter wheat breeding using high-throughput phenotyping and deep learning. Front Plant Sci 15:1410249. https://doi.org/10.3389/fpls.2024.1410249

Kumar S, Chagné D, Bink MCAM et al (2012) Genomic selection for fruit quality traits in apple (Malus×domestica Borkh.). PLoS One 7:e36674. https://doi.org/10.1371/journal.pone.0036674

Lado B, Vázquez D, Quincke M et al (2018) Resource allocation optimization with multi-trait genomic prediction for bread wheat (Triticum aestivum L.) baking quality. Theor Appl Genet 131:2719–2731. https://doi.org/10.1007/s00122-018-3186-3

Li Y, Kaur S, Pembleton LW et al (2022) Strategies of preserving genetic diversity while maximizing genetic response from implementing genomic selection in pulse breeding programs. Theor Appl Genet 135:1813–1828. https://doi.org/10.1007/s00122-022-04071-6

Liu L, Li Y, Li S et al (2012) Comparison of next-generation sequencing systems. J Biomed Biotechnol 2012:1–11. https://doi.org/10.1155/2012/251364

Liu L, Jeffers D, Zhang Y et al (2015) Introgression of the crtRB1 gene into quality protein maize inbred lines using molecular markers. Mol Breeding 35:154. https://doi.org/10.1007/s11032-015-0349-7

Lopez-Cruz M, Crossa J, Bonnett D et al (2015) Increased prediction accuracy in wheat breeding trials using a marker × environment interaction genomic selection model. G3 Genes|Genomes|Genetics 5:569–582. https://doi.org/10.1534/g3.114.016097

Lorenz AJ, Chao S, Asoro FG et al (2011) Genomic selection in plant breeding. In: Advances in agronomy. Elsevier, pp 77–123

Martin MJ, Chicaiza O, Caffarel JC et al (2020) Development of barley introgression lines carrying the leaf rust resistance genes *Rph1* to *Rph15*. Crop Sci 60:282–302. https://doi.org/10.1002/csc2.20057

Messina CD, Technow F, Tang T et al (2018) Leveraging biological insight and environmental variation to improve phenotypic prediction: integrating crop growth models (CGM) with whole genome prediction (WGP). Eur J Agron 100:151–162. https://doi.org/10.1016/j.eja.2018.01.007

Meuwissen THE, Hayes BJ, Goddard ME (2001) Prediction of Total genetic value using genome-wide dense marker maps. Genetics 157:1819–1829. https://doi.org/10.1093/genetics/157.4.1819

Mickelbart MV, Hasegawa PM, Bailey-Serres J (2015) Genetic mechanisms of abiotic stress tolerance that translate to crop yield stability. Nat Rev Genet 16:237–251. https://doi.org/10.1038/nrg3901

Montesinos-López OA, Montesinos-López A, Crossa J et al (2016) A genomic Bayesian multi-trait and multi-environment model. G3 Genes|Genomes|Genetics 6:2725–2744. https://doi.org/10.1534/g3.116.032359

Montgomery DC, Peck EA, Vining GG (2020) Introduction to linear regression analysis, 5th edn. Wiley, Hoboken

Mrode RA, Thompson R (eds) (2005) Linear models for the prediction of animal breeding values, 2nd edn. CABI Publishing, UK

Muleta KT, Bulli P, Zhang Z et al (2017) Unlocking diversity in germplasm collections via genomic selection: a case study based on quantitative adult plant resistance to stripe rust in spring wheat. The plant. Genome 10:plantgenome2016.12.0124. https://doi.org/10.3835/plantgenome2016.12.0124

Muvunyi BP, Zou W, Zhan J et al (2022) Multi-trait genomic prediction models enhance the predictive ability of grain trace elements in rice. Front Genet 13:883853. https://doi.org/10.3389/fgene.2022.883853

Nascimento M, Nascimento ACC, Azevedo CF et al (2024) Enhancing genomic prediction with stacking ensemble learning in Arabica coffee. Front Plant Sci 15:1373318. https://doi.org/10.3389/fpls.2024.1373318

Nawade B, Mishra GP, Radhakrishnan T et al (2019) Development of high oleic peanut lines through marker-assisted introgression of mutant ahFAD2 alleles and its fatty acid profiles under open-field and controlled conditions. 3 Biotech 9:243. https://doi.org/10.1007/s13205-019-1774-9

Ogutu JO, Piepho H-P, Schulz-Streeck T (2011) A comparison of random forests, boosting and support vector machines for genomic selection. BMC Proc 5:S11. https://doi.org/10.1186/1753-6561-5-S3-S11

Olivoto T, Lúcio AD (2020) metan: an R package for multi-environment trial analysis. Methods Ecol Evol 11:783–789. https://doi.org/10.1111/2041-210X.13384

Park T, Casella G (2008) The Bayesian Lasso. J Am Stat Assoc 103:681–686. https://doi.org/10.1198/016214508000000337

Parveen R, Kumar M, Swapnil et al (2023) Understanding the genomic selection for crop improvement: current progress and future prospects. Mol Gen Genomics 298:813–821. https://doi.org/10.1007/s00438-023-02026-0

Persa R, Bernardeli A, Jarquin D (2020) Prediction strategies for leveraging information of associated traits under single- and multi-trait approaches in soybeans. Agriculture 10:308. https://doi.org/10.3390/agriculture10080308

Persa R, Canella Vieira C, Rios E et al (2023) Improving predictive ability in sparse testing designs in soybean populations. Front Genet 14:1269255. https://doi.org/10.3389/fgene.2023.1269255

Poland JA, Rife TW (2012) Genotyping-by-sequencing for plant breeding and genetics. Plant Genome 5:plantgenome2012.05.0005. https://doi.org/10.3835/plantgenome2012.05.0005

R Core Team (2024) R: a language and environment for statistical computing. R Foundation for Statistical Computing, Vienna, Austria

Raftery AE, Lewis SM (1992) [Practical Markov Chain Monte Carlo]: comment: one long run with diagnostics: implementation strategies for Markov Chain Monte Carlo. Stat Sci 7. https://doi.org/10.1214/ss/1177011143

Ray DK, Mueller ND, West PC, Foley JA (2013) Yield trends are insufficient to double global crop production by 2050. PLoS One 8:e66428. https://doi.org/10.1371/journal.pone.0066428

Rutkoski JE, Crain J, Poland J, Sorrells ME (2017) Genomic selection for small grain improvement. In: Varshney RK, Roorkiwal M, Sorrells ME (eds) Genomic selection for crop improvement. Springer International Publishing, Cham, pp 99–130

Rutkoski JE, Krause MR, Sorrells ME (2022) Breeding methods: population improvement and selection methods. In: Reynolds MP, Braun H-J (eds) Wheat improvement. Springer International Publishing, Cham, pp 83–96

Saavedra LM, Caixeta ET, Barka GD et al (2023) Marker-assisted recurrent selection for pyramiding leaf rust and coffee berry disease resistance alleles in Coffea arabica L. Genes 14: 189. https://doi.org/10.3390/genes14010189

Sanger F, Nicklen S, Coulson AR (1977) DNA sequencing with chain-terminating inhibitors. Proc Natl Acad Sci USA 74:5463–5467. https://doi.org/10.1073/pnas.74.12.5463

Satam H, Joshi K, Mangrolia U et al (2023) Next-generation sequencing technology: current trends and advancements. Biology 12:997. https://doi.org/10.3390/biology12070997

Schölkopf B, Smola AJ (2002) Learning with kernels: support vector machines, regularization, optimization, and beyond. MIT Press, Cambridge, Mass, Reprint

Silva ÉDBD, Xavier A, Faria MV (2021) Impact of genomic prediction model, selection intensity, and breeding strategy on the long-term genetic gain and genetic erosion in soybean breeding. Front Genet 12:637133. https://doi.org/10.3389/fgene.2021.637133

Sousa ICD, Nascimento M, Silva GN et al (2021) Genomic prediction of leaf rust resistance to Arabica coffee using machine learning algorithms. Sci Agric (Piracicaba, Braz) 78:e20200021. https://doi.org/10.1590/1678-992x-2020-0021

Sun J, Rutkoski JE, Poland JA et al (2017) Multitrait, random regression, or simple repeatability model in high-throughput phenotyping data improve genomic prediction for wheat grain yield. The plant. Genome 10:plantgenome2016.11.0111. https://doi.org/10.3835/plantgenome2016.11.0111

Technow F, Messina CD, Totir LR, Cooper M (2015) Integrating crop growth models with whole genome prediction through approximate Bayesian computation. PLoS One 10:e0130855. https://doi.org/10.1371/journal.pone.0130855

Tibshirani R (1996) Regression shrinkage and selection via the Lasso. J R Stat Soc Series B: Stat Methodol 58:267–288. https://doi.org/10.1111/j.2517-6161.1996.tb02080.x

Tuberosa R (2012) Phenotyping for drought tolerance of crops in the genomics era. Front Physiol 3. https://doi.org/10.3389/fphys.2012.00347

Van Eeuwijk FA, Bustos-Korts DV, Malosetti M (2016) What should students in plant breeding know about the statistical aspects of genotype × environment interactions? Crop Sci 56:2119–2140. https://doi.org/10.2135/cropsci2015.06.0375

VanRaden PM (2008) Efficient methods to compute genomic predictions. J Dairy Sci 91:4414–4423. https://doi.org/10.3168/jds.2007-0980

Varshney RK, Bohra A, Roorkiwal M et al (2021) Fast-forward breeding for a food-secure world. Trends Genet 37:1124–1136. https://doi.org/10.1016/j.tig.2021.08.002

Volpato L, Alves RS, Teodoro PE et al (2019) Multi-trait multi-environment models in the genetic selection of segregating soybean progeny. PLoS One 14:e0215315. https://doi.org/10.1371/journal.pone.0215315

Voss-Fels KP, Cooper M, Hayes BJ (2019) Accelerating crop genetic gains with genomic selection. Theor Appl Genet 132:669–686. https://doi.org/10.1007/s00122-018-3270-8

Wang C, Hu S, Gardner C, Lübberstedt T (2017) Emerging avenues for utilization of exotic germplasm. Trends Plant Sci 22:624–637. https://doi.org/10.1016/j.tplants.2017.04.002

Wang X, Xu Y, Hu Z, Xu C (2018) Genomic selection methods for crop improvement: current status and prospects. Crop J 6:330–340. https://doi.org/10.1016/j.cj.2018.03.001

Washburn JD, Burch MB, Franco JAV (2020) Predictive breeding for maize: making use of molecular phenotypes, machine learning, and physiological crop models. Crop Sci 60:622–638. https://doi.org/10.1002/csc2.20052

Weckwerth W, Ghatak A, Bellaire A et al (2020) PANOMICS meets germplasm. Plant Biotechnol J 18:1507–1525. https://doi.org/10.1111/pbi.13372

Wilson S, Zheng C, Maliepaard C et al (2024) Multienvironment genomic prediction in tetraploid potato. G3: genes, genomes. Genetics 14:jkae011. https://doi.org/10.1093/g3journal/jkae011

Xu Y, Li P, Zou C et al (2017) Enhancing genetic gain in the era of molecular breeding. J Exp Bot 68:2641–2666. https://doi.org/10.1093/jxb/erx135

Xu Y, Liu X, Fu J et al (2020) Enhancing genetic gain through genomic selection: from livestock to plants. Plant Commun 1:100005. https://doi.org/10.1016/j.xplc.2019.100005

Yang H, Li C, Lam H-M et al (2015) Sequencing consolidates molecular markers with plant breeding practice. Theor Appl Genet 128:779–795. https://doi.org/10.1007/s00122-015-2499-8

Zhang X, Pérez-Rodríguez P, Semagn K et al (2015) Genomic prediction in biparental tropical maize populations in water-stressed and well-watered environments using low-density and GBS SNPs. Heredity 114:291–299. https://doi.org/10.1038/hdy.2014.99

Zhang J, Liu F, Reif JC, Jiang Y (2021) On the use of GBLUP and its extension for GWAS with additive and epistatic effects. G3 Genes|Genomes|Genetics 11:jkab122. https://doi.org/10.1093/g3journal/jkab122

Chapter 10
Integrated Omics Approaches to Accelerate New Strategies for Crop Improvement

Qamar U. Zaman, Shen Shuangqian, and Jie Luo

Abstract Developing genomic databases and integrating the next-generation sequencing tools could advance the efforts of crop improvement and assist in harvesting the actual potential of plant genetic resources. Under changing environmental conditions, exploring the wild and cultivated genetic resources and adjusting their multi-agronomic traits could achieve sustainable solutions for food security challenges. Next-generation sequencing technologies have advanced phenotyping strategies by enabling more accurate assessments under diverse environmental conditions and facilitating the detection of underlying genetic mechanisms. Recently, many omics approaches have been formulated with precise linkage from genome to phenome for biotic and abiotic stress tolerance for the generation of sustainable crop production. Creating crops that support nutritional security requires alternative methods and breeding approaches to diversify the crops at both genetic and phenotypic levels. Here, we demonstrate that integrating the genomics-assisted tools with the 3D chromatin structure of the plant genomes could predict the precise target loci regulating the phenotypic traits of the plants and create an opportunity to improve the crops against the looming challenges of a changing climate and food scarcity. These integrated genomic approaches could assist us in understanding, optimising, and innovating plant breeding strategies under changing climatic conditions.

Keywords OMICS · 3D genomes · TADs · Compartmentalisation · New breeding strategies

Q. U. Zaman · S. Shuangqian
Yazouwan National Laboratory, Sanya, China

J. Luo (✉)
Yazouwan National Laboratory, Sanya, China

School of Breeding and Multiplication (Sanya Institute of Breeding and Multiplication), Hainan Yazhou-Bay Seed Laboratory, Hainan University, Sanya, China
e-mail: luojie@yzwlab.cn

V. Garg et al. (eds.), *DNA of Sustainability*, Sustainability Sciences in Asia and Africa,
https://doi.org/10.1007/978-981-95-5109-5_10

10.1 Introduction

The current trajectory of crop improvement through genomic approaches paves the way for inducing specific variations at desired loci to produce stress tolerance, generate climate-smart crops, and nourish the world's growing population. Greater and consistent crop production has remained a key element throughout the history of plant breeding for economic gains and global food security. Over the decades, traditional plant breeding methods have been used to develop new varieties with desirable traits by crossing individuals of the same species (Chao et al. 2023). However, these methods proved time-consuming, labour-intensive, and often limited to genetic variations in plant species. In addition to these limitations, climate change, lack of arable land, water scarcity, and a continuously growing population require constant crop improvement to meet food security (Steinwand and Ronald 2020; Zhang et al. 2022). Among these, extensive population growth and climate change are the major challenges and put pressure on the efforts of the sustainable food supply chain and global crop production, respectively (Lenaerts et al. 2019). In recent decades, after the green revolution, there has been impressive growth in crop yield after producing short-stature varieties, disease-resistant cultivars, and increased use of chemical fertilisers. However, population pressure continues to imbalance the food supply chain, land-water management, resource degradation, deforestation, rural-urban migration, and environmental waste-stacking. In addition, climate change is a majorly contributor to the threat to agriculture due to changes in temperature and precipitation (Xiong et al. 2022). Environmental fluctuation negatively affect food productivity, pest and disease dynamics, and soil nutrient properties.

Conventional plant breeding contributed a lot to feeding the masses in the last several decades by producing high-yielding, low-stature, and high-stress-tolerant varieties (Xiong et al. 2022). The challenge to conventional plant breeding is the slow process of variety development of desired traits. In this twenty-first century, longer time variety development processes, longer breeding cycles, breeding based on phenotype, and time-consuming practices are hindering the efforts of researchers to improve crop traits under uncertain climate conditions across the globe. The impact of climate change and growing global demands due to increasing population pressure needed an integrative and multidisciplinary response using modern technology. Integrating modern breeding and genomics tools for crop improvement can deliver precise and efficient results to meet the requirements of the masses and cope with environmental challenges.

The advances in next-generation sequencing (NGS) technology have provided bases for crop improvement strategies and led to the new era of genomics-assisted breeding (GAB) (Varshney et al. 2021). The genomic approaches provided an efficient tool for locating the desirable loci regulating the phenotype of the plants. Using a large amount of genomics data (transcriptomics, epigenomics, proteomics, and metabolomics) assisted the researchers in finding the genetic loci regulating the phenotype of the specific traits (Zhang et al. 2022). These GAB tools helped to modernise plant breeding by effectively exploiting the variation in available germplasm for cultivar development. Currently, these genomics tools meet the crops' 3-dimensional genomic landscape (high-throughput chromatin capture-Hi–C). The eukaryotic genome is hierarchically folded and packed in chromatin loops, A & B compartments, and topologically associated domains (TADs) (Zaman et al. 2025). Chromatin folding plays a key role in gene expression, phenotypic regulation, and transcriptional control. It is also associated with physical contact between the target gene and the respective enhancer-promoter sequences (Kempfer and Pombo 2020). Transcription factors inside the TAD boundaries also play a key role as a regulator and repressor at once and mechanistically regulate the genes within the TADs (Zaman et al. 2025). TAD boundaries enriched by active gene expression and euchromatic epigenetic marks posit that transcription factors (*TCP1* and *bZIP*) bridge the different genes to form a transcriptional hub. In a recent study, Dong and colleagues (2017) found that plant chromatin loops were formed between gene islands beyond the repressive domains and closely associated with active compartments. Besides spatial chromatin packing modules than the mammals, plant TADs are associated with transcription factor activities. In addition, long-range chromatin interactions are important for transcriptional regulation of genes. Genes with higher chromatin interactions showed higher expression levels than those without interactions (Li et al. 2019a). By combining the Hi–C and genomics-assisted tools, similar to Liu et al. (2023) did in metazoans (Liu et al. 2023), the potential association of genes with a trait can be identified at a spatial-temporal stage in specific varieties which is being cultivated at specific agro-climatic regions. This method can assist in identifying the candidate genes from the specific cultivars of one species grown in specific regions or temporal climates. This integrated strategy to cope with the changing climatic challenges can ensure sustainability in crop improvement by developing new and better crops.

10.2 Role of Genome Sequencing Technologies

The strategies that decoded the genetic codes of life made a paradigm revolution in revealing the basic information of the central dogma. Initially, DNA sequencing was performed using the Sanger sequencing method, which is based on a chain termination system (Sanger et al. 1977). However, integrating Sanger's sequencing methods, capillary electrophoresis (CE) based automatic sequencing approaches, and data analyser tools of bioinformatics laid the foundation of improvements in sequencing strategies (Ashraf et al. 2022). Over the last 2 decades, next-generation sequencing technologies provided opportunities to fast-forward genetic studies using high-throughput sequencing methods (Wheeler et al. 2008) and develop millions of novel markers, genes regulating desired traits, major phenotypic regulatory mechanisms, and identification of variations within the different varieties of the same species. Since then, genome sequencing technology has become the norm of the day and changed it into a big enterprise. Third-generation sequencing potentially generates 5-30 kb read lengths and produces a considerable overlapping read length to avoid amplification bias in genome assembly. Third-generation long-read sequencing platforms are comprised of single-molecule real-time (SMRT) sequencing (Clark et al. 2011), Oxford-Nanopore platform (Jain et al. 2016; Wang et al. 2021b), and electron microscopy to generate long-read data (Peterson et al. 2010).

Sequencing technologies are the functional bases of genomics, and whole genome assemblies of various plant species have been characterised using these platforms. High-quality reference genome assemblies of various diploid and complex polyploid species provided the basic genetic information, biological niche, genetic diversity, and variation between inter- and intra-species (Montenegro et al. 2017; Zhao et al. 2018; Khan et al. 2024; Zaman et al. 2025). Exploitation of variations within the germplasm resources is the basis of the trait improvement and variety development programme. Advancements in NGS led to the development of high-density genetic maps in various crops to predict the linear sequence of markers based on the recombination frequency of the population developed through crosses between genetically diverse parents (Wang et al. 2019). By labelling the traits with high resolution, linkage groups help locate the scaffolds on the map and substitute association mapping for QTL mapping (Thottathil et al. 2016; Pei et al. 2021b). The availability of reference genomes of many important crops made it possible to discover targeted regions of the genomes regulating agronomic components (Sasaki and International Rice Genome Sequencing 2005; Zaman et al. 2025). It enhanced the precision of conventional plant breeding strategies. Many important crop varieties were developed using marker-assisted selection methods. NGS tools assisted in detecting the variation in the genome, structural variations including deletions, insertions, and translocations compared to the reference genomes within a short period at a low cost (Krishna et al. 2023). Identifying all sets of genes and comparing genomic information provides a way forward in understanding the inheritance of agronomic components throughout the evolutionary process (Edwards and Batley 2010).

10.3 Omics Role in Revealing the Genetic Flow of Central Dogma to Accelerate the Crop Improvement

Integrating the omics data, complex genetic regulatory mechanisms controlling agronomic components can be predicted to design new crops that could produce higher yields, better quality, and more resilience against biotic and abiotic stresses. These integrated approaches have enabled to identification of the candidate genes that can be manipulated for precision breeding (Li and Yan 2020; Raza et al. 2024) and elaborated on how plants interact with the environment in response to climate change and different stresses. In addition, plant omics databases and platforms, including the genome size in the *Asteraceae* database-GSAD (Garnatje et al. 2011), Ensemble (Monaco et al. 2013), Brassica.info (http://www.brassica.info/

tools/databases.html), Brassica database-BRAD (Chen et al. 2021), NCBI, plant GARDEN (Ichihara et al. 2023), Phytozome (Goodstein et al. 2011), plant DNA C-values database (Pellicer and Leitch 2020), plant rDNA database (Garcia et al. 2014), plant GDB genome browser (Duvick et al. 2008), BnaOmics (Cui et al. 2023), and ArecaceaeMDB (Yang et al. 2023b) are providing valuable insights and resources for crop improvement. Integrated omics tools significantly enhanced crop improvement strategies and remain key in ensuring scientific progress to secure global food security (Purugganan and Jackson 2021) by leveraging the data from various approaches, including genomics, transcriptomics, epigenomics, metabolomics, proteomics, and phenomics.

10.3.1 Transcriptomics

RNA sequencing is a high-resolution technique that uses the NGS tool to analyse the complete set of transcripts (mRNA, noncoding RNAs, and small RNAs) generated by the genome under specific time and space (Satam et al. 2023). Generally, a population of RNA (total or fractionated) is added to the library of fragmented cDNAs with adaptors attached to one or both sides. Each molecule, with or without amplification, is sequenced using many high-throughput sequencing technologies, including Illumina IG (Nagalakshmi et al. 2008), Applied Biosystem SOLiD (Cloonan et al. 2008), and Roche 454 Life Science (Barbazuk et al. 2007; Emrich et al. 2007) to get single-end sequencing or pair-end sequencing with the size of 30–400 bp. The resulting reads are normally aligned to the reference genome or reference transcripts to generate the genome-scale transcription map that profiles gene and transcript-level expression (Srivastava et al. 2020). Short reads from RNA-sequence provide information to reveal the connectivity, how two exons are connected, while longer reads or pair-end short reads reveal connectivity between multiple exons (Morin et al. 2008). The key aim of transcriptomics is to determine the transcriptional structure of the genes, quantify the gene expression, start sites, 5′ and 3′ ends, splicing patterns, gene fusion, mutation, and changes in the gene expression level of each transcript during development or changes in environmental conditions (Conesa et al. 2016). This technique significantly contributed to identifying the genes and linking them with functional analysis that reveals underlying mechanisms of biological processes, molecular functions, and cellular components (Tyagi et al. 2022). In addition, it enables to compare the gene expression under different environmental conditions, the study of alternative splice variants, alternative genome sequencing, the discovery of novel genes, and noncoding RNAs (Yoo et al. 2018).

These advancements in RNA-seq have revolutionised the field of genomics by improving gene annotation methods, building transcriptome databases (CNCB-NGDC Members and Partner 2021; Katz et al. 2022; Lim et al. 2022; Liu et al. 2022; Yu et al. 2022), and presenting their applications to uncover the underlying mechanism of plant phenotype (Wang et al. 2024a). Researchers compare the differential expression of different tissues to identify the differences in gene expression levels at the developmental stage of plants. In addition, it can assist in analysing the stress tolerance (Khokhar et al. 2024) and elucidate the scavenging mechanism of plants against the biotic and abiotic stresses (Zaman et al. 2024a). The origin and evolution of many species were analysed using the transcriptome of 1173 plants, and a database was established preceding the origins of green plants, land plants, vascular plants, and patterns of flowering plants and ferns (Leebens-Mack et al. 2019). A vast collection of transcriptomic data from the model plant Arabidopsis (Berardini et al. 2015), and many important cultivated plant species, including wheat (Pearce et al. 2015), cotton (Dai et al. 2022; Yang et al. 2023a), maize (Lawrence et al. 2004), sugarcane (Healey et al. 2024), tomato (Fei et al. 2011; Kudo et al. 2017), barley (Milne et al. 2021; Li et al. 2023a), *Brassica napus* (Cui et al. 2023), and rice (Kawahara et al. 2013; Sato et al. 2013; Yu et al. 2023), provided a way forward to carry out experiments and reveal the complex underlying genetic regulatory mechanism.

10.3.2 Epigenomics

Epigenomics is an important approach to understand gene regulatory mechanisms, which is associated with protein modification, including histone tails (phosphorylation, ubiquitination, acetylation, and deacetylation) and DNA methylation (Wang and Chang 2018; Choi et al. 2021). Epigenetic regulations are attained with the changes in gene expression that are heritable due to DNA methylation and post-translational histone modifications without altering the underlying DNA sequence (Kapazoglou et al. 2018). In addition, the action of noncoding RNAs led the genome to be in a state of open chromatin (euchromatin) or closed chromatin (heterochromatin), which is directly associated with gene activation or gene silencing, respectively (Kapazoglou et al. 2018; Gupta and Salgotra 2022). Epigenomics provides insights into the complex regulatory mechanisms and how these mechanisms regulate the phenotype via the gene expression system. In the last 2 decades, advances in high-throughput sequencing technology have assisted the researchers in quantifying and analysing the DNA, RNA, and histone modifications with various quantitative, high-resolution, whole-genome/transcriptome approaches that specifically employ the regulatory mechanism of cell differentiation and develop the unique phenotype (Chen et al. 2023). This epigenetic regulation reflects the mode of development in plants and evolutionary patterns. To perform the epigenetic sequencing, DNA methylation is transformed into a quantitative signal to screen it throughout the genome. There are various approaches (Chen et al. 2023) to quantify the epigenetic marks on the genome including methylation sequencing (detect 5-methylcytosine-5mC) (Rauluseviciute et al. 2019), PacBio methylation sequencing (reveal the diverse array of DNA modifications of 6-methyladenine, 4-methylcytosine) (Ni et al. 2023), Assay for Transposase Accessible Chromatin-ATAC-sequencing (employ Tn5 transposase to identify the open chromatin) (Buenrostro et al. 2015),

Chip-Sequencing (Identify protein–DNA interactions and histone modifications) (Park 2009) and an improved single-cell NanoATAC sequencing (identify genome-wide chromatin accessibility by profiling single-cell chromatin structure and DNA binding proteins that regulate gene expression) (Hu et al. 2023). Open chromatin regions constitute nucleosome-depleted regions of chromatin and contain cis-regulatory elements that provide a binding platform for transcription factors and other regulatory components. Chromatin of promoters and enhancers often exists in an open conformation to facilitate transcription (Kundaje et al. 2015). Epigenetic modifications mediate transcriptional regulation in many biological processes in plants (Liang et al. 2020). Histone acetylation is one of the most characterised and generally associated with transcription by removing acetyl moieties from the acetylated lysines of histones, which are crucial players in plant growth and development as well as response to stress stimuli (Chen et al. 2016b).

In plants, 5-cytosine is the main site of DNA methylation, which occurs in two symmetric forms, also called "CG islands" (CH and CHG), and one asymmetric context (CHH). In addition to DNA methylation, modification of histones and nuclear proteins, which are involved in the packaging of DNA strands in the nucleus, play a role in transcription, and translation for genetic regulation (Brukhin and Albertini 2021). Using the epigenomics approaches, epigenetic variants have been characterised in many model and field crops (Pecinka et al. 2019) including rice (drought tolerance) (Zheng et al. 2017), Arabidopsis (seed dormancy) (Zheng et al. 2017), Brachypodium (drought tolerance) (Song et al. 2019), tomato (fruit ripening) (Chen et al. 2024), and soybean (reduce the seed size) (Wang et al. 2024b).

10.3.3 Proteomics

Within the diverse field of 'OMICS', proteomics plays a vital role in the post-genomics era and has become a key tool to identify protein quantification, posttranslational modification, protein structure, protein function, and protein-protein interactions in cells (Baginsky 2009; Cui et al. 2022). Protein sequencing is a foundational molecular biology and biochemistry technique that determines the precise order of amino acids (also known as building blocks of proteins) within the protein molecule. One of the earliest methods of protein sequencing was Edman degradation, which selectively removes and identifies the N-terminal amino acids from a purified peptide or protein (Masoodi et al. 2021). However, conducting sequencing from this method was not feasible because it requires large quantities of highly purified peptides and challenges with repetitive sequences (Alfaro et al. 2021). In the last 2 decades, mass spectrometry has emerged as a powerful tool for protein sequencing. It ionises protein fragments by specific enzymatic or chemical hydrolysis, measures the mass-to-charge ratio of each peptide, and identifies the peptide spectrum data for known proteins (Coon et al. 2005; Wang and Wilson 2013). In addition to these protein sequencing approaches, NGS also offers high-throughput capabilities for protein sequencing indirectly by relying on mRNA sequences. These protein sequencing methods mainly follow a database search approach and a de novo sequencing approach (Wang and Wilson 2013). Various methods have been developed to identify protein sequences by developing theoretical spectra in silico and comparing them with a given protein sequence database, including the open mass spectrometry search algorithm (OMSSA) (Geer et al. 2004), TANDEM mass spectra (Craig and Beavis 2004), scoring algorithm-based computer program (Mascot) (Perkins et al. 1999), liquid chromatography using tandem mass spectrometry (Nakashima et al. 2018), NGS protein sequencing platform (mass spectrometry, protein pathway array, next-generation tissue microarray, and Luminex) (Callahan et al. 2020), and single-molecule protein sequencing technologies (Alfaro et al. 2021; Ren et al. 2021).

Proteomics necessarily generates more complexity than genomics by collecting more proteins than the total number of genes in a single organism (Schad et al. 2011). Protein directs biological function in all organisms and does not function independently, but interacts in a highly complex regulatory mechanism that exhibits extensive dynamic range and difficult-to-model (Aslam et al. 2017). The role of protein is crucial in the growth and development of plants. The dynamic regulation of proteins in a complex pathway and crosstalk between multiple plant signals generate an effective response to cellular mechanisms. Various proteins are likely to function together and play multiple roles in stress response (Liu et al. 2019). During stresses, plants initiate a sophisticated immune response pattern recognition receptors (PRRs), which are bio-synthesised in the endoplasmic reticulum and moved to the plasma membrane to identify the danger associated with the molecular pattern and generate an effective pattern-triggered-immunity (PTI) (Jones and Dangl 2006; Gust et al. 2017). The advent of high-throughput approaches, investigations of posttranscriptional modifications, and protein-protein interactions provided new insightful data that revealed how plant cells decide and integrate several signals to generate an effective coping strategy.

10.3.4 Metabolomics

Metabolomics is an emerging field of study and a set of techniques to comprehensively measure the entire set of phytochemicals (known as metabolites in a biological specimen) by integrating various disciplines, including analytical chemistry, statistics, and biochemistry (Shen et al. 2023). These small molecules comprise a wide range of compounds, including sugars, amino acids, organic acids, alkaloids, flavonoids, and lipids. Metabolites have complex structures, are chemically diverse in nature, and pose challenges to visualising and interpreting (Hall et al. 2022). It's a favourable approach to decode the interconnected metabolic processes to regulate the growth, physiological components, biotic, and abiotic stress tolerance throughout the life span. It's a standard approach to quantifying the metabolites based on liquid/gas chromatography, mass spectrometry, and nuclear magnetic resonance. Nuclear magnetic resonance (NMR) and mass spectrometry (MS) are the most used methods to quantify plant metabolomics. NMR provides a flexible platform to cover the broad range of metabolites, a holistic picture of biological samples and elucidate the structures of unknown samples (Selegato et al. 2019). However, MS coupled with liquid or gas chromatography (Liu and Rochfort 2013; Yang et al. 2022) is a preferred approach because of its higher sensitivity and requires fewer samples. LC-MS can analyse a much wider range of chemical species than GC-MS (Liu and Rochfort 2014). GC-MS can also determine many polar metabolites, but it cannot detect oligosaccharides. In addition, the formation of isomers during the derivatisation of different metabolites increases the complexity of the data analysis (Liu and Rochfort 2013).

The quality and derivative products of the plants, including the taste, fragrance, colour, and storage duration, are all directly dependent on the concentration of metabolites (Jamaloddin et al. 2021). In the plant kingdom, it is estimated that 100,000–200,000 metabolites are present in diverse forms in cell membranes as lipids, simple ions, acidic and basic chemicals for cell elements, stable oxidisation systems, and structures (Gupta et al. 2021). These metabolites can be used to track the gene expression and response to stress tolerance in plants. Metabolomics tracks the specific metabolites of plant species which could help us to characterise them with the agronomic components, including regeneration of plants from callus, plant growth, tillering, plant height, fruit development, grain colour and size (Chen et al. 2016a), shelf life, food aroma and storage. Metabolomics studies have identified specific metabolites of glucosinolates synthesis in Brassicaceae (Zang et al. 2009), feruloylserotonin and l-asparagine content variation in rice (Li et al. 2019b), cis-zeatin glucosyltransferase activity (Chen et al. 2018), regulation of biosynthetic pathways (De Geyter et al. 2012), metabolic spatial distribution in the presence of photosynthates in shoots and absence in roots (Gaudinier et al. 2015), metabolic quantitative trait loci regulating morphological traits in rice (Gong et al. 2013), secondary cell wall differentiation during xylem vessel formation (Yamaguchi et al. 2010; Yamaguchi et al. 2011), presence of metabolites in fruit serve to increase the attractiveness and protect it from biotic and abiotic stresses (Pott et al. 2019) and cell death (Ohashi-Ito et al. 2010).

10.4 Current Developments in Omics Integration for Predicting Precise Chromatin Interactions and Improving Plant Breeding Strategies

The activities of genomic components at the molecular level form the complex traits through a series of biological processes in plants. Recent biotechnological tools have been harnessed to understand genome profiling, transcriptomes, epigenomes, proteomes, metabolomes, and various other OMIC components. These high-throughput approaches have emerged as cutting-edge tools to advance crop improvement and contend with the challenges of food security. These technologies have enabled researchers to identify the candidate genes that play a key role in underlying complex pathways to regulate the phenotype of plants. Integration of these datasets is crucial in providing comprehensive knowledge of agronomic traits and genetic interactions (Raza et al. 2024). New biotechnological tools and next-generation sequencing have provided novel strategies to introduce elite foundation material from wild species to cultivated crops to provide gains in productivity, environmental adaptations, and other beneficial traits.

10.4.1 3D Genomic Architecture and Emerging Role of TAD and TADs Boundaries

Currently, various bioinformatics tools have been developed to link the datasets of the central dogma via computational methods with existing omics approaches. In recent decades, with the advent of 3-dimensional genome structure (using Hi–C approaches) (Lieberman-Aiden et al. 2009), it has become easier to map the spatial chromatin conformation in plants and investigate the possible role of genomic components by uncovering the hierarchical organisations of plant genomes including chromosomal territories, compartmentalisation, TADs, sub-TADs, TAD boundaries, chromatin loops, detecting statistically significant chromatin interactions (Zaman et al. 2025), promoters, and cis-regulatory DNA elements (enhancers and silencers) (Consortium EP 2012). Chromatin status of the genome is directly correlated with DNA replication (Dekker and Heard 2015; Pei et al. 2021a), cell type-specific gene expression, and other biological functions (Dekker et al. 2013). 3D spatial organisation of the genome can be partitioned into two types of chromatins, referred to as active (A) and inactive (B) compartments (Ouyang et al. 2020). Several high-throughput methods, including chromatin interaction analysis by paired-end tag sequencing (ChIA-PET), In-Situ Hi–C (Padmarasu et al. 2019), Hi–C, and chromatin immunoprecipitation (HiChIP) (Wang et al. 2021a) have been explored in mammals and plants (Sun et al. 2024; Zaman et al. 2025). Technological optimisation of Hi–C, Hi–C 2.0, and Hi–C 3.0 has substantially enhanced the resolution from the megabase scale to the kilobase scale. Chromatin architecture differs between plants in mammals. In mammals, loop extrusion and compartmentalisation suggest that both operate jointly to maintain the spatial organisation of the genome. However, TAD-like domains and compartmentalisation have also been observed in plants, but the mechanism of formation largely remains unknown (Doğan and Liu 2018). There are open questions concerning their structures, dynamics, and transcriptional regulatory function in plant species in the absence of *CTCF* proteins. This also indicates that the fundamental architecture of TADs can exist and perform its function without the presence of *CTCF* proteins.

Transposable elements (TEs) also play a crucial role in shaping the 3D genome organisation and can significantly influence TADs in plants. In *Verticillium dahlia*, it was suggested that TEs contribute to genome dynamics by promoting structural variations such as insertions, deletions, and rearrangements, which can disrupt or create new TAD boundaries (Torres et al. 2021). These TE-induced changes can reorganise chromatin interactions, altering the spatial proximity of regulatory elements and genes, thereby impacting gene expression. Such reorganisation not only drives genomic plasticity but may also underpin adaptive responses to environmental stresses by modulating expression patterns of stress-responsive genes. Thus, TEs serve as dynamic elements that link genome architecture with evolutionary potential in plants.

In a conserved feature of genome organisation, the presence of TADs (Szabo et al. 2019) suggests a role in shaping functional chromosomal organisation and defining DNA-dependent regulatory landscape (Lieberman-Aiden et al. 2009; Beagan and Phillips-Cremins 2020). Mammalian TADs are contiguous 3D chromatin structures formed by active extrusion of chromatin loops by cohesion at a sub-chromosomal scale and vary in size significantly between species, from a few kb to Mb scale. Most of the mammalian TADs are enriched with conserved CCCTC binding proteins (CTCF). This again differs in plants without CTCF plant genome enrichment mediates domain formation (Dong et al. 2017). Plant TADs might play different functions due to regulatory contact between enhancers and promoters appearing around the TAD boundaries. Wang and colleagues (2015) called the TADs ‘positive strips’ which show special structural features and distinct epigenetic marks on the Arabidopsis genome packed in a 2 kb resolution Hi–C map. These positive strips exhibited kbp-sized segments representing higher intrachromosomal interaction rates compared to neighbouring chromatin regions and genes mapped in genome regions of positive strips showed lower expression levels. The presence of TADs is unlikely to depend on genome size in plants; TADs have been observed in Marchantia (~290 Mb). TAD size varied greatly across plant species. The determination of TAD sizes is influenced by experimental and technical conditions. TADs are typically detected by Hi–C, and the absolute number in the genome depends on their resolution. It’s easier to get high resolution of Hi–C maps in small genomes, whereas it is more challenging in large genomes (Tourdot and Grob 2023).

Liu et al. (2017) found that TAD boundaries were enriched with a plant-specific family of *TCP* TF, which was positively correlated with higher than average gene expression (Liu et al. 2017). In pepper, mammal-like TAD domain structures were identified. In total, 60% of the pepper genome corresponded to the repressed regions that were enriched by repetitive sequences and heterochromatin marks (*H3K9me2*). Pepper TAD-like domains were enriched by genes that may act to connect genes via gene-to-gene loops, leading to the spatial clustering of active genes (Liao et al. 2022). In yellow (*Selenicereus megalanthus*) and red-peel white pulp pitaya (*Selenicereus undatus*), Zaman and colleagues (2025) found 2031 and 3376 TADs, respectively (Zaman et al. 2025). In red-peel pitaya, 24,384 genes were distributed in 3376 TADs, while in yellow pitaya, 2031 TADs were comprised of 24,775 genes. However, 4045 and 2162 genes were located in the TAD boundaries of red-peel and yellow pitaya, respectively. In red-peel pitaya, TAD boundaries were significantly enriched by *PHYPADRAFT_64121*, *RAV1*, *AT3G24120*, *WRKY60*, and *WRKY18* motifs. In yellow pitaya, TAD boundaries were enriched with

PHYPADRAFT_64121, *RAV1*, *WRKY75*, *WRKY60*, and *MYB3* motifs. Zaman and colleagues further identified the role of genes expressed to regulate two contrasting phenotypes in pitaya species (red peel and yellow peel pitaya). Both phenotypes of pitaya are regulated by the betalain biosynthesis pathway. Higher expression of the *SmeADH1* gene in the TAD region (135480000-135520000) of yellow-peel pitaya and lower expression of *HuADH1* gene in the TAD region (87160000-87200000) of red-peel pitaya created the phenotypic differences. In addition, the homoeolog copy of *CYP76AD1* that produces the enzymes to oxidase the tyrosine into L-DOPA and generate betacyanins in red-peel pitaya showed two-fold higher expression as compared to the genes in yellow pitaya. Expression of the genes regulating the betalain pathway was highly correlated with motif count in TAD boundaries of both pitaya species. It was observed that the genes exhibited higher expression within the TADs that comprised a lower number of TFs in TAD boundaries. Intensive functional verifications are required to characterise the potential role of proteins at TAD boundaries, which might be shaping plant traits. The substantial variation of TAD boundaries in different plant species suggests that chromatin organisation might be more diverse in plants, being affected more by gene density, linear distribution of genes, and transposable elements (Stam et al. 2019) than in multicellular metazoans.

Genes within the TADs have a more conserved expression pattern, and any alteration or disruption in TADs by evolutionary changes causes a change in expression profile (Krefting et al. 2018). Long et al. (2022) found that single genes that occupy a TAD alone are statistically more constrained compared to the genes in TADs with multiple genes while investigating the relationship between gene evolution and TADs. These results suggested that genes occupying TADs singly are functionally enriched and exhibited a strong association with the developmental and evolutionary process (Long et al. 2022). In addition, Doebley et al. (1995) identified that *TEOSINTE BRANCHED 1* was involved in the teosinte plant's response to the local environment to produce long or short branches and under the domestication process that maize evolution involved changes at this locus that produced short branches under all environments (Doebley et al. 1995). *TEOSINTE BRANCHED 1* protein is a *TCP* transcription factor that might assist in understanding the connection between higher-order, chromatin-level epigenomics and chromosome rearrangement during the evolutionary process. Genomic rearrangements could affect gene expression, which causes instability in TADs. In *Gossypium* species, epigenomic and expression correlate with the nonrandom chromosomal rearrangements and 3D architecture during plant evolution (Li et al. 2023b). By categorising the TADs into two groups, including evolutionary rearranged and conserved TADs, it was found that the number of conserved TADs between paired species (*Gossypium kikii* vs. *Gossypium arboreum*, *Gossypium kirkii* vs. *Gossypium raimondii*, *Gossypium arboretum* vs. *Gossypium raimondii*) exhibited association with their phylogenetic distance. The number of conserved TADs was decreased with an increase in divergence time, which suggests that TAD stabilisation acts antagonistically in increasing and decreasing expression divergence (Li et al. 2023b). By comparing the two species of pitaya that carry two contrasting phenotypes (red and yellow peel), Zaman and colleagues found that the *CYP36A12* genes in red-peel pitaya exhibited higher expression and were located inside the TAD region (compartment A of the genome) and might be directly coordinated with the transcription factors positioned in respective TAD boundaries. However, a homologous copy of the *CYP36A12* gene in yellow pitaya showed lower expression and was located in the open euchromatin and condensed heterochromatin (compartments A and B) part of the yellow pitaya genome. Lower gene expression in the yellow pitaya species might lead to the loss of capacity to oxidase tyrosine into *L-DOPA* and produce betaxanthin instead of betacyanin (Zaman et al. 2025).

10.4.2 Integration of Omics Approaches and Way Forward for the Development of New Crops

The development of new breeding technologies has improved crop varieties very rapidly and facilitated the beginning of a new era of crop domestication by enhancing the precise genetic variations required for sustainable agroecosystem and food system diversity (Krug et al. 2023). Improving cultivated crops and introducing new diverse materials for achieving the sustainable development goals, uninterrupted supply chain, zero hunger, and improved nutrition requires a major transformation in the global food system (Qaim and Policy 2020). Plant breeding strategies need to effectively address the challenges associated with climate change and a growing population. However, most modern breeding tools do not effectively assist in precisely selecting the putative candidate genes in plants. For instance, quantitative trait locus (QTL) mapping indicates a potential and allows the breeder to associate the genes with the phenotype of the plant. However, these molecular tools do not replace agronomic trials under on-farm conditions and more precise selection tools are needed.

The development of Hi–C sequencing strategies offers a good opportunity to accelerate the identification of genes regulating at the compartment level. In recent studies, the genome has been shown to have conserved TADs in closely related species, as 8.23% of foxtail millet TADs exhibited syntenic conservation features also found in the sorghum genome. However, none of the sorghum TADs were found in the maize genome (Dong et al. 2017). Other than plant species, TADs showed 30–40% conservation in the genomes of *D. melanogaster* and *D. pseudoobscura*. In addition, 43% of the human genome has syntenic relations with the closest chimpanzees (Eres et al. 2019). plant genomes have clear 3D chromatin organisation and TAD structure, determining the gene functionality and fine-tuning of physiological processes for appropriate cellular responses (Zhang and Wang 2021). In cotton, by comparing the sub-genome of the tetraploid, it was found that allo-polyploidisation has significantly contributed to a switch and reorganisation of the A/B compartments and TADs. Karaaslan et al. (2020) have found that mutations in *TCP1-rich* TAD boundary lines explained significant changes in the expression of the genes located within the TAD (Karaaslan et al. 2020). Interaction between proximal and distal regulatory regions further suggests that chromatin interactions are important for transcriptional regulation of genes associated with plant traits. In a recent study, Dong and colleagues (2017) found that plant chromatin loops were formed between gene islands beyond the repressive domains and closely associated with active compartments.

Genes that have higher chromatin interactions showed higher expression levels in maize compared to those without interactions (Li et al. 2019a). Chromatin interaction network (ChIN) and analysed gene ontology terms for the 50 network components. Previously, it was known that *Dwarf8* regulates plant height in maize. However, ChIN analysis predicted that the *Dwarf8* gene, along with nine other genes, was associated with flowering traits. By identifying genes in active TADs and combining the analysis of ChIN, complex regulatory pathways can be identified for quantitative and qualitative traits of plants (Li et al. 2019a). Though plant TAD structures and metazoan TADs are entirely different, some common concepts could be used to conduct investigations, as the research is lagging in plants. Recently, Liu et al. (2023) combined Hi–C and RNA-sequencing to explore the functional relationships between the 3D genome and transcriptome during mouse embryonic lineage specification. It was observed that two cell types closer in developmental stages showed more conserved genome compartmentalisation and exhibited changes with the progress of development. The same pattern was observed in the formation of TAD boundaries (Liu et al. 2023) and chromatin structures that were rearranged extensively in various tissues, cell types and displayed strong tissue-specific interactions (Schmitt et al. 2016).

In the future, the TAD-based trait improvement methods could complement and enhance marker-assisted breeding approaches, as the TADs can be used to manipulate specific targets that regulate agronomic traits in plants and can ensure the expression of the untapped potential of functional and cryptic genes in the specific variety. In the recent past, plant breeders used marker-assisted selection (MAS), which allows the faster identification of specific genes and the selection of desirable traits. However, markers developed for one plant population may not be transferable to other populations because of a lack of markers or the absence of marker trait association. Often, if the trait can be transferred, then the F1 generations need to be backcrossed for many generations to clear the undesirable genetic background from the population developed by marker-assisted breeding. In principle, TAD-based plant improvement strategies start from selecting the cultivated or wild progenitor to identify the genome compartments, TADs, sub-TADs, and genes inside the TADs associated with the specific desired plant trait. By combining Hi–C and scRNA-seq data, similar to the work Liu et al. (2023) did in metazoans, the potential association of genes with the trait can be identified at a spatial-temporal stage in specific varieties that are being cultivated at specific agro-climatic regions. This method can assist in identifying the candidate genes from a specific cultivar of one species grown in specific regions or temporal climates. Using the CRISPR system (Zaman et al. 2023a, b), mutagenesis can be induced directly in associated genes or motifs in TAD-like boundaries in plants to alter the expression of the genes significantly, similar to research carried out in *Marchantia* plants (Karaaslan et al. 2020).

10.4.3 TAD-Based Trait Improvement Strategies

The domestication of new crops would promote agricultural diversity, adaptation to environmental changes, and associated problems with intensive agriculture (Østerberg et al. 2017). New breeding techniques are emerging rapidly due to advances in genomics research. This enables the researcher to identify the target precisely and make reliable changes in the genome for crop improvement. The impact of genomics is increasing the rate of improvement of crops by breeding, especially in the staple crops (Bevan and Uauy 2013). Creating crops that support nutritional resilience requires alternative methods and breeding approaches to diversify the crops at both genetic and phenotypic levels. Rapid development in genome sequencing technologies provides a powerful tool to identify the genes that played their role in domestication in crop plants and their wild relatives on a larger scale (Henry 2011).

TAD-based breeding methods can facilitate the crop domestication process multifold as marker-assisted breeding strategies are no longer robust as other modern techniques to map the candidate genes. Alteration in gene expression may be influenced by perturbations at TAD-like boundaries in crops, which are being cultivated in different climate zones and carrying spatial–temporal variations. Most of the agronomic and quantitative traits have complex genetic pathways that are regulated by multiple genes. TAD-based breeding methods can directly pick the active TADs, and ChIN can facilitate identifying the whole network of regulatory pathways that are affecting the phenotype. These active TADs represent promising regulatory units that could be targeted to potentially accelerate the domestication related trait improvement and the development of elite crop varieties. Genes within the TADs and enhancer–promoter regions were exploited in *Marchantia* species, which significantly changed the expression of the genes within the TADs. By linking the TADs with tissue-specific expression of the plants, genes, and their chromatin interaction network can be linked to the phenotype of the plant precisely. Identifying the putative candidate genes regulating desired traits in plants can be exploited for TAD-based breeding programmes, and potential biological questions can be solved from the perspective of CGVs and TADs. The specific variety that is cultivated in spatial–temporal proximity can be used to transfer the improved traits to other cultivated varieties grown in other climatic regions. This TAD-based breeding can complement and refine or might replace the marker-assisted breeding method, which is laborious and takes years for mapping, and most of the time predicts many genes at a single QTL (Fig. 10.1).

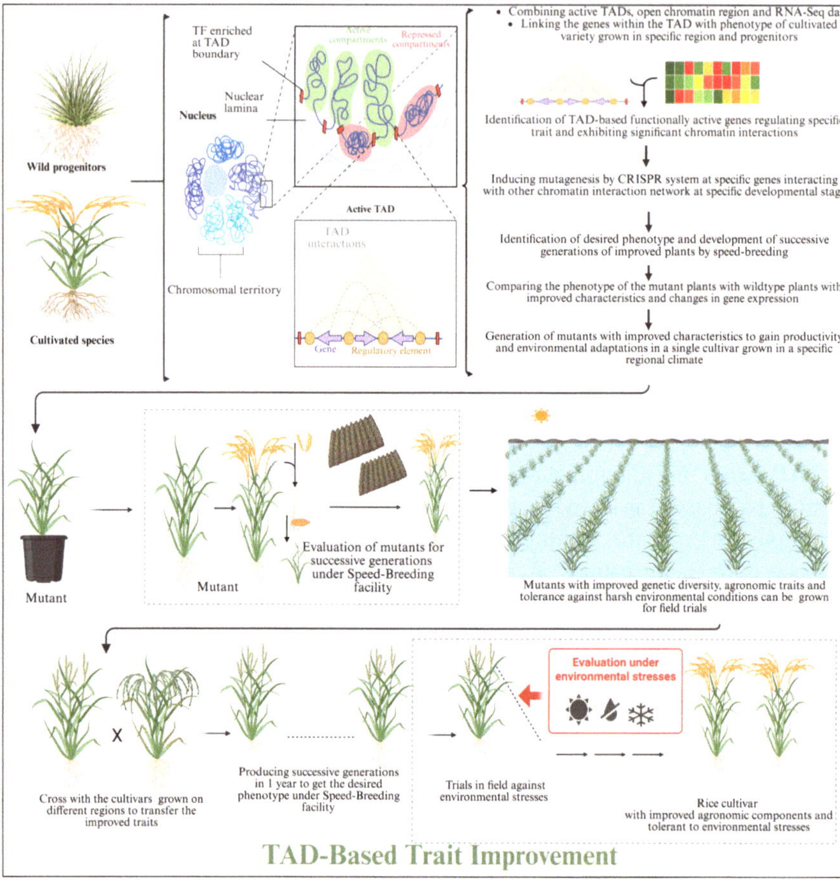

Fig. 10.1 TAD based trait improvement in plants. The genome of cultivated variety grown in a specific region and wild progenitor species can be differentiated into active (**a**) and inactive (**b**) compartments. TADs and scRNA-seq data can be linked to the phenotype of plants. Integration of multi-omics approaches to identify the candidate genes and regulatory elements associated with the agronomic traits. Perturbation and functional validation of these targets, followed by speed breeding and field evaluation under diverse environmental conditions, might facilitate the development of improved crop varieties

10.5 Concluding Remarks

Plant growth is subjected to complex molecular interactions of genes that rapidly respond to the transition of spatiotemporal variations and are directly correlated with phenotypic diversification. During the evolutionary and domestication periods, plants have had to pass through a range of environmental conditions and significant changes in genome architecture. However, the domestication process resulted in a narrow genetic base of the cultivated species, which is prone to climate change and other stresses. Wild relatives of the cultivated species typically exhibit high levels of genetic diversity and phenotypic variations. The sequencing of large and complex genomes by advanced sequencing technologies has facilitated the development of contiguous and chromosome-scale assemblies in many crops. Recent advances provide new opportunities to complement conventional plant breeding approaches because TAD like domains show varying degree of conservation and can be identified using high-resolution chromatin conformation approaches. Candidate genes and ChIN associated with the regulation of the desired phenotype enable more refined prioritization of candidate genes compared with other genomic approaches. Specific genes regulating specific traits within the TADs may serve as potential targets for genome editing. Mutations in transcription factors in TAD boundaries may greatly alter the expression of the genes within the TAD.

References

Alfaro JA, Bohländer P, Dai M, Filius M, Howard CJ et al (2021) The emerging landscape of single-molecule protein sequencing technologies. Nat Methods 18(6):604–617

Ashraf MF, Hou D, Hussain Q, Imran M, Pei J et al (2022) Entailing the next-generation sequencing and metabolome for sustainable agriculture by improving plant tolerance. Int J Mol Sci 23(2): 651

Aslam B, Basit M, Nisar MA, Khurshid M, Rasool MH (2017) Proteomics: technologies and their applications. J Chromatogr Sci 55(2):182–196

Baginsky S (2009) Plant proteomics: concepts, applications, and novel strategies for data interpretation. Mass Spectrom Rev 28(1):93–120

Barbazuk WB, Emrich SJ, Chen HD, Li L, Schnable PS (2007) SNP discovery via 454 transcriptome sequencing. Plant J 51(5):910–918

Beagan JA, Phillips-Cremins JE (2020) On the existence and functionality of topologically associating domains. Nat Genet 52(1):8–16

Berardini TZ, Reiser L, Li D, Mezheritsky Y, Muller R, Strait E, Huala E (2015) The Arabidopsis information resource: making and mining the "gold standard" annotated reference plant genome. Genesis 53(8):474–485

Bevan MW, Uauy C (2013) Genomics reveals new landscapes for crop improvement. Genome Biol 14(6):206

Brukhin V, Albertini E (2021) Epigenetic modifications in plant development and reproduction. Epigenomes 5(4):25

Buenrostro JD, Wu B, Chang HY, Greenleaf WJ (2015) ATAC-seq: a method for assaying chromatin accessibility genome-wide. Curr Protoc Mol Biol 109:21

Callahan N, Tullman J, Kelman Z, Marino J (2020) Strategies for development of a next-generation protein sequencing platform. Trends Biochem Sci 45(1):76–89

Chao H, Zhang S, Hu Y, Ni Q, Xin S et al (2023) Integrating omics databases for enhanced crop breeding. J Integr Bioinform 20(4):20230012

Chen W, Wang W, Peng M, Gong L, Gao Y et al (2016a) Comparative and parallel genome-wide association studies for metabolic and agronomic traits in cereals. Nat Commun 7(1):12767

Chen X, Lu L, Mayer KS, Scalf M, Qian S et al (2016b) POWERDRESS interacts with HISTONE DEACETYLASE 9 to promote aging in Arabidopsis. elife 5:e17214

Chen J, Wang J, Chen W, Sun W, Peng M et al (2018) Metabolome analysis of multi-connected biparental chromosome segment substitution line populations. Plant Physiol 178(2):612–625

Chen H, Wang T, He X, Cai X, Lin R et al (2021) BRAD V3.0: an upgraded Brassicaceae database. Nucleic Acids Res 50(D1):D1432–D1441
Chen X, Xu H, Shu X, Song C-X (2023) Mapping epigenetic modifications by sequencing technologies. Cell Death Differ 32:56–65
Chen S, Han J, Wu S, Guo S, Tang Y et al (2024) From non-coding RNAs to histone modification: the epigenetic mechanisms in tomato fruit ripening and quality regulation. Plant Physiol Biochem 215:109070
Choi MY, Fritzler MJ, Mahler M (2021) Chapter 8 – Development of multi-omics approach in autoimmune diseases. In: Mahler M (ed) Precision medicine and artificial intelligence. Academic, pp 189–201
Clark TA, Spittle KE, Turner SW, Korlach J (2011) Direct detection and sequencing of damaged DNA bases. Genome Integr 2:10
Cloonan N, Forrest AR, Kolle G, Gardiner BB, Faulkner GJ et al (2008) Stem cell transcriptome profiling via massive-scale mRNA sequencing. Nat Methods 5(7):613–619
CNCB-NGDC Members and Partners (2021) Database resources of the National Genomics Data Center, China National Center for Bioinformation in 2021. Nucleic Acids Res 49(D1):D18–D28
Conesa A, Madrigal P, Tarazona S, Gomez-Cabrero D, Cervera A et al (2016) A survey of best practices for RNA-seq data analysis. Genome Biol 17(1):13
Consortium EP (2012) An integrated encyclopedia of DNA elements in the human genome. Nature 489(7414):57
Coon JJ, Syka JEP, Shabanowitz J, Hunt DF (2005) Tandem mass spectrometry for peptide and protein sequence analysis. Biotechniques 38(4):519–523
Craig R, Beavis RC (2004) TANDEM: matching proteins with tandem mass spectra. Bioinformatics 20(9):1466–1467
Cui M, Cheng C, Zhang L (2022) High-throughput proteomics: a methodological mini-review. Lab Investig 102(11):1170–1181
Cui X, Hu M, Yao S, Zhang Y, Tang M, Liu L, Cheng X, Tong C, Liu S (2023) BnaOmics: a comprehensive platform combining pan-genome and multi-omics data from Brassica napus. Plant Commun 4(5):100609
Dai F, Chen J, Zhang Z, Liu F, Li J, Zhao T, Hu Y, Zhang T, Fang L (2022) COTTONOMICS: a comprehensive cotton multi-omics database. Database 2022:baac080
De Geyter N, Gholami A, Goormachtig S, Goossens A (2012) Transcriptional machineries in jasmonate-elicited plant secondary metabolism. Trends Plant Sci 17(6):349–359
Dekker J, Heard E (2015) Structural and functional diversity of topologically associating domains. FEBS Lett 589(20PartA):2877–2884
Dekker J, Marti-Renom MA, Mirny LA (2013) Exploring the three-dimensional organization of genomes: interpreting chromatin interaction data. Nat Rev Genet 14(6):390–403
Doebley J, Stec A, Gustus C (1995) Teosinte branched1 and the origin of maize: evidence for epistasis and the evolution of dominance. Genetics 141(1):333–346
Doğan ES, Liu C (2018) Three-dimensional chromatin packing and positioning of plant genomes. Nat Plants 4(8):521–529
Dong P, Tu X, Chu PY, Lü P, Zhu N, Grierson D, Du B, Li P, Zhong S (2017) 3D chromatin architecture of large plant genomes determined by local A/B compartments. Mol Plant 10(12): 1497–1509
Duvick J, Fu A, Muppirala U, Sabharwal M, Wilkerson MD, Lawrence CJ, Lushbough C, Brendel V (2008) PlantGDB: a resource for comparative plant genomics. Nucleic Acids Res 36(Database issue):D959–D965
Edwards D, Batley J (2010) Plant genome sequencing: applications for crop improvement. Plant Biotechnol J 8(1):2–9
Emrich SJ, Barbazuk WB, Li L, Schnable PS (2007) Gene discovery and annotation using LCM-454 transcriptome sequencing. Genome Res 17(1):69–73
Eres IE, Luo K, Hsiao CJ, Blake LE, Gilad Y (2019) Reorganization of 3D genome structure may contribute to gene regulatory evolution in primates. PLoS Genet 15(7):e1008278

Fei Z, Joung JG, Tang X, Zheng Y, Huang M, Lee JM, McQuinn R, Tieman DM, Alba R, Klee HJ, Giovannoni JJ (2011) Tomato functional genomics database: a comprehensive resource and analysis package for tomato functional genomics. Nucleic Acids Res 39(Database issue): D1156–D1163

Garcia S, Gálvez F, Gras A, Kovařík A, Garnatje T (2014) Plant rDNA database: update and new features. Database (Oxford) 2014:bau063

Garnatje T, Canela MÁ, Garcia S, Hidalgo O, Pellicer J et al (2011) GSAD: a genome size in the Asteraceae database. Cytometry A 79A(6):401–404

Gaudinier A, Tang M, Kliebenstein DJ (2015) Transcriptional networks governing plant metabolism. Curr Plant Biol 3-4:56–64

Geer LY, Markey SP, Kowalak JA, Wagner L, Xu M (2004) Open mass spectrometry search algorithm. J Proteome Res 3(5):958–964

Gong L, Chen W, Gao Y, Liu X, Zhang H et al (2013) Genetic analysis of the metabolome exemplified using a rice population. Proc Natl Acad Sci 110(50):20320–20325

Goodstein DM, Shu S, Howson R, Neupane R, Hayes RD et al (2011) Phytozome: a comparative platform for green plant genomics. Nucleic Acids Res 40(D1):D1178–D1186

Gupta C, Salgotra RK (2022) Epigenetics and its role in effecting agronomical traits. Front Plant Sci 13:925688

Gupta V, Kumar S, Kamboj D, Mishra CN, Singh C, Singh G, Singh GP (2021) Recent advances in OMICS technologies. In: Kumar A, Kumar R, Shukla P, Pandey MK (eds) Omics technologies for sustainable agriculture and global food security volume 1. Springer Singapore, pp 1–18

Gust AA, Pruitt R, Nürnberger T (2017) Sensing danger: key to activating plant immunity. Trends Plant Sci 22(9):779–791

Hall RD, D'Auria JC, Silva Ferreira AC, Gibon Y, Kruszka D et al (2022) High-throughput plant phenotyping: a role for metabolomics? Trends Plant Sci 27(6):549–563

Healey AL, Garsmeur O, Lovell JT, Shengquiang S, Sreedasyam A, Jenkins J et al (2024) The complex polyploid genome architecture of sugarcane. Nature 628(8009):804–810

Henry RJ (2011) Next-generation sequencing for understanding and accelerating crop domestication. Brief Funct Genomics 11(1):51–56

Hu Y, Jiang Z, Chen K, Zhou Z, Zhou X et al (2023) scNanoATAC-seq: a long-read single-cell ATAC sequencing method to detect chromatin accessibility and genetic variants simultaneously within an individual cell. Cell Res 33(1):83–86

Ichihara H, Yamada M, Kohara M, Hirakawa H, Ghelfi A et al (2023) Plant GARDEN: a portal website for cross-searching between different types of genomic and genetic resources in a wide variety of plant species. BMC Plant Biol 23(1):391

Jain M, Olsen HE, Paten B, Akeson M (2016) The Oxford Nanopore MinION: delivery of nanopore sequencing to the genomics community. Genome Biol 17(1):239

Jamaloddin M, Maliha A, Gokulan CG, Gaur N, Patel HK (2021) Metabolomics-assisted breeding for crop improvement: an emerging approach. In: Kumar A, Kumar R, Shukla P, Pandey MK (eds) Omics technologies for sustainable agriculture and global food security volume 1. Springer Singapore, pp 241–279

Jones JD, Dangl JL (2006) The plant immune system. Nature 444(7117):323–329

Kapazoglou A, Ganopoulos I, Tani E, Tsaftaris A (2018) Chapter nine – epigenetics, epigenomics and crop improvement. In: Kuntz M (ed) Advances in botanical research. Academic, pp 287–324

Karaaslan ES, Wang N, Faiß N, Liang Y, Montgomery SA et al (2020) Marchantia TCP transcription factor activity correlates with three-dimensional chromatin structure. Nat Plants 6(10): 1250–1261

Katz K, Shutov O, Lapoint R, Kimelman M, Brister JR, O'Sullivan C (2022) The sequence read archive: a decade more of explosive growth. Nucleic Acids Res 50(D1):D387–D390

Kawahara Y, de la Bastide M, Hamilton JP, Kanamori H, McCombie WR et al (2013) Improvement of the Oryza sativa Nipponbare reference genome using next generation sequence and optical map data. Rice 6(1):4

Kempfer R, Pombo A (2020) Methods for mapping 3D chromosome architecture. Nat Rev Genet 21(4):207–226

Khan AW, Garg V, Sun S, Gupta S, Dudchenko O, Roorkiwal M et al (2024) Cicer super-pangenome provides insights into species evolution and agronomic trait loci for crop improvement in chickpea. Nat Genet 56(6):1225–1234

Khokhar AA, Hui L, Khan D, You Z, Zaman QU, Usman B, Wang HF (2024) Transcriptome profiles reveal key regulatory networks during single and multifactorial stresses coupled with melatonin treatment in pitaya (Selenicereus undatus L.). Int J Mol Sci 25(16):8901

Krefting J, Andrade-Navarro MA, Ibn-Salem. (2018) Evolutionary stability of topologically associating domains is associated with conserved gene regulation. BMC Biol 16:1–12

Krishna TPA, Veeramuthu D, Maharajan T, Soosaimanickam M (2023) The era of plant breeding: conventional breeding to genomics-assisted breeding for crop improvement. Curr Genomics 24(1):24–35

Krug AS, B. M, Drummond E, Van Tassel DL, Warschefsky EJ (2023) The next era of crop domestication starts now. Proc Natl Acad Sci 120(14):e2205769120

Kudo T, Kobayashi M, Terashima S, Katayama M, Ozaki S, Kanno M, Saito M, Yokoyama K, Ohyanagi H, Aoki K, Kubo Y, Yano K (2017) TOMATOMICS: a web database for integrated omics information in tomato. Plant Cell Physiol 58(1):e8

Kundaje A, Meuleman W, Ernst J, Bilenky M, Yen A, Heravi-Moussavi A, et al (2015) Integrative analysis of 111 reference human epigenomes. Nature 518(7539):317–330

Lawrence CJ, Dong Q, Polacco ML, Seigfried TE, Brendel V (2004) MaizeGDB, the community database for maize genetics and genomics. Nucleic Acids Res 32(Database issue):D393–D397

Leebens-Mack JH, Barker MS, Carpenter EJ, Deyholos MK, Gitzendanner MA, et al, One Thousand Plant Transcriptomes I (2019) One thousand plant transcriptomes and the phylogenomics of green plants. Nature 574(7780):679–685

Lenaerts B, Collard BCY, Demont M (2019) Review: improving global food security through accelerated plant breeding. Plant Sci 287:110207

Li Q, Yan J (2020) Sustainable agriculture in the era of omics: knowledge-driven crop breeding. Genome Biol 21(1):154

Li E, Liu H, Huang L, Zhang X, Dong X, Song W, Zhao H, Lai (2019a) Long-range interactions between proximal and distal regulatory regions in maize. Nat Commun 10(1):2633

Li K, Wang D, Gong L, Lyu Y, Guo H, Chen W, Jin C, Liu X, Fang C, Luo J (2019b) Comparative analysis of metabolome of rice seeds at three developmental stages using a recombinant inbred line population. Plant J 100(5):908–922

Li T, Li Y, Shangguan H, Bian J, Luo R, Tian Y, Li Z, Nie X, Cui L (2023a) BarleyExpDB: an integrative gene expression database for barley. BMC Plant Biol 23(1):170

Li X, Wang J, Yu Y, Li G, Wang J, Li C et al (2023b) Genomic rearrangements and evolutionary changes in 3D chromatin topologies in the cotton tribe (Gossypieae). BMC Biol 21(1):56

Liang Z, Riaz A, Chachar S, Ding Y, Du H, Gu X (2020) Epigenetic modifications of mRNA and DNA in plants. Mol Plant 13(1):14–30

Liao Y, Wang J, Zhu Z, Liu Y, Chen J et al (2022) The 3D architecture of the pepper genome and its relationship to function and evolution. Nat Commun 13(1):3479

Lieberman-Aiden E, van Berkum NL, Williams L, Imakaev M, Ragoczy T, Telling A et al (2009) Comprehensive mapping of long-range interactions reveals folding principles of the human genome. Science 326(5950):289–293

Lim PK, Zheng X, Goh JC, Mutwil M (2022) Exploiting plant transcriptomic databases: resources, tools, and approaches. Plant Commun 3(4):100323

Liu Z, Rochfort S (2013) A fast liquid chromatography–mass spectrometry (LC–MS) method for quantification of major polar metabolites in plants. J Chromatogr B 912:8–15

Liu Z, Rochfort S (2014) Recent progress in polar metabolite quantification in plants using liquid chromatography–mass spectrometry. J Integr Plant Biol 56(9):816–825

Liu C, Cheng Y-J, Wang J-W, Weigel D (2017) Prominent topologically associated domains differentiate global chromatin packing in rice from Arabidopsis. Nat Plants 3(9):742–748

Liu Y, Lu S, Liu K, Wang S, Huang L, Guo L (2019) Proteomics: a powerful tool to study plant responses to biotic stress. Plant Methods 15(1):135
Liu J, Zhang Y, Zheng Y, Zhu Y, Shi Y, Guan Z, Lang K, Shen D, Huang W, Dou D (2022) PlantExp: a platform for exploration of gene expression and alternative splicing based on public plant RNA-seq samples. Nucleic Acids Res 51(D1):D1483–D1491
Liu Z, Chen Y, Xia Q, Liu M, Xu H, Chi Y, Deng Y, Xing D (2023) Linking genome structures to functions by simultaneous single-cell Hi-C and RNA-seq. Science 380(6649):1070–1076
Long HS, Greenaway S, Powell G, Mallon A-M, Lindgren CM, Simon MM (2022) Making sense of the linear genome, gene function and TADs. Epigenetics Chromatin 15(1):4
Masoodi KZ, Lone SM, Rasool RS (2021) Chapter 16 – Dansyl-Edman method of peptide sequencing. In: Masoodi KZ, Lone SM, Rasool RS (eds) Advanced methods in molecular biology and biotechnology. Academic, pp 89–92
Milne L, Bayer M, Rapazote-Flores P, Mayer C-D, Waugh R, Simpson CG (2021) EORNA, a barley gene and transcript abundance database. Sci Data 8(1):90
Monaco MK, Stein J, Naithani S, Wei S, Dharmawardhana P, Kumari S (2013) Gramene 2013: comparative plant genomics resources. Nucleic Acids Res 42(D1):D1193–D1199
Montenegro JD, Golicz AA, Bayer PE, Hurgobin B, Lee H (2017) The pangenome of hexaploid bread wheat. Plant J 90(5):1007–1013
Morin R, Bainbridge M, Fejes A, Hirst M, Krzywinski M (2008) Profiling the HeLa S3 transcriptome using randomly primed cDNA and massively parallel short-read sequencing. Biotechniques 45(1):81–94
Nagalakshmi U, Wang Z, Waern K, Shou C, Raha D, Gerstein M, Snyder M (2008) The transcriptional landscape of the yeast genome defined by RNA sequencing. Science 320(5881):1344–1349
Nakashima Y, Nahar S, Miyagi-Shiohira C, Kinjo T, Toyoda Z, Kobayashi N, Saitoh I, Watanabe M, Fujita J, Noguchi H (2018) A liquid chromatography with tandem mass spectrometry-based proteomic analysis of the proteins secreted by human adipose-derived mesenchymal stem cells. Cell Transplant 27(10):1469–1494
Ni P, Nie F, Zhong Z, Xu J, Huang N, Zhang J, Zhao H, Zou Y, Huang Y, Li J, Xiao C-L, Luo F, Wang J (2023) DNA 5-methylcytosine detection and methylation phasing using PacBio circular consensus sequencing. Nat Commun 14(1):4054
Ohashi-Ito K, Oda Y, Fukuda H (2010) Arabidopsis VASCULAR-RELATED NAC-DOMAIN6 directly regulates the genes that govern programmed cell death and secondary wall formation during xylem differentiation. Plant Cell 22(10):3461–3473
Østerberg JT, Xiang W, Olsen LI, Edenbrandt AK, Vedel SE (2017) Accelerating the domestication of new crops: feasibility and approaches. Trends Plant Sci 22(5):373–384
Ouyang W, Xiong D, Li G, Li X (2020) Unraveling the 3D genome architecture in plants: present and future. Mol Plant 13(12):1676–1693
Padmarasu S, Himmelbach A, Mascher M, Stein N (2019) In situ Hi-C for plants: an improved method to detect long-range chromatin interactions. Methods Mol Biol 1933:441–472
Park PJ (2009) ChIP–seq: advantages and challenges of a maturing technology. Nat Rev Genet 10(10):669–680
Pearce S, Vazquez-Gross H, Herin SY, Hane D, Wang Y, Gu YQ, Dubcovsky J (2015) WheatExp: an RNA-seq expression database for polyploid wheat. BMC Plant Biol 15(1):299
Pecinka A, Chevalier C, Colas I, Kalantidis K, Varotto S, Krugman T, Michailidis C, Vallés M-P, Muñoz A, Pradillo M (2019) Chromatin dynamics during interphase and cell division: similarities and differences between model and crop plants. J Exp Bot 71(17):5205–5222
Pei L, Li G, Lindsey K, Zhang X, Wang M (2021a) Plant 3D genomics: the exploration and application of chromatin organization. New Phytol 230(5):1772–1786
Pei S, Liu T, Ren X, Li W, Chen C, Xie Z (2021b) Benchmarking variant callers in next-generation and third-generation sequencing analysis. Brief Bioinform 22(3):bbaa148
Pellicer J, Leitch IJ (2020) The plant DNA C-values database (release 7.1): an updated online repository of plant genome size data for comparative studies. New Phytol 226(2):301–305

Perkins DN, Pappin DJ, Creasy DM, Cottrell JS (1999) Probability-based protein identification by searching sequence databases using mass spectrometry data. Electrophoresis 20(18):3551–3567
Peterson TW, Nam SJ, Darby A (2010) Next gen sequencing survey, North America equity research. New York, NY: JP Morgan Chase & Co.
Pott DM, Osorio S, Vallarino JG (2019) From central to specialized metabolism: an overview of some secondary compounds derived from the primary metabolism for their role in conferring nutritional and organoleptic characteristics to fruit. Front Plant Sci 10:835
Purugganan MD, Jackson SA (2021) Advancing crop genomics from lab to field. Nat Genet 53(5): 595–601
Qaim MJAEP, Policy (2020) Role of new plant breeding technologies for food security and sustainable agricultural development. Appl Econ Perspect Policy 42(2):129–150
Rauluseviciute I, Drabløs F, Rye MB (2019) DNA methylation data by sequencing: experimental approaches and recommendations for tools and pipelines for data analysis. Clin Epigenetics 11(1):193
Raza A, Bashir S, Khare T, Karikari B, Copeland RGR, Jamla M, Abbas S, Charagh S, Nayak SN, Djalovic I, Rivero RM, Siddique KHM, Varshney RK (2024) Temperature-smart plants: a new horizon with omics-driven plant breeding. Physiol Plant 176(1):e14188
Ren AH, Diamandis EP, Kulasingam V (2021) Uncovering the depths of the human proteome: antibody-based technologies for ultrasensitive multiplexed protein detection and quantification. Mol Cell Proteomics 20:100155
Sanger F, Nicklen S, Coulson AR (1977) DNA sequencing with chain-terminating inhibitors. Proc Natl Acad Sci USA 74(12):5463–5467
Sasaki T, International Rice Genome Sequencing P (2005) The map-based sequence of the rice genome. Nature 436(7052):793–800
Satam H, Joshi K, Mangrolia U, Waghoo S, Zaidi G, Rawool S, Thakare RP, Banday S, Mishra AK, Das G, Malonia SK (2023) Next-generation sequencing technology: current trends and advancements. Biology 12(7):997
Sato Y, Takehisa H, Kamatsuki K, Minami H, Namiki N, Ikawa H, Ohyanagi H, Sugimoto K, Antonio BA, Nagamura Y (2013) RiceXPro version 3.0: expanding the informatics resource for rice transcriptome. Nucleic Acids Res 41(D1):D1206–D1213
Schad E, Tompa P, Hegyi H (2011) The relationship between proteome size, structural disorder and organism complexity. Genome Biol 12(12):R120
Schmitt AD, Hu M, Jung I, Xu Z, Qiu Y, Tan CL, Li Y, Lin S, Lin Y, Barr CL, Ren B (2016) A compendium of Chromatin contact maps reveals spatially active regions in the human genome. Cell Rep 17(8):2042–2059
Selegato DM, Pilon AC, Carnevale Neto F (2019) Plant metabolomics using NMR spectroscopy. In: Gowda GAN, Raftery D (eds) NMR-based metabolomics: methods and protocols. Springer, New York, pp 345–362
Shen S, Zhan C, Yang C, Fernie AR, Luo J (2023) Metabolomics-centered mining of plant metabolic diversity and function: past decade and future perspectives. Mol Plant 16(1):43–63
Song J, Henry HAL, Tian L (2019) Brachypodium histone deacetylase BdHD1 positively regulates ABA and drought stress responses. Plant Sci 283:355–365
Srivastava A, Malik L, Sarkar H, Zakeri M, Almodaresi F, Soneson C, Love MI, Kingsford C, Patro R (2020) Alignment and mapping methodology influence transcript abundance estimation. Genome Biol 21(1):239
Stam M, Tark-Dame M, Fransz P (2019) 3D genome organization: a role for phase separation and loop extrusion? Curr Opin Plant Biol 48:36–46
Steinwand MA, Ronald PC (2020) Crop biotechnology and the future of food. Nat Food 1(5): 273–283
Sun L, Zhou J, Xu X, Liu Y, Ma N, Liu Y, Nie W, Zou L, Deng XW, He H (2024) Mapping nucleosome-resolution chromatin organization and enhancer-promoter loops in plants using Micro-C-XL. Nat Commun 15(1):35

Szabo Q, Bantignies F, Cavalli G (2019) Principles of genome folding into topologically associating domains. Sci Adv 5(4):eaaw1668
Thottathil GP, Jayasekaran K, Othman AS (2016) Sequencing crop genomes: a gateway to improve tropical agriculture. Trop Life Sci Res 27(1):93–114
Torres DE, Thomma BPHJ, Seidl MF (2021) Transposable elements contribute to genome dyanamics and gene expression variation in the fungal plant patogen *Verticillium dahliae*. Genome Biol Evol 13(7):evab135
Tourdot E, Grob S (2023) Three-dimensional chromatin architecture in plants – —general features and novelties. Eur J Cell Biol 102(4):151344
Tyagi P, Singh D, Mathur S, Singh A, Ranjan R (2022) Upcoming progress of transcriptomics studies on plants: an overview. Front Plant Sci 13:1030890
Varshney RK, Bohra A, Yu J, Graner A, Zhang Q, Sorrells ME (2021) Designing future crops: genomics-assisted breeding comes of age. Trends Plant Sci 26(6):631–649
Wang KC, Chang HY (2018) Epigenomics: technologies and applications. Circ Res 122(9): 1191–1199
Wang P, Wilson SR (2013) Mass spectrometry-based protein identification by integrating de novo sequencing with database searching. BMC Bioinformatics 14 Suppl 2(Suppl 2):S24
Wang C, Liu C, Roqueiro D, Grimm D, Schwab R, Becker C, Lanz C, Weigel D (2015) Genome-wide analysis of local chromatin packing in Arabidopsis thaliana. Genome Res 25(2):246–256
Wang H, Zaman QU, Huang W, Mei D, Liu J, Wang W, Ding B, Hao M, Fu L, Cheng H, Hu Q (2019) QTL and candidate gene identification for silique length based on high-dense genetic map in Brassica napus L. Front Plant Sci 10:1579
Wang P, Feng Y, Zhu K, Chai H, Chang Y-t, Yang X, Liu X, Shen C, Gega E, Lee B, Kim M, Ruan X, Ruan Y (2021a) In situ chromatin interaction analysis using paired-end tag sequencing. Curr Protoc 1(8):e174
Wang Y, Zhao Y, Bollas A, Wang Y, Au KF (2021b) Nanopore sequencing technology, bioinformatics and applications. Nat Biotechnol 39(11):1348–1365
Wang H, Xu Y, Zhang Z, Zhang G, Tan C, Ye L (2024a) Development and application of transcriptomics technologies in plant science. Crop Design 3(2):100057
Wang W, Zhang T, Liu C, Liu C, Jiang Z, Zhang Z, Ali S, Li Z, Wang J, Sun S, Chen Q, Zhang Q, Xie L (2024b) A DNA demethylase reduces seed size by decreasing the DNA methylation of AT-rich transposable elements in soybean. Commun Biol 7(1):613
Wheeler DA, Srinivasan M, Egholm M, Shen Y, Chen L et al (2008) The complete genome of an individual by massively parallel DNA sequencing. Nature 452(7189):872–876
Xiong W, Reynolds M, Xu Y (2022) Climate change challenges plant breeding. Curr Opin Plant Biol 70:102308
Yamaguchi M, Goué N, Igarashi H, Ohtani M, Nakano Y, Mortimer JC, Nishikubo N, Kubo M, Katayama Y, Kakegawa K (2010) VASCULAR-RELATED NAC-DOMAIN6 and VASCULAR-RELATED NAC-DOMAIN7 effectively induce transdifferentiation into xylem vessel elements under control of an induction system. Plant Physiol 153(3):906–914
Yamaguchi M, Mitsuda N, Ohtani M, Ohme-Takagi M, Kato K, Demura T (2011) VASCULAR-RELATED NAC-DOMAIN 7 directly regulates the expression of a broad range of genes for xylem vessel formation. Plant J 66(4):579–590
Yang C, Shen S, Zhou S, Li Y, Mao Y, Zhou J et al (2022) Rice metabolic regulatory network spanning the entire life cycle. Mol Plant 15(2):258–275
Yang Z, Wang J, Huang Y, Wang S, Wei L, Liu D, Weng Y, Xiang J, Zhu Q, Yang Z, Nie X, Yu Y, Yang Z, Yang Q-Y (2023a) CottonMD: a multi-omics database for cotton biological study. Nucleic Acids Res 51(D1):D1446–D1456
Yang Z, Liu Z, Xu H, Li Y, Huang S, Cao G et al (2023b) ArecaceaeMDB: a comprehensive multi-omics database for Arecaceae breeding and functional genomics studies. Plant Biotechnol J 21(1):11–13
Yoo BC, Kim K-H, Woo SM, Myung JK (2018) Clinical multi-omics strategies for the effective cancer management. J Proteome 188:97–106

Yu Y, Zhang H, Long Y, Shu Y, Zhai J (2022) Plant public RNA-seq database: a comprehensive online database for expression analysis of ~45 000 plant public RNA-Seq libraries. Plant Biotechnol J 20(5):806–808

Yu Z, Chen Y, Zhou Y, Zhang Y, Li M, Ouyang Y et al (2023) Rice gene index: a comprehensive pan-genome database for comparative and functional genomics of Asian rice. Mol Plant 16(5): 798–801

Zaman QU, Raza A, Gill RA, Hussain MA, Wang HF, Varshney RK (2023a) New possibilities for trait improvement via mobile CRISPR-RNA. Trends Biotechnol 41:1335

Zaman QU, Raza A, Lozano-Juste J, Chao L, Jones MGK, Wang HF, Varshney RK (2023b) Engineering plants using diverse CRISPR-associated proteins and deregulation of genome-edited crops. Trends Biotechnol 42:560–574

Zaman QU, Garg V, Raza A, Nazir MF, Hui L, Khan D, Khokhar AA, Hussain MA, Wang HF, Varshney RK (2024a) Unique regulatory network of dragon fruit simultaneously mitigates the effect of vanadium pollutant and environmental factors. Physiol Plant 176(4):e14416

Zaman QU, Raza A, Hui L, Nazir MF, Garg V et al (2025) Chromosome-level genome assembly of the autotetraploid yellow pitaya provides novel insights into the evolution of trait patterning in pitaya species with different ploidy. Genome Biol 26:234 https://doi.org/10.1186/s13059-025-03695-3

Zang YX, Kim HU, Kim JA, Lim MH, Jin M, Lee SC, Kwon SJ, Lee SI, Hong JK, Park TH, Mun JH, Seol YJ, Hong SB, Park BS (2009) Genome-wide identification of glucosinolate synthesis genes in Brassica rapa. FEBS J 276(13):3559–3574

Zhang X, Wang T (2021) Plant 3D chromatin organization: important insights from chromosome conformation capture analyses of the last 10 years. Plant Cell Physiol 62(11):1648–1661

Zhang R, Zhang C, Yu C, Dong J, Hu J (2022) Integration of multi-omics technologies for crop improvement: status and prospects. Front Bioinform 2:1027457

Zhao Q, Feng Q, Lu H, Li Y, Wang A et al (2018) Pan-genome analysis highlights the extent of genomic variation in cultivated and wild rice. Nat Genet 50(2):278–284

Zheng X, Chen L, Xia H, Wei H, Lou Q, Li M, Li T, Luo L (2017) Transgenerational epimutations induced by multi-generation drought imposition mediate rice plant's adaptation to drought condition. Sci Rep 7:39843

Chapter 11
Advances and Applications of Single-Cell and Spatial Omics for Crop Improvement

Saurabh Gupta, Aamir W. Khan, and Mohit Verma

Abstract Crop improvement is essential for ensuring global food security and sustainability amidst a growing population and changing climate. Traditional breeding methods, while effective, often fall short of addressing complex traits and rapid environmental changes. Single-cell omics technologies offer a transformative approach, providing unprecedented resolution into the cellular heterogeneity of plants. This chapter explores the application of single-cell genomics, transcriptomics, epigenomics, proteomics, and metabolomics in enhancing crop resilience, yield, and nutritional value. By dissecting plant biology at the single-cell level, we can identify key genes, regulatory pathways, and metabolic processes that govern crucial agronomic traits. This chapter highlights successful case studies in rice, maize, and other crops, demonstrating the power of single-cell omics in unravelling complex biological mechanisms. We also discuss the technical challenges, data analysis hurdles, and future directions of this emerging field, including its integration with artificial intelligence, gene editing technologies, and synthetic biology. Overcoming these challenges will pave the way for developing resilient, high-yielding crops tailored to specific environments, ultimately contributing to global food security.

Keywords Single-cell omics · Spatial omics · Cellular heterogeneity · Stress resilience

S. Gupta (✉)
Curtin Medical Research Institute, Curtin Medical School, Curtin University, Perth, WA, Australia

UWA Centre for Medical Research, University of Western Australia, Perth, WA, Australia
e-mail: saurabh.gupta@curtin.edu.au

A. W. Khan
Division of Plant Science and Technology, University of Missouri, Columbia, MO, USA

M. Verma
Department of Biomedical Informatics, Biostatistics, and Medical Epidemiology (BBME), University of Missouri, Columbia, MO, USA

V. Garg et al. (eds.), *DNA of Sustainability*, Sustainability Sciences in Asia and Africa,
https://doi.org/10.1007/978-981-95-5109-5_11

11.1 Introduction

Crop improvement is essential for ensuring food security, sustainability, and climate resilience. With the global population projected to reach nearly 10 billion by 2050 (Kenny and Gehan 2023; Hickey et al. 2019), agriculture must adapt to rising food demands, environmental fluctuations, and resource limitations. While traditional breeding and molecular approaches have enhanced productivity, future progress requires precision breeding strategies that leverage advanced biotechnological tools to develop resilient, high-yielding crops.

Plant processes rely on coordinated signalling across cells, tissues, and organs. Traditional bulk omics methods, such as transcriptomics, proteomics, and metabolomics, provide valuable insights but average data across many cells, masking cellular heterogeneity. Single-cell omics overcomes this limitation by capturing molecular profiles at the individual cell level, providing deeper insights into plant development and stress responses. Advances in protoplast isolation, sequencing, and spatial technologies have made single-cell and spatial studies feasible in plants, revolutionising the ability to explore cellular diversity in both isolated cells and intact tissues.

Single-cell omics is an emerging and powerful technology that provides unprecedented resolution for studying biological processes at the level of individual cells (Kolodziejczyk et al. 2015). Unlike traditional tissue-level analyses, which average signals from millions of cells and obscure cellular heterogeneity, single-cell omics enables exploration of the intricate diversity within a cell population. This technology has transformed our ability to uncover rare cell types or transient cellular states that play pivotal roles in plant growth, development, and adaptation to environmental conditions (Luo et al. 2020).

This rapidly evolving field leverages single-cell transcriptomics, proteomics, and metabolomics to uncover molecular mechanisms at a cellular level. Single-cell transcriptomics reveals cell-type-specific gene expression patterns linked to environmental stress and development (Efroni and Birnbaum 2016), while proteomics and metabolomics uncover functional proteins and metabolites involved in stress adaptation, nutrient absorption, and defence mechanisms (Yan et al. 2022; Clark et al. 2021; Barkla and Vera-Estrella 2015). Single-cell genomics further enables the detection of somatic mutations, structural variations, and epigenetic modifications that drive plant differentiation and adaptation (Goel et al. 2024). Additionally, spatial omics complements single-cell techniques by preserving the tissue context, allowing gene expression and molecular interactions to be mapped within plant organs. Spatial transcriptomics, for instance, enables the visualisation of gene activity across different tissue layers, revealing localised responses to biotic and abiotic stressors (Yu et al. 2023). Integrating single-cell and spatial omics provides a comprehensive view of plant physiology, bridging the gap between cellular diversity and tissue organisation.

These approaches offer critical insights into cellular responses to environmental stressors such as drought, heat, and salinity factors that are becoming increasingly prevalent due to climate change (Li et al. 2023; Berrío et al. 2025; Sun et al. 2022). Understanding how individual cells respond to these stressors enables the identification of genes and regulatory pathways that confer stress tolerance, facilitating precision breeding and genetic engineering of resilient crops.

Single-cell omics also enhance our understanding of plant immunity, nutrient use efficiency, and gene regulation. Identifying cell-type-specific immune responses may lead to novel resistance genes for disease-resistant crops (Chen et al. 2024; Yan et al. 2025). Single-cell ATAC-seq (Assay for Transposase-Accessible Chromatin using sequencing) maps chromatin accessibility, linking gene expression with epigenetic regulation to refine breeding strategies (Feng et al. 2022). Additionally, single-cell analyses of meristematic cells can reveal early indicators of desirable agronomic traits, accelerating breeding programmes (Liao and Wang 2023; Zhu et al. 2023a).

In summary, single-cell and spatial omics represent a paradigm shift in agricultural research, providing unprecedented insights into cellular diversity, functional genomics, and adaptive mechanisms in crops. By integrating cellular and spatial data, these technologies hold transformative potential for developing stress-resilient, high-yielding crops and optimising agricultural practices to meet global food challenges. Addressing technical and regulatory barriers will be crucial to fully harnessing the power of single-cell and spatial omics in crop improvement.

11.2 Single-Cell Omics Approaches

Single-cell omics encompasses a range of techniques that provide comprehensive insights into cellular heterogeneity by analysing gene expression, epigenetic modifications, protein abundance, and metabolic profiles at the individual cell level. The different single-cell omics approaches are discussed below and summarised in Fig. 11.1.

11.2.1 Single-Cell Transcriptomics

Single-cell transcriptomics, primarily through single-cell RNA sequencing (scRNA-seq), is a transformative technology that provides an unparalleled view of gene expression patterns within individual plant cells, revealing the intricate heterogeneity of plant tissues. Unlike bulk RNA sequencing, which averages gene expression across a population of cells, scRNA-seq captures the unique transcriptional profiles of individual cells, offering insights into cell-type-specific functions, developmental trajectories, and responses to environmental cues (Fig. 11.2). This level of resolution is crucial for understanding how diverse cell populations contribute to overall plant physiology and adaptation (Clark et al. 2021). In plant biology, scRNA-seq has become indispensable for dissecting complex developmental processes and stress

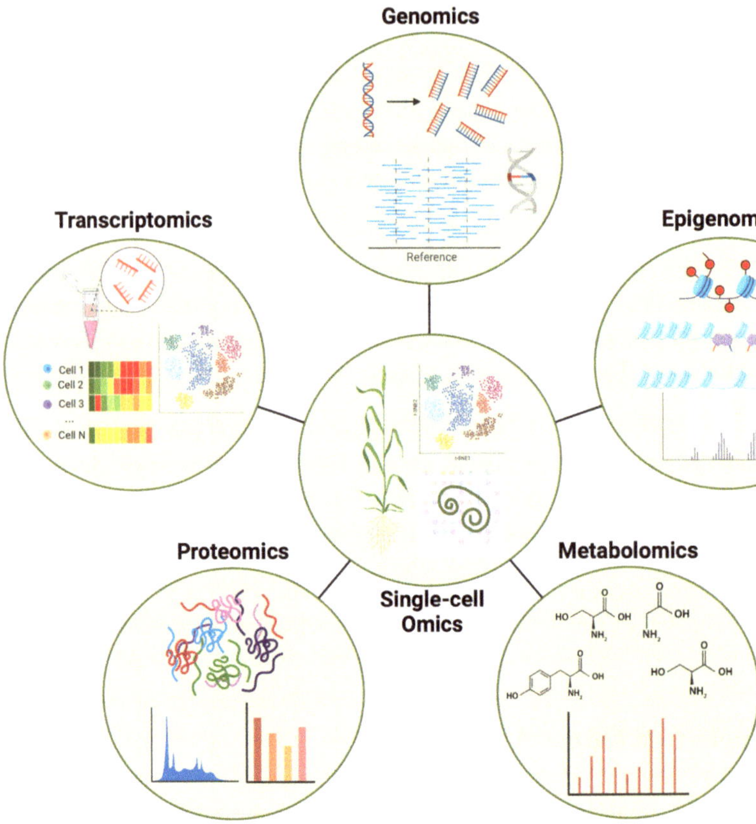

Fig. 11.1 Overview of different single-cell omics approaches. This figure illustrates the various single-cell omics approaches used to study cellular heterogeneity at multiple molecular levels. Genomics focuses on sequencing and analysing DNA to identify genetic variations and structural features. Epigenomics examines modifications such as DNA methylation and histone modifications, which regulate gene expression without altering the DNA sequence. Transcriptomics captures RNA expression profiles at the single-cell level, enabling the identification of cell-type-specific gene expression patterns. Proteomics investigates protein composition, modifications, and interactions, providing insights into cellular functions. Metabolomics analyses small molecules and metabolic pathways, reflecting the biochemical state of individual cells. Together, these approaches enable a comprehensive understanding of cellular diversity, regulatory mechanisms, and functional dynamics in complex biological systems

responses, as it overcomes the limitations of bulk RNA sequencing by enabling the identification of rare cell types, the delineation of cell lineages, and the mapping of gene expression dynamics during development and stress. Techniques like droplet-based scRNA-seq enable the high-throughput analysis of thousands of individual cells, providing a comprehensive view of cellular diversity within plant tissues.

The process of scRNA-seq typically begins with the isolation of individual cells, often through enzymatic digestion of plant cell walls to create protoplasts or by isolating nuclei. Once isolated, the RNA from each cell is captured,

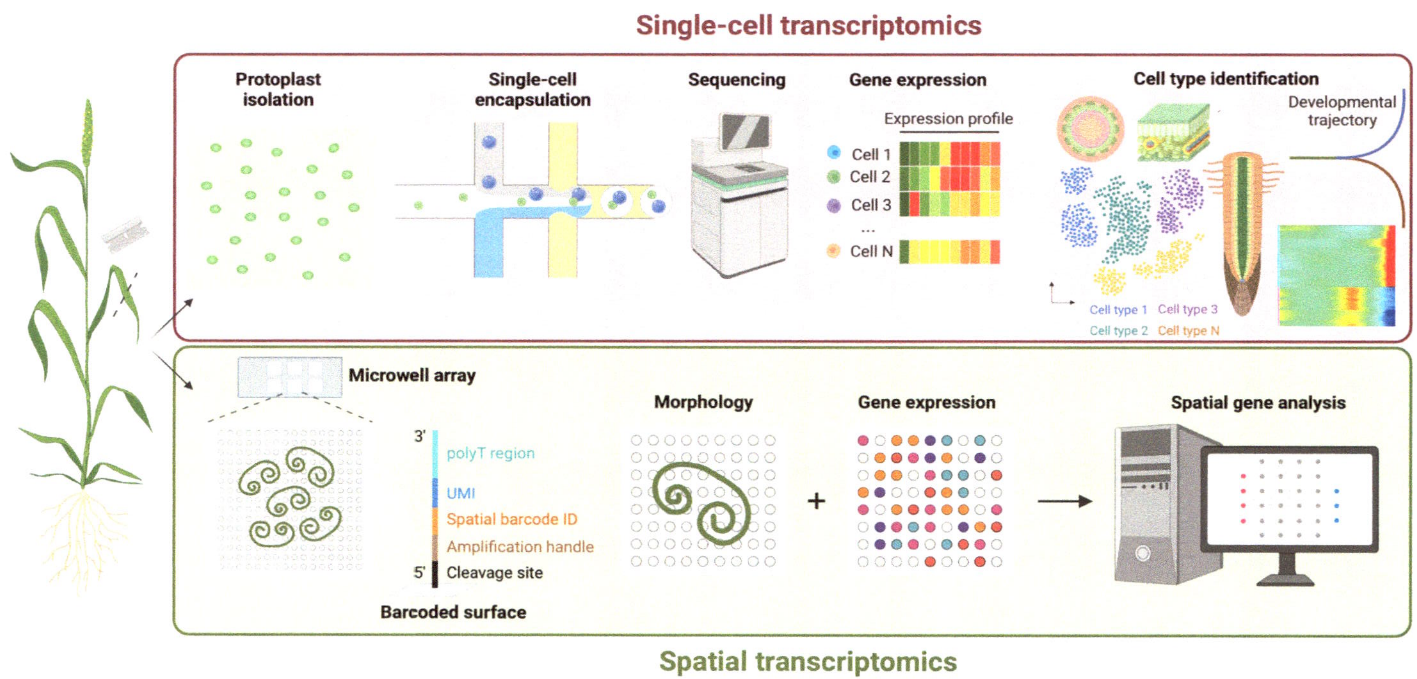

Fig. 11.2 An overview of single-cell and spatial transcriptomics. The single-cell transcriptomics workflow follows isolation of protoplasts, followed by single-cell encapsulation, sequencing, and quantification of gene expression for each cell. The cells are then clustered based on their transcriptome profiles, leading to annotation of cell types based on the expression of marker genes in each cluster. The downstream analysis includes predicting developmental trajectories and identifying genes that determine various cell fates. A typical spatial transcriptomics workflow begins with preserving tissue architecture while capturing gene expression. The process involves sectioning and fixing tissue onto specialised slides, followed by in situ hybridisation or barcoded mRNA capture to spatially tagged transcripts. Sequencing and computational analysis then map gene expression back to its original spatial context, enabling insights into tissue organisation and cellular interactions

reverse-transcribed into cDNA, and amplified. Unique barcodes are added to each cDNA molecule, allowing for the identification of the cell of origin during sequencing. Next-generation sequencing (NGS) is then performed, and the resulting sequence data are analysed using bioinformatics tools to reconstruct the transcriptome of each individual cell. This allows for the identification of distinct cell clusters based on their gene expression profiles, revealing cell-type-specific gene expression patterns and regulatory networks.

Spatial transcriptomics further enriches our understanding by integrating gene expression data with spatial coordinates, effectively mapping the transcriptome within the native architecture of plant tissues (Fig. 11.2; Yu et al. 2023; Giacomello et al. 2017). Unlike scRNA-seq, which necessitates the dissociation of cells and thus loses spatial context, spatial transcriptomics preserves the positional information of cells, providing insights into how gene activity is organised and regulated within tissues and organs. This technology allows the visualisation of gene expression patterns across different tissue layers, identifies localised responses to stimuli, and explores cell–cell interactions within their native environment.

Recent advances clearly demonstrate that scRNA-seq and spatial transcriptomics together form powerful complementary strategies for interrogating plant responses to environmental perturbations and developmental transitions. For example, during drought or heat stress, specific cell populations exhibit distinct transcriptional changes, reflecting their specialised roles in stress adaptation. Single-cell and spatial approaches allow for the identification of these responsive cell types and the characterisation of their unique gene expression signatures, providing insights into the molecular mechanisms underlying stress tolerance (Wang et al. 2024). Moreover, these techniques are crucial for understanding plant development, revealing the dynamic changes in gene expression that drive cell differentiation and tissue formation. By mapping developmental trajectories and identifying cell-type-specific gene activity patterns within tissues, key regulatory genes and signalling pathways that control plant growth and morphogenesis can be identified. Furthermore, when combined, the gene expression data can be mapped back to the tissue, allowing for a more complete understanding of cellular interactions and how different cell types coordinate their activities in space.

11.2.2 Single-Cell Epigenomics

Single-cell epigenomics is an emerging field that enables the investigation of chromatin dynamics and regulatory mechanisms at the single-cell level, providing new insights into plant development, stress responses, and cellular heterogeneity. Unlike traditional genomics, which focuses on genetic information, epigenomics examines the modifications to DNA and histone proteins that regulate gene expression without altering the underlying DNA sequence. These modifications, including DNA methylation, histone modifications, and chromatin accessibility, are crucial for cell differentiation, developmental patterning, and adaptation to environmental changes (Dorrity et al. 2021; Marand et al. 2021; Zhang et al. 2021a).

In plants, epigenetic regulation plays a pivotal role in controlling developmental processes such as flowering, root development, and responses to abiotic stresses like drought or salinity. However, traditional bulk epigenetic profiling methods have limitations in resolving the fine-scale heterogeneity of epigenetic states within populations of cells. Single-cell epigenomics overcomes this limitation by enabling the profiling of epigenetic modifications in individual cells, thus revealing the diversity of epigenetic states across different cell types and developmental stages.

Advances in technologies such as single-cell bisulfite sequencing for DNA methylation, single-cell ATAC-seq for chromatin accessibility, and single-cell ChIP-seq (chromatin immunoprecipitation sequencing) for histone modifications have opened up new possibilities for studying epigenetic regulation at an unprecedented resolution. For instance, single-cell bisulfite sequencing allows for the analysis of DNA methylation patterns at a single-cell level, revealing how methylation landscapes vary between cell types and how these patterns correlate with gene expression and developmental processes (Smallwood et al. 2014). Similarly, single-cell ATAC-seq provides insights into chromatin accessibility, highlighting regions of the genome that are actively engaged in transcriptional regulation and how this varies across different cell types, tissues, or developmental stages (Dorrity et al. 2021; Marand et al. 2021).

Single-cell epigenomics is helpful for identifying cell-specific changes that occur during stress and development. For example, during abiotic stress responses, such as heat or drought, certain cell populations exhibit epigenetic reprogramming, which leads to altered gene expression that enhances stress tolerance. These epigenetic changes can be inherited or transient, influencing how plants adapt to changing environments. Single-cell approaches allow for the identification of specific cell types that undergo such reprogramming, providing valuable insights into the molecular mechanisms behind stress adaptation (Zhao et al. 2020).

Moreover, single-cell epigenomics is becoming increasingly relevant for understanding plant developmental plasticity. By linking epigenetic modifications to cellular fate decisions and gene expression programmes, researchers can explore how plants generate diverse cell types and tissues from a common genome, a process central to growth and development. For instance, during somatic embryogenesis or organogenesis, epigenetic modifications can regulate key transcriptional networks, influencing which genes are activated or silenced in a cell-specific manner (Huang et al. 2021).

11.2.3 Single-Cell Proteomics and Metabolomics

Single-cell proteomics and metabolomics extend beyond gene expression to provide a direct measure of cellular function, revealing the dynamic interplay between proteins and metabolites at the individual cell level. Unlike transcriptomics, which reflects potential gene activity, proteomics and metabolomics capture the actual functional outputs of cells, offering insights into protein modifications, metabolic pathways, and cellular physiology (Marx 2019; Zenobi 2013; Montes et al. 2024). These technologies are essential for understanding how cells respond to environmental cues, nutrient availability, and developmental signals, providing a more complete picture of cellular states.

In plants, single-cell proteomics and metabolomics are emerging fields that hold immense potential for unravelling the complexity of cellular processes. Traditional bulk proteomics and metabolomics provide averaged data, obscuring the heterogeneity of protein and metabolite profiles across cell populations. Single-cell approaches overcome this limitation by enabling the analysis of protein and metabolite levels in individual cells, revealing cell-type-specific functions and metabolic states. Techniques like mass spectrometry-based proteomics and metabolomics allow for the identification and quantification of thousands of proteins and metabolites in individual cells, providing a comprehensive view of cellular biochemistry (Grobei et al. 2009; Wada et al. 2020; Vu et al. 2024).

Specifically, single-cell proteomics utilises advanced mass spectrometry techniques to identify and quantify proteins within individual cells, revealing post-translational modifications, protein interactions, and dynamic changes in protein abundance. This level of detail is crucial for understanding how proteins contribute to cellular functions and responses to stimuli. Similarly, single-cell metabolomics employs techniques like mass spectrometry and nuclear magnetic resonance (NMR) spectroscopy to analyse the small molecules within single cells, providing a snapshot of cellular metabolic pathways and physiological states. This allows for the detection of subtle changes in metabolite levels that reflect cellular responses to environmental stressors, developmental cues, or nutrient availability. For example, during nutrient deficiency or pathogen attack, specific cell populations exhibit distinct changes in protein and metabolite profiles, reflecting their specialised roles in stress adaptation and defence. Single-cell approaches allow for the identification of these responsive cell types and the characterisation of their unique protein and metabolite signatures, providing insights into the molecular mechanisms underlying stress tolerance and immunity. Moreover, these technologies are crucial for understanding plant development, revealing the dynamic changes in protein and metabolite levels that drive cell differentiation and tissue formation. By mapping metabolic pathways and protein interactions, researchers can identify key regulatory networks that control plant growth and morphogenesis. The integration of proteomics and metabolomics with transcriptomics and epigenomics offers a holistic view of plant function, revealing the interconnectedness of gene regulation, protein activity, and metabolic pathways at the single-cell level.

11.2.4 Single-Cell Genomics

Single-cell genomics enables the comprehensive analysis of the DNA sequence within individual plant cells, revealing genetic variation and mutations that are often masked in bulk sequencing. This technique captures the unique genetic makeup of individual cells, offering insights into somatic mutations, structural variations, and copy number variations. This level of resolution is crucial for understanding how genetic diversity contributes to plant development, adaptation, and evolution.

In plant biology, single-cell genomics is vital for dissecting complex genetic phenomena, such as polyploidy, genetic mosaicism, and the evolution of plant tissues. Traditional bulk genomics often fails to capture the subtle genetic differences between cell types, limiting our ability to understand the precise roles of specific genetic variations. Single-cell genomics overcomes this limitation by enabling the analysis of the complete DNA sequence in individual cells, revealing cell-type-specific genetic profiles. Techniques like single-cell whole-genome sequencing allow for the high-throughput analysis of thousands of individual cells, providing a comprehensive view of genetic diversity within plant tissues.

The process of single-cell genomics typically involves isolating individual cells, often through microfluidic devices or laser capture microdissection. Once isolated, the DNA from each cell is amplified using techniques like multiple displacement amplification (MDA) or polymerase chain reaction (PCR). The amplified DNA is then subjected to high-throughput sequencing, generating vast amounts of sequence data. Bioinformatics tools are subsequently used to analyse the resulting sequence data, identifying single nucleotide variations (SNVs), insertions/deletions (indels), structural variations, and copy number variations. This allows for the construction of detailed genetic maps of individual cells, revealing the genetic diversity within plant tissues.

Single-cell genomics has proven invaluable in studying plant responses to environmental stressors and developmental processes. For example, during stress responses, such as UV radiation or chemical exposure, specific cell populations may accumulate somatic mutations that contribute to adaptation or resistance. Single-cell approaches allow for the identification of these mutated cell types and the characterisation of their unique genetic profiles, providing insights into the molecular mechanisms underlying stress-induced mutagenesis. Moreover, single-cell genomics is crucial for understanding plant development, revealing the dynamic changes in genome structure that drive cell differentiation and tissue formation. By mapping clonal lineages and identifying cell-type-specific genetic variations, researchers can identify key genetic events that control plant growth and morphogenesis (Goel et al. 2024). Furthermore, single-cell genomics is vital for understanding polyploidy, allowing for the observation of subgenome dominance and variation, which is key in understanding the evolution and breeding of many crop plants.

11.3 Workflow for Single-Cell Omics Experiment

11.3.1 Sample Acquisition and Preparation

The initial phase involves procuring the biological material of interest. For plants, this often necessitates meticulous tissue dissection and, frequently, enzymatic digestion to remove cell walls, yielding protoplasts (Ren and Wang 2023). Alternatively, especially for tissues with robust cell walls or those sensitive to protoplasting, nuclei isolation is employed (Thibivilliers and Libault 2021). The objective is to generate a homogenous suspension of individual cells or nuclei, minimising cell clumping and ensuring viability. This stage demands careful optimisation, as plant tissues vary significantly in their composition and fragility (Jeong et al. 2021). Spatial omics requires a different strategy, emphasising the preservation of cellular spatial distribution through tissue embedding and sectioning. This can be accomplished using formalin-fixed paraffin-embedded (FFPE) or fresh-frozen tissue samples, both designed to maintain tissue architecture.

11.3.2 Single-Cell/Nuclei Isolation and Capture

Once a single-cell or single-nucleus suspension is achieved, or the tissue section is prepared, the next step is to isolate and capture individual cells or nuclei. This is accomplished through various techniques, including droplet-based microfluidics (e.g. 10x Chromium, Drop-Seq, and DNBelab C4 system), microwell systems (e.g. BD Rhapsody), fluorescence-activated cell sorting (FACS), or spatial capture areas (e.g. 10X Visium, Visium HD, and Stereo-Seq). The chosen method depends on the specific omics application and experimental requirements. The goal is to isolate individual cells or nuclei into discrete compartments, where they can be processed independently. In contrast, spatial omics captures molecular data within the tissue's original spatial context, instead of isolating individual cells.

11.3.3 Molecular Tagging and Indexing

Following capture, the molecular components of interest (RNA, DNA, proteins, or metabolites) are tagged or indexed with unique identifiers. This step is essential for distinguishing the molecular profiles of individual cells. For transcriptomics, RNA molecules are barcoded; for epigenomics, DNA fragments are tagged; for proteomics, proteins may be labelled with chemical tags; and for metabolomics, metabolites may be derivatised. Spatial omics uses barcoded probes to capture the molecules in their location.

11.3.4 Library Preparation and High-Throughput Data Generation

The tagged molecular components are then processed into a sequencing or analysis library. This typically involves reverse transcription (for RNA), DNA fragmentation and adapter ligation (for DNA), protein digestion and peptide labelling (for proteins), or metabolite extraction and derivatisation (for metabolites). Amplification steps are often included to increase the abundance of target molecules, ensuring sufficient signal for downstream analysis.

The prepared library is then subjected to high-throughput analysis, usually involving NGS for transcriptomics and epigenomics, mass spectrometry for proteomics and metabolomics, or imaging-based methods for spatial omics. This step generates vast amounts of data, capturing the molecular profiles of thousands of individual cells or spatial locations.

11.3.5 Data Processing and Analysis

The raw data generated from high-throughput analysis is then processed using specialised bioinformatics tools/pipelines. This involves aligning sequencing reads to a reference genome (for transcriptomics and epigenomics), identifying and quantifying proteins or metabolites (for proteomics and metabolomics), or reconstructing spatial gene expression patterns. Downstream analysis includes clustering cells based on their molecular profiles, identifying cell types and states, and exploring gene regulatory networks, protein interactions, or metabolic pathways. Several bioinformatics tools and pipelines are available for analysing single-cell data, from quality control to downstream functional analysis (Table 11.1).

11.3.6 Integration and Interpretation

Finally, the results obtained from single-cell/spatial omics experiments are integrated and interpreted in the context of the biological question being addressed. This may involve combining data from multiple omics approaches, integrating single-cell data with other datasets, and visualising results in a biologically meaningful way. The ultimate goal is to gain a deeper understanding of cellular heterogeneity, function, and interactions within complex biological systems.

Table 11.1 List of commonly used tools for analysing single-cell and spatial omics data

Tool	Language	Description	Reference
Preprocessing and quality control			
CellRanger	Rust	A widely used pipeline for processing raw sequencing data from 10x genomics platforms, including barcode assignment, UMI counting, and gene expression quantification.	https://github.com/10XGenomics/cellranger
Scater	R	Tools for quality control, normalisation, and variance modelling of single-cell RNA sequencing (scRNA-seq) data.	McCarthy et al. (2017)
DropletUtils	R	Provides methods for handling droplet-based scRNA-seq datasets, including empty droplet detection.	Lun et al. (2019)
Dimensionality reduction and clustering			
Seurat	R	One of the most popular packages for analysing single-cell and spatial omics data, supporting normalisation, clustering, differential expression analysis, and integration of multiple datasets.	Hao et al. (2024)
Scanpy	Python	A scalable and efficient toolkit for analysing large single-cell datasets, offering dimensionality reduction, clustering, and trajectory inference.	Wolf et al. (2018)
Monocle	R	Used for pseudotime trajectory analysis to infer differentiation pathways.	Trapnell et al. (2014)
Cell type annotation			
CellTypist	Python	A machine-learning-based tool for automated cell type annotation, supporting large-scale datasets.	Domínguez Conde et al. (2022)
SingleR	R	Assigns cell identities based on reference transcriptomic datasets.	Aran et al. (2019)
Garnett	R	Uses marker-based classification for cell type identification in single-cell data.	Pliner et al. (2019)
Integration and batch effect correction			
Harmony	R	A powerful tool for integrating single-cell datasets from different experiments while correcting batch effects.	Korsunsky et al. (2019)
BBKNN	Python/R	A batch correction tool designed for Scanpy workflows	Polański et al. (2020)
LIGER	R	Uses factorisation-based methods for integrating multiple single-cell datasets, including multi-species comparisons.	Liu et al. (2020)
Gene regulatory network inference			

SCENIC	Python/R	A computational framework for inferring gene regulatory networks and identifying transcription factors.	Aibar et al. (2017)
GRNBoost2	Python	Based on tree-based regression models, useful for large-scale single-cell gene regulatory network reconstruction.	Moerman et al. (2019)
Spatial transcriptomics			
SpaceRanger	Rust	Pipeline developed by 10x genomics for processing and analysing spatial transcriptomics data, mapping gene expression to tissue sections while preserving spatial context.	https://github.com/10XGenomics/spaceranger
SPOTlight	R	Deconvolves spatial transcriptomics data using single-cell references to infer cell composition.	Elosua-Bayes et al. (2021)
STUtility	R	Provides visualisation and integration tools for spatial transcriptomics data.	Bergenstråhle et al. (2020)
Multi-omics integration			
MOFA+	Python/R	Integrates multi-omics single-cell data, such as scRNA-seq and scATAC-seq.	Argelaguet et al. (2020)
SCVI	Python	Uses deep generative models for multi-modal single-cell integration.	Lopez et al. (2018)
Signac	R	Developed for single-cell chromatin accessibility (scATAC-seq) data analysis and integration with transcriptomic data.	Stuart et al. (2021)
Functional annotation and pathway analysis			
GSVA	R	Gene set variation analysis to assess pathway activity at a single-cell level.	Hänzelmann et al. (2013)
GSEA	R	Identifies enriched pathways in different cell populations.	Franchini et al. (2023)
CellChat	R	Infers intercellular communication networks from single-cell expression data.	Jin et al. (2025)

11.4 Applications of Single-Cell Omics

The utility of single-cell omics transcends fundamental research, providing tangible strategies for crop enhancement through applications such as precision breeding, stress response characterisation, and targeted gene modification. In the following sections, we will examine a selection of these applications.

11.4.1 Understanding Cellular Heterogeneity

Plants consist of diverse cell types, each with specific roles in development, growth, and adaptation. Single-cell omics enable the analysis of this cellular diversity, revealing how distinct cell types in tissues such as roots, leaves, and reproductive organs contribute to plant development (Apelt et al. 2022; Turco et al. 2019; Shulse et al. 2019; Zhang et al. 2021b; Liu et al. 2024; Marand et al. 2021). By understanding the unique functions of each cell type, researchers can better comprehend how plants grow, differentiate, and respond to environmental stimuli like nutrient availability, pathogens, and abiotic stresses (e.g. drought and heat) (Cao et al. 2023; Wang et al. 2025; Li et al. 2023). This deeper understanding of cellular heterogeneity provides the foundation for more targeted approaches to crop improvement.

For example, single-cell analyses of root systems can reveal which cells are most effective at nutrient uptake, providing insights that can be applied to enhance nutrient use efficiency in crops (Lhamo and Luan 2021; Li et al. 2022). Similarly, studying leaf tissue at the cellular level can help identify specific cells responsible for photosynthetic efficiency, which could lead to innovations that improve crop productivity.

11.4.2 Precision Breeding

Single-cell omics also play a crucial role in precision breeding by identifying key molecular markers at the cellular level. Traditional breeding methods rely on bulk tissue analysis, which can obscure rare but important cellular variations. In contrast, single-cell omics can uncover specific gene expressions, protein functions, or metabolite activities linked to desirable traits such as higher yield, disease resistance, or stress tolerance (Jean-Baptiste et al. 2019; Wang et al. 2021; Sun et al. 2022).

By pinpointing these molecular markers, breeders can create more tailored crops, accelerating the development of varieties optimised for specific traits. For instance, cells responsible for drought tolerance can be isolated and studied, allowing breeders to develop crop varieties with enhanced resilience to water scarcity. The ability to focus on individual cells also means breeders can optimise plants for other desired characteristics, such as faster growth or increased reproductive success.

11.4.3 Unveiling Plant Developmental Mechanisms

Understanding plant development requires detailed knowledge of how cells differentiate and form tissues during growth and reproduction. Single-cell omics provide a high-resolution view of cellular differentiation, helping to elucidate key processes involved in tissue formation and organ development (Zhu et al. 2022; Nelms and Walbot 2019; Satterlee et al. 2020; Lopez-Anido et al. 2021; Serrano-Ron et al. 2021). In particular, studying meristematic cells—undifferentiated plant cells responsible for growth—can provide insights into how plants regulate stem cells to control growth, reproduction, and yield (Satterlee et al. 2020; Liao and Wang 2023; Xu et al. 2021).

This technology can also reveal how environmental cues influence plant development at the cellular level, allowing researchers to manipulate these signals to improve crop growth and productivity. By analysing the regulatory networks involved in plant development, single-cell omics can inform strategies to enhance traits like faster growth, improved fruit set, or optimised flowering times (Hou et al. 2021).

11.4.4 Understanding Stress Responses

Plants constantly encounter biotic (pests and pathogens) and abiotic (drought and temperature) stresses, which can negatively affect growth and yield (Cramer et al. 2011; Pandey et al. 2017). Single-cell omics offer an unparalleled view of how individual cells respond to these stressors, identifying the specific genes, proteins, and metabolites involved in stress resilience. By studying how certain cells activate defence mechanisms or adapt their metabolic processes under stress, researchers can uncover pathways that could be harnessed to develop crops with improved stress tolerance (Berrío and Dubois 2024).

For example, single-cell transcriptomics can reveal how specific cells in a plant's roots or leaves respond to drought by activating drought-responsive genes, providing targets for engineering drought-resistant crops (Berrío et al. 2024). Additionally, understanding cellular responses to pests and pathogens can lead to the identification of new genes for disease resistance, helping to safeguard crops against agricultural pests and diseases (Delannoy et al. 2023; Zhu et al. 2023b; Nobori et al. 2023).

By integrating single-cell data into crop breeding programmes, it becomes possible to engineer plants that are more resilient to climate change, ensuring food production remains stable in the face of increasing environmental challenges.

11.4.5 Integration with CRISPR and Gene Editing

The integration of single-cell omics with CRISPR and other gene editing technologies has the potential to significantly advance plant biodesign and crop improvement. By identifying key genes and regulatory elements in specific cell types, single-cell omics enable precise genetic modifications tailored to enhance desired traits such as drought tolerance, disease resistance, and nutrient use efficiency (Islam et al. 2024).

Beyond conventional applications, single-cell omics facilitate the discovery and design of cell-type-specific promoters, which are essential for targeted gene expression while minimising unintended pleiotropic effects. These promoters, identified through scRNA-seq analysis, enable cell-type-specific CRISPR/Cas-based gene editing, metabolic pathway engineering, and synthetic symbiosis between plants and microbes. For example, cell-type-specific promoters have been used to drive CRISPR/Cas9 expression in targeted cell types, optimising traits like lignin biosynthesis while avoiding growth defects (Liang et al. 2019). Additionally, single-cell approaches aid in identifying cis-regulatory elements (CREs) and microRNAs, further refining the precision of genetic interventions.

Together with chromatin profiling methods such as single-cell ATAC-seq, single-cell transcriptomics provides a powerful framework for designing synthetic cell-type-specific promoters, enabling spatially controlled gene expression. The application of these approaches in plant synthetic biology holds immense potential for engineering crops with improved resilience, productivity, and sustainability, paving the way for future innovations in agriculture.

11.5 Status of Single-Cell Omics in Different Crops

Single-cell omics is revolutionising plant research by providing unprecedented insights into cellular functions, developmental processes, and stress responses at a resolution previously unattainable. These technologies are being applied across various plant species to unravel the complexities of plant biology, offering new avenues for crop improvement (Table 11.2). In this section, we highlight key examples of single-cell omics studies across different crops, showcasing their impact on advancing plant science.

11.5.1 Rice

Rice, crucial for global food security, has been deeply explored through single-cell omics to understand its responses to environmental stresses and developmental biology. scRNA-seq has been pivotal in dissecting how rice cells react to salinity

Table 11.2 List of selected single-cell studies in different plant species

Species	Technology	Tissue/organ studied	Stress/condition	Reference
Arabidopsis thaliana	Drop-Seq	Root	–	Apelt et al. (2022)
Arabidopsis thaliana	Drop-Seq	Root	No sucrose	Shulse et al. (2019)
Arabidopsis thaliana	10X chromium	Seedling	–	Han et al. (2023)
Arabidopsis thaliana	10X chromium	Root	Heat	Jean-Baptiste et al. (2019)
Arabidopsis thaliana	10X chromium	Rosette	Drought	Illouz-Eliaz et al. (2023)
Arabidopsis thaliana	10X chromium	Root	–	Nolan et al. (2023)
Arabidopsis thaliana	DNBelab C4 system	Callus	–	Yin et al. (2024)
Barley (*Hordeum vulgare*)	10X chromium, Visium	Grain, seed	–	Peirats-Llobet et al. (2023)
Cassava (*Manihot esculenta*)	10X chromium	Leaf	–	Zang et al. (2023)
Chinese cabbage (*Brassica rapa* ssp. pekinensis)	10X chromium	3rd true leaf	Heat	Sun et al. (2022)
Cotton (*Gossypium arboreum*)	10X chromium	Lateral root tips	Salt	Li et al. (2023)
Cotton (*Gossypium hirsutum*)	10X chromium	Hypocotyl	–	Zhu et al. (2023a)
Lotus japonicus	Stereo-seq	Nodules	–	Ye et al. (2024)
Maize (*Zea mays*)	10X chromium	Root	Heat	Wang et al. (2025)
Maize (*Zea mays*)	10X chromium	Maize ears	–	Xu et al. (2021)
Maize (*Zea mays*)	DNBelab C4 system	Leaf	*Puccinia polysora*	Yan et al. (2025)

(continued)

Table 11.2 (continued)

Species	Technology	Tissue/organ studied	Stress/condition	Reference
Maize (*Zea mays*)	10X chromium	Seedlings, axillary buds, staminate, embryonic root tips, and pistillate inflorescence.	–	Marand et al. (2021)
Maize (*Zea mays*)	10X chromium	Seedlings	Nitrate deficient	Li et al. (2022)
Medicago truncatula	10X chromium	Root	–	Liu et al. (2023a)
Pea (*Pisum sativum*)	10X chromium	Shoot apex	Boron deficient	Chen et al. (2024)
Peanut (*Arachis hypogaea*)	10X chromium	Pegs/pods	–	Cui et al. (2024)
Peanut (*Arachis hypogaea*)	10X chromium	Leaf	–	Liu et al. (2024)
Peanut (*Arachis hypogaea*)	10X chromium	Leaf	Dark	Deng et al. (2024)
Pitaya (*Hylocereus undatus*)	10X chromium, Visium	Pericarp	Senescence	Pang et al. (2024)
Pitaya (*Hylocereus undatus*)	10X chromium, Visium	Pericarp and peel	Senescence	Li et al. (2024)
Rice (*Oryza sativa*)	10X chromium	Radicles	–	Zhang et al. (2021a)
Rice (*Oryza sativa*)	BD rhapsody	Floret and inflorescence	–	Zong et al. (2022)
Rice (*Oryza sativa*)	10X chromium	Leaf	*Nilaparvata lugens*	Zha et al. (2023)
Rice (*Oryza sativa*)	10X chromium	Seedlings	Salinity, low nitrogen, and iron deficiency.	Wang et al. (2021)
Rice (*Oryza sativa*)	10X chromium	Root tips	–	Feng et al. (2022)
Soybean (*Glycine max*)	10X chromium, Visium	Leaf, young pod, nodule, hypocotyl, and root	–	Zhang et al. (2025)
Soybean (*Glycine max*)	10X chromium and stereo-seq	Nodules	*Bradyrhizobium japonicum* USDA 110	Liu et al. (2023b)
Strawberry (*Fragaria vesca*)	10X chromium	Leaf	*Botrytis cinerea*	Bai et al. (2022)

(continued)

Table 11.2 (continued)

Species	Technology	Tissue/organ studied	Stress/condition	Reference
Tomato (*Solanum lycopersicum*)	10X chromium	Shoot apex	–	Tian et al. (2020)
Wheat (*Triticum aestivum*)	Visium	Developing grain	Heat	Wang et al. (2024)
Wheat (*Triticum aestivum*)	10X chromium	Root tip	–	Zhang et al. (2023)
Wheat (*Triticum aestivum*)	DNBelab C4 system	Shoot apex	–	Liu et al. (2025)

stress. Studies have shown that in rice roots, specific cell types express genes managing ion transport and osmotic regulation, aiding adaptation to high salinity (Zha et al. 2023). This approach has revealed the diversity in cellular responses, with some cells specialising in ion homeostasis, while others focus on osmotic adjustment, showcasing the intricate adaptation strategies within the root system.

Moreover, single-cell transcriptomics has provided insights into rice grain filling. Gene expression patterns linked to starch biosynthesis have been mapped across different cell types during grain development, shedding light on their roles in determining grain size and quality. This work has highlighted the complex cellular mechanisms involved, offering potential for guiding genetic engineering to enhance rice yield and nutritional content (Wang et al. 2021). These advancements underscore the transformative potential of single-cell omics in both stress adaptation and developmental biology of rice, paving the way for more resilient and nutritious varieties.

11.5.2 Maize

Maize, a globally important crop, has been extensively studied using single-cell omics technologies to enhance its resilience, yield, and nutritional quality. Recent advances in scRNA-seq and spatial transcriptomics have provided unprecedented insights into how individual maize cells respond to environmental stressors and regulate key developmental processes. For instance, scRNA-seq has been used to characterise maize roots under heat stress, identifying distinct cell-type-specific responses, including the activation of heat shock proteins and other protective mechanisms, which are critical for improving heat tolerance (Wang et al. 2025). Similarly, single-cell transcriptomics has been applied to understand maize's response to drought and nutrient deficiencies, revealing regulatory genes that enhance water and nutrient uptake, which could inform breeding strategies for stress-resilient maize varieties (Yan et al. 2025). In addition to stress responses, single-cell omics have been instrumental in unravelling the molecular basis of maize grain development. By profiling gene expression during endosperm differentiation, researchers have mapped the regulatory networks controlling kernel formation, grain size, and composition, providing key targets for genetic improvement (Marand et al. 2021). Moreover, spatial transcriptomics has shed light on the dynamics of sucrose transport and allocation during grain filling, identifying genes involved in carbohydrate metabolism that influence kernel nutritional quality and yield (Li et al. 2022). Furthermore, integrating single-cell and spatial omics has provided insights into maize root development and nutrient uptake efficiency, helping to optimise breeding strategies for higher-yielding and more resilient varieties (Xu et al. 2021). These studies collectively highlight the power of single-cell and spatial omics technologies in advancing maize breeding, enabling targeted genetic improvements to enhance stress tolerance, yield potential, and nutritional value.

11.5.3 Wheat

Wheat, a staple crop worldwide, has benefited from significant advancements through single-cell omics technologies, paving the way for improving its resilience, yield, and nutritional quality. Studies employing scRNA-seq and spatial transcriptomics have provided detailed insights into cellular responses to environmental stressors and developmental processes. For instance, single-cell multi-omics analysis has revealed asymmetric gene expression and cell-type-specific regulatory networks in bread wheat roots. This study identified 22 distinct cell types and reconstructed cell differentiation trajectories, highlighting key transcription factors involved in root development and stress responses, providing a detailed transcriptional landscape of wheat root organisation at single-cell resolution (Zhang et al. 2023). Beyond stress responses, single-cell omics have contributed significantly to understanding wheat inflorescence development. By profiling gene expression during spikelet and floret formation, researchers have mapped the regulatory networks governing these processes, identifying key targets for genetic improvement (Liu et al. 2025). These discoveries emphasise the crucial role of single-cell and spatial omics in shaping the future of wheat breeding, contributing to the development of superior varieties.

11.5.4 Other Crops

In Chinese cabbage (*Brassica rapa* ssp. pekinensis), scRNA-seq revealed that genes in the least fractioned subgenome were predominantly expressed across cell types and that heat stress alters gene expression and subgenome dominance in a cell type-specific manner. These findings enhance our understanding of leaf development and transcriptional regulation under stress, aiding efforts to improve yield and climate resilience (Sun et al. 2022). Similarly, soybean (*Glycine max*) research has leveraged single-nucleus and spatial transcriptomics to study nodule maturation, capturing transitional states in nitrogen fixation. These insights enable the development of soybean varieties with enhanced nitrogen-fixing capabilities, reducing reliance on synthetic fertilisers (Zhang et al. 2025). Cassava (*Manihot esculenta*), a staple crop in tropical regions, has also benefited from single-cell technology, leading to a deeper understanding of its structural and photosynthetic characteristics. A single-cell transcriptome atlas of cassava leaves has revealed key insights into leaf cell types, including mesophyll, epidermis, and vascular tissues, along with their distinctive metabolic activities. By identifying 120 candidate marker genes and constructing a leaf cell development trajectory, researchers have uncovered genes involved in cell differentiation fate. This knowledge provides a theoretical foundation for enhancing cassava yield and nutritional quality, contributing to its resilience against environmental challenges (Zang et al. 2023).

In addition to these staple crops, single-nucleus RNA-seq has been applied to tomato (*Solanum lycopersicum*) to resolve spatiotemporal developmental trajectories in the shoot apex. This approach allowed researchers to overcome challenges associated with plant cell walls that typically hinder single-cell RNA sequencing. The resulting high-resolution gene expression atlas identified numerous distinct cell types across major shoot tissues and developmental stages. By delineating the developmental trajectories of mesophyll, vasculature, epidermal, and trichome cells, the study uncovered key regulatory networks that control tomato shoot growth and differentiation. These findings provide crucial insights into hierarchical cellular relationships and developmental regulation in crops, demonstrating the power of snRNA-seq in advancing plant research (Tian et al. 2020).

These advancements highlight the power of single-cell omics in unravelling complex cellular processes that govern crop development and stress adaptation. Future research could leverage these findings to develop targeted breeding and biotechnology strategies, ensuring global food security while promoting sustainability.

11.6 Challenges and Limitations

11.6.1 Technical Challenges

The application of single-cell omics to crop improvement faces several technical challenges. One of the primary obstacles is the difficulty of isolating single plant cells. Unlike animal cells, plant cells are enclosed by rigid cell walls, which makes their extraction and manipulation more complex (Shulse et al. 2019). Technologies such as protoplasting (the enzymatic removal of the cell wall) have been developed to address this, but they are not universally applicable to all plant species or tissue types and may introduce cell stress or altered gene expression (Long et al. 2021).

Alternative approaches, such as single-nucleus RNA-seq, are gaining traction, but they also have limitations, including lower RNA capture efficiency.

Additionally, the structure of plant tissues, such as roots, stems, and leaves, presents further challenges. Variability in tissue composition affects cell dissociation efficiency, leading to potential biases in single-cell datasets. This makes it difficult to achieve comprehensive cellular coverage across different plant organs.

11.6.2 Low Abundance Cells

A significant challenge in single-cell omics is the accurate detection and characterisation of low-abundance cell types. These rare cells can play crucial roles in plant development and stress responses but are often underrepresented in single-cell datasets (Cuperus 2021). The difficulty in capturing sufficient numbers of these cells can lead to incomplete or biased representations of tissue heterogeneity. Techniques such as enrichment strategies, optimised dissociation protocols, and deeper sequencing are being explored to overcome this limitation. However, further methodological advancements are needed to ensure the reliable identification and analysis of these critical cell populations.

11.6.3 Cost and Accessibility

The high cost of single-cell omics technologies remains a major barrier to their widespread adoption. Single-cell RNA sequencing, proteomics, and metabolomics require specialised instrumentation, expensive reagents, and significant computational resources, limiting accessibility, especially in regions where agricultural research funding is constrained. Moreover, the vast amount of data generated by these techniques creates additional hurdles in terms of storage, management, and computational analysis (Lähnemann et al. 2020).

Efforts to lower costs include the miniaturisation of sequencing platforms, the development of cost-effective library preparation methods, and the leveraging of cloud-based bioinformatics solutions to reduce computational infrastructure requirements. However, further innovations are needed to make single-cell omics more widely available for crop research.

11.6.4 Computational Challenges and Data Integration

The large-scale datasets produced by single-cell omics present substantial computational challenges. A single experiment can generate thousands to millions of data points, requiring sophisticated bioinformatics pipelines for processing, normalisation, and interpretation. Single-cell transcriptomics, for instance, produces extensive gene expression data that must be carefully analysed to extract biologically meaningful patterns.

Integrating multiple layers of omics data—such as genomic, transcriptomic, proteomic, and metabolomic datasets adds another layer of complexity. Each omics layer provides unique insights, but combining them into a coherent framework demands advanced machine-learning algorithms and statistical models. For example, linking gene expression patterns from single-cell transcriptomics with protein abundance data from single-cell proteomics can reveal regulatory networks, but challenges such as differences in data formats, resolutions, and scales must be addressed (Lähnemann et al. 2020). Additionally, data integration increases the risk of false positives and missing critical interactions, necessitating robust validation strategies.

11.6.5 Cell Annotation Issues

Accurate cell annotation, or the assignment of cell identities based on their molecular profiles, is a critical step in single-cell omics analysis. However, this process is often challenging due to the lack of comprehensive reference datasets and standardised annotation pipelines, especially for non-model plant species (Cuperus 2021). A significant hurdle is the absence of well-defined, universally applicable molecular markers for many plant cell types, making it difficult to confidently assign cell identities based solely on transcriptomic data. The reliability of cell annotation directly impacts the interpretation of single-cell data and the generation of biologically meaningful insights. Issues such as batch effects, variability in cell capture, and the presence of transitional cell states can further complicate cell annotation. To address these challenges, in situ validation techniques, such as spatial transcriptomics or in situ hybridisation, are essential for confirming the localisation and identity of annotated cell types. These methods provide crucial spatial context, allowing researchers to verify the distribution and characteristics of specific cell populations within plant tissues.

Standardising single-cell protocols across plant species is another ongoing challenge. Unlike well-established model systems in human and animal research, plant single-cell workflows require customisation for different species and tissue types. Establishing universal guidelines for data generation, processing, and sharing will be critical to advancing the field.

11.7 Future Directions

11.7.1 Integration with AI and Machine Learning

One of the most promising directions for single-cell omics in crop improvement lies in the integration of advanced computational techniques, particularly artificial intelligence (AI) and machine learning (ML). The vast datasets generated by single-cell omics are often too complex for traditional data analysis approaches, but AI and ML offer the potential to analyse and interpret these datasets more efficiently and accurately. By leveraging these technologies, researchers can uncover previously hidden patterns and interactions within cellular data, facilitating a more detailed understanding of plant biology at the single-cell level (Chau et al. 2024).

For instance, AI algorithms can be used to predict phenotypic outcomes, such as yield, disease resistance, or drought tolerance-based on single-cell transcriptomic, proteomic, and metabolomic data. ML models can identify correlations between gene expression in specific cell types and the development of desired traits, providing a powerful tool for plant breeders to make more informed decisions during the breeding process. Additionally, predictive modelling using AI could help forecast how crops will respond to environmental changes or stresses at the cellular level, enabling the design of more resilient plant varieties (Farooq et al. 2024).

11.7.2 Gene Editing and Synthetic Biology

The integration of single-cell omics with gene editing technologies like CRISPR is poised to revolutionise crop improvement by enabling precision breeding at an unprecedented scale. Single-cell omics provide a detailed map of gene function and regulation in individual cells, which can be used to identify specific genetic targets for modification. CRISPR technology allows for the precise editing of these targets, resulting in crops with highly specific improvements, such as enhanced drought tolerance or faster growth.

Beyond gene editing, synthetic biology offers exciting new opportunities for designing crops with novel traits based on single-cell omics insights. By understanding the molecular pathways at the single-cell level, scientists can engineer new metabolic pathways or introduce synthetic genes that confer desirable traits such as improved nutrient use efficiency or enhanced pest resistance. For example, single-cell omics data can inform the design of crops that produce more biomass or are more resistant to extreme environmental conditions, pushing the boundaries of what is possible with traditional breeding methods.

A major frontier in synthetic biology is the design of 'smart crops' that dynamically adjust their physiological responses to environmental conditions in real time. For example, by leveraging single-cell transcriptomics, scientists could engineer crops that can modulate their water uptake, nutrient allocation, or immune responses in response to environmental cues, thereby increasing resilience to fluctuating climate conditions. Additionally, synthetic biology-driven metabolic engineering could create crops with enhanced photosynthetic efficiency, thereby increasing productivity while minimising resource inputs.

11.7.3 Scaling and Field Application

While the potential of single-cell omics in the laboratory is immense, one of the major challenges for the future is scaling these discoveries for large-scale field applications. Bridging the gap between laboratory-based single-cell research and practical, field-ready solutions requires the development of robust strategies that can translate cellular-level insights into crop improvement on a global scale.

One approach is integrating single-cell omics with traditional breeding programmes, allowing breeders to harness cellular-level data to select for desired traits in a more targeted and efficient manner. Large-scale genotyping and phenotyping platforms, coupled with single-cell data, will enable breeders to develop new crop varieties faster and with greater precision. Additionally, the cost of single-cell omics technologies is expected to decrease as they become more widely adopted, making them more accessible for breeding programmes around the world.

Another key challenge is ensuring that crops developed using single-cell omics and gene editing technologies can be successfully cultivated in diverse environmental conditions. Field trials and pilot programmes will be crucial in testing the performance of these crops in real-world settings, ensuring that the traits developed in the lab can be expressed and maintained in varying climates, soils, and agricultural systems. Overcoming these challenges will be essential for the large-scale adoption of single-cell omics in agriculture.

Furthermore, regulatory acceptance of omics-driven crop improvements will be crucial for their deployment at scale. Policies that support the responsible use of gene editing and single-cell technologies while ensuring environmental and food safety will be essential in facilitating widespread adoption. Public engagement and education efforts should accompany technological advancements to improve societal acceptance and trust in single-cell omics applications in agriculture.

11.8 Conclusions

The advent of single-cell omics has transformed the field of crop improvement by providing unprecedented insights into plant biology at the cellular level. These technologies—spanning genomics, transcriptomics, proteomics, and metabolomics—have enabled scientists to explore the cellular heterogeneity in plants, shedding light on how different cell types contribute to growth, development, and stress responses. By dissecting plant biology at the single-cell resolution, researchers have identified key molecular markers and regulatory pathways that play critical roles in traits such as drought tolerance, nutrient uptake, and disease resistance. From rice and maize to wheat and soybean, single-cell omics have revealed the complexity of plant systems, offering new avenues for precision breeding and genetic engineering.

Moreover, integrating these insights with cutting-edge technologies like CRISPR and AI-based data analytics has provided novel solutions for crop improvement. CRISPR, for instance, has enabled the precise editing of genes identified through single-cell analyses, pushing the boundaries of crop design. Similarly, AI and machine learning have facilitated the analysis of large single-cell datasets, predicting phenotypic outcomes with greater accuracy. These advancements are helping to address global challenges in agriculture, including food security and climate resilience.

Looking ahead, single-cell technologies will play an increasingly central role in shaping the future of agriculture. As the cost of single-cell omics technologies decreases and the scalability of these tools improves, their integration into routine breeding programmes will become more accessible. In the future, crop science will likely see a convergence of single-cell omics, gene editing, and computational biology, enabling a new era of precision agriculture where crops are tailored to specific environments, resistances, and nutritional needs.

However, realising the full potential of single-cell omics in agriculture will require a collaborative, interdisciplinary approach. Strong partnerships between plant scientists, bioinformaticians, and agriculture breeders will be necessary to translate omics-driven discoveries into practical, field-ready solutions. As these fields converge, the next generation of crops will not only be more resilient and productive but also more adaptable to the rapidly changing environmental conditions brought on by climate change.

In conclusion, single-cell omics represent a paradigm shift in agricultural science, offering a powerful toolkit for enhancing crop resilience, optimising resource use, and sustaining global food production in the face of future challenges. By embracing technological advancements and fostering interdisciplinary collaboration, the agricultural community can harness the full potential of single-cell omics to drive sustainable and resilient crop innovation.

References

Aibar S et al (2017) SCENIC: single-cell regulatory network inference and clustering. Nat Methods 14:1083–1086

Apelt F et al (2022) Shoot and root single cell sequencing reveals tissue- and daytime-specific transcriptome profiles. Plant Physiol 188:861–878

Aran D et al (2019) Reference-based analysis of lung single-cell sequencing reveals a transitional profibrotic macrophage. Nat Immunol 20:163–172

Argelaguet R et al (2020) MOFA+: a statistical framework for comprehensive integration of multi-modal single-cell data. Genome Biol 21:111

Bai Y et al (2022) Development of a single-cell atlas for woodland strawberry (*Fragaria Vesca*) leaves during early *Botrytis cinerea* infection using single cell RNA-seq. Hortic Res 9:uhab055

Barkla BJ, Vera-Estrella R (2015) Single cell-type comparative metabolomics of epidermal bladder cells from the halophyte *Mesembryanthemum crystallinum*. Front Plant Sci 6:435

Bergenstråhle J et al (2020) Seamless integration of image and molecular analysis for spatial transcriptomics workflows. BMC Genomics 21:482

Berrío RT, Dubois M (2024) Single-cell transcriptomics reveals heterogeneity in plant responses to the environment: a focus on biotic and abiotic interactions. J Exp Bot 75:5188–5203

Berrío RT et al (2024) Dual and spatially resolved drought responses in the Arabidopsis leaf mesophyll revealed by single-cell transcriptomics. Nw Phytol 246(3):840–858. https://nph.onlinelibrary.wiley.com/doi/10.1111/nph.20446

Cao Y et al (2023) Single-cell RNA sequencing profiles reveal cell type-specific transcriptional regulation networks conditioning fungal invasion in maize roots. Plant Biotechnol J 21:1839–1859

Chau TN et al (2024) Advancing plant single-cell genomics with foundation models. Curr Opin Plant Biol 82:102666

Chen X et al (2024) Single-cell transcriptomic analysis of pea shoot development and cell-type-specific responses to boron deficiency. Plant J 117:302–322

Clark NM, Elmore JM, Walley JW (2021) To the proteome and beyond: advances in single-cell omics profiling for plant systems. Plant Physiol 188:726–737

Cramer GR et al (2011) Effects of abiotic stress on plants: a systems biology perspective. BMC Plant Biol 11:163

Cui Y et al (2024) Single-nucleus RNA and ATAC sequencing analyses provide molecular insights into early pod development of peanut fruit. Plant Commun 5:100979

Cuperus JT (2021) Single-cell genomics in plants: current state, future directions, and hurdles to overcome. Plant Physiol 188:749–755

Delannoy E et al (2023) Cell specialization and coordination in Arabidopsis leaves upon pathogenic attack revealed by scRNA-Seq. Plant Commun 4:100676

Deng Q et al (2024) scRNA-Seq reveals dark- and light-induced differentially expressed gene atlases of seedling leaves in *Arachis hypogaea* L. Plant Biotechnol J 22:1848–1866

Domínguez Conde C et al (2022) Cross-tissue immune cell analysis reveals tissue-specific features in humans. Science 376:eabl5197

Dorrity MW et al (2021) The regulatory landscape of *Arabidopsis thaliana* roots at single-cell resolution. Nat Commun 12:3334

Efroni I, Birnbaum KD (2016) The potential of single-cell profiling in plants. Genome Biol 17:65

Elosua-Bayes M et al (2021) SPOTlight: seeded NMF regression to deconvolute spatial transcriptomics spots with single-cell transcriptomes. Nucleic Acids Res 49:e50

Farooq MA et al (2024) Artificial intelligence in plant breeding. Trends Genet 40:891–908

Feng D et al (2022) Chromatin accessibility illuminates single-cell regulatory dynamics of rice root tips. BMC Biol 20:274

Franchini M et al (2023) Single-cell gene set enrichment analysis and transfer learning for functional annotation of scRNA-seq data. NAR Genom Bioinform 5:lqad024

Giacomello S et al (2017) Spatially resolved transcriptome profiling in model plant species. Nat Plants 3:17061

Goel M et al (2024) The vast majority of somatic mutations in plants are layer-specific. Genome Biol 25:194

Grobei MA et al (2009) Deterministic protein inference for shotgun proteomics data provides new insights into Arabidopsis pollen development and function. Genome Res 19:1786–1800

Han X et al (2023) Time series single-cell transcriptional atlases reveal cell fate differentiation driven by light in Arabidopsis seedlings. Nat Plants 9:2095–2109

Hänzelmann S et al (2013) GSVA: gene set variation analysis for microarray and RNA-seq data. BMC Bioinform 14:7
Hao Y et al (2024) Dictionary learning for integrative, multimodal and scalable single-cell analysis. Nat Biotechnol 42:293–304
Hickey LT et al (2019) Breeding crops to feed 10 billion. Nat Biotechnol 37:744–754
Hou Z et al (2021) High-throughput single-cell transcriptomics reveals the female germline differentiation trajectory in *Arabidopsis thaliana*. Commun Biol 4:1149
Huang C-Y et al (2021) The chromatin-remodeling protein BAF60/SWP73A regulates the plant immune receptor NLRs. Cell Host Microbe 29:425–434.e4
Illouz-Eliaz N et al (2023) Stress recovery triggers rapid transcriptional reprogramming and activation of immunity in plants. bioRxiv. https://doi.org/10.1101/2023.02.27.530256
Islam MT et al (2024) Advances in the application of single-cell transcriptomics in plant systems and synthetic biology. Biodes Res 6:0029
Jean-Baptiste K et al (2019) Dynamics of gene expression in single root cells of *Arabidopsis thaliana*. Plant Cell 31(5):993–1011
Jeong YY et al (2021) Optimization of protoplast regeneration in the model plant *Arabidopsis thaliana*. Plant Methods 17:21
Jin S et al (2025) CellChat for systematic analysis of cell-cell communication from single-cell transcriptomics. Nat Protoc 20:180–219
Kenny C, Gehan Z (2023) Scenarios for future global growth to 2050. Working Papers, March. https://ideas.repec.org//p/cgd/wpaper/634.html
Kolodziejczyk AA et al (2015) The technology and biology of single-cell RNA sequencing. Mol Cell 58(4):610–620
Korsunsky I et al (2019) Fast, sensitive and accurate integration of single-cell data with Harmony. Nat Methods 16:1289–1296
Lähnemann D et al (2020) Eleven grand challenges in single-cell data science. Genome Biol 21:31
Lhamo D, Luan S (2021) Potential networks of nitrogen-phosphorus-potassium channels and transporters in Arabidopsis roots at a single-cell resolution. Front Plant Sci 12:689545
Li X et al (2022) Single-cell RNA sequencing reveals the landscape of maize root tips and assists in identification of cell type-specific nitrate-response genes. Crop J 10:1589–1600
Li P et al (2023) Transcriptional landscape of cotton roots in response to salt stress at single-cell resolution. Plant Commun 5:100740
Li X et al (2024) Single-cell and spatial RNA sequencing reveal the spatiotemporal trajectories of fruit senescence. Nat Commun 15:3108
Liang Y et al (2019) A screening method to identify efficient sgRNAs in Arabidopsis, used in conjunction with cell-specific lignin reduction. Biotechnol Biofuels 12:130
Liao RY, Wang JW (2023) Analysis of meristems and plant regeneration at single-cell resolution. Curr Opin Plant Biol 74:102378
Liu J et al (2020) Jointly defining cell types from multiple single-cell datasets using LIGER. Nat Protoc 15:3632–3662
Liu Z et al (2023a) Single-nucleus transcriptomes reveal spatiotemporal symbiotic perception and early response in Medicago. Nat Plants 9:1734–1748
Liu Z et al (2023b) Integrated single-nucleus and spatial transcriptomics captures transitional states in soybean nodule maturation. Nat Plants 9:515–524
Liu H et al (2024) A single-nucleus resolution atlas of transcriptome and chromatin accessibility for peanut (*Arachis hypogaea* L.) leaves. Adv Biol 8:e2300410
Liu X et al (2025) Decoding cellular transcriptional regulatory networks governing wheat inflorescence development. *bioRxiv*. https://doi.org/10.1101/2025.01.18.633750
Long Y et al (2021) FlsnRNA-Seq: Protoplasting-free full-length single-nucleus RNA profiling in plants. Genome Biol 22:66
Lopez R et al (2018) Deep generative modeling for single-cell transcriptomics. Nat Methods 15: 1053–1058

Lopez-Anido CB et al (2021) Single-cell resolution of lineage trajectories in the Arabidopsis stomatal lineage and developing leaf. Dev Cell 56:1043–1055.e4
Lun ATL et al (2019) EmptyDrops: distinguishing cells from empty droplets in droplet-based single-cell RNA sequencing data. Genome Biol 20:63
Luo C et al (2020) Single-cell genomics and epigenomics: technologies and applications in plants. Trends Plant Sci 25:1030–1040
Marand AP et al (2021) A cis-regulatory atlas in maize at single-cell resolution. Cell 184:3041–3055.e21
Marx V (2019) A dream of single-cell proteomics. Nat Methods 16:809–812
McCarthy DJ et al (2017) Scater: pre-processing, quality control, normalization and visualization of single-cell RNA-seq data in R. Bioinformatics 33:1179–1186
Moerman T et al (2019) GRNBoost2 and Arboreto: efficient and scalable inference of gene regulatory networks. Bioinformatics 35:2159–2161
Montes C et al (2024) Single-cell proteomics differentiates Arabidopsis root cell types. New Phytol 244:1750–1759
Nelms B, Walbot V (2019) Defining the developmental program leading to meiosis in maize. Science 364:52–56
Nobori T et al (2023) Time-resolved single-cell and spatial gene regulatory atlas of plants under pathogen attack. bioRxiv. https://doi.org/10.1101/2023.04.10.536170
Nolan TM et al (2023) Brassinosteroid gene regulatory networks at cellular resolution in the Arabidopsis root. Science 379:eadf4721
Pandey P et al (2017) Impact of combined abiotic and biotic stresses on plant growth and avenues for crop improvement by exploiting physio-morphological traits. Front Plant Sci 8:537
Pang X et al (2024) Integration of single-cell and spatial RNA sequencing uncovers spatiotemporal transition of fruit senescence trajectory from exocarp to mesocarp in pitaya (*Hylocereus undatus*). Postharvest Biol Technol 213:112954
Peirats-Llobet M et al (2023) Spatially resolved transcriptomic analysis of the germinating barley grain. Nucleic Acids Res 51:7798–7819
Pliner HA et al (2019) Supervised classification enables rapid annotation of cell atlases. bioRxiv. https://doi.org/10.1101/538652
Polański K et al (2020) BBKNN: fast batch alignment of single cell transcriptomes. Bioinformatics 36:964–965
Ren S, Wang Y (2023) Protoplast isolation for plant single-cell RNA-Seq. Methods Mol Biol (Clifton, NJ) 2686:301–305
Satterlee JW et al (2020) Plant stem-cell organization and differentiation at single-cell resolution. Proc Natl Acad Sci 117(33689–33):699
Serrano-Ron L et al (2021) Reconstruction of lateral root formation through single-cell RNA sequencing reveals order of tissue initiation. Mol Plant 14:1362–1378
Shulse CN et al (2019) High-throughput single-cell transcriptome profiling of plant cell types. Cell Rep 27:2241–2247.e4
Smallwood SA et al (2014) Single-cell genome-wide bisulfite sequencing for assessing epigenetic heterogeneity. Nat Methods 11:817–820
Stuart T et al (2021) Single-cell chromatin state analysis with Signac. Nat Methods 18:1333–1341
Sun X et al (2022) Single-cell transcriptome reveals dominant subgenome expression and transcriptional response to heat stress in Chinese cabbage. Genome Biol 23:262
Thibivilliers S, Libault M (2021) Plant single-cell multiomics: cracking the molecular profiles of plant cells. Trends Plant Sci 26:662–663
Tian C et al (2020) Single-nucleus RNA-seq resolves spatiotemporal developmental trajectories in the tomato shoot apex. bioRxiv. https://doi.org/10.1101/2020.09.20.305029
Trapnell C et al (2014) The dynamics and regulators of cell fate decisions are revealed by pseudotemporal ordering of single cells. Nat Biotechnol 32:381–386
Turco GM et al (2019) Molecular mechanisms driving switch behavior in xylem cell differentiation. Cell Rep 28:342–351.e4

Vu AH et al (2024) Quantitative single-cell mass spectrometry provides a highly resolved analysis of natural product biosynthesis partitioning in plants. J Am Chem Soc 146(23891–23):900
Wada H et al (2020) On-site single pollen metabolomics reveals varietal differences in phosphatidylinositol synthesis under heat stress conditions in rice. Sci Rep 10:2013
Wang Y et al (2021) Single-cell transcriptome atlas of the leaf and root of rice seedlings. J Genet Genom 48:881–898
Wang J et al (2024) Spatial transcriptomics uncover coordinated cellular responses to heat stress in developing wheat grains. Res Sq. https://doi.org/10.21203/rs.3.rs-4253930/v1
Wang T et al (2025) Single-cell transcriptomes reveal spatiotemporal heat stress response in maize roots. Nat Commun 16:177
Wolf FA et al (2018) SCANPY: large-scale single-cell gene expression data analysis. Genome Biol 19:15
Xu X et al (2021) Single-cell RNA sequencing of developing maize ears facilitates functional analysis and trait candidate gene discovery. Dev Cell 56:557–568.e6
Yan S et al (2022) Recent advances in proteomics and metabolomics in plants. Mol Hortic 2:17
Yan X-C et al (2025) Single-cell transcriptomic profiling of maize cell heterogeneity and systemic immune responses against *Puccinia polysora* Underw. Plant Biotechnol J 23:549–563
Ye K et al (2024) Mapping the molecular landscape of *Lotus japonicus* nodule organogenesis through spatiotemporal transcriptomics. Nat Commun 15:6387
Yin R et al (2024) A single-cell transcriptome atlas reveals the trajectory of early cell fate transition during callus induction in Arabidopsis. Plant Commun 5:100941
Yu X et al (2023) Single-cell and spatial multi-omics in the plant sciences: technical advances, applications, and perspectives. Plant Commun 4:100508
Zang Y et al (2023) Single-cell RNA-sequencing profiles reveal the developmental landscape of the *Manihot esculenta* Crantz leaves. Plant Physiol 194:456–474
Zenobi R (2013) Single-cell metabolomics: analytical and biological perspectives. Science 342: 1243259
Zha W et al (2023) Single-cell RNA sequencing of leaf sheath cells reveals the mechanism of rice resistance to brown planthopper (*Nilaparvata lugens*). Front Plant Sci 14:1200014
Zhang T-Q et al (2021a) Single-cell transcriptome atlas and chromatin accessibility landscape reveal differentiation trajectories in the rice root. Nat Commun 12:2053
Zhang T-Q et al (2021b) A single-cell analysis of the Arabidopsis vegetative shoot apex. Dev Cell 56:1056–1074.e8
Zhang L et al (2023) Asymmetric gene expression and cell-type-specific regulatory networks in the root of bread wheat revealed by single-cell multiomics analysis. Genome Biol 24:65
Zhang X et al (2025) A spatially resolved multi-omic single-cell atlas of soybean development. Cell 188:550–567.e19
Zhao L et al (2020) Integrative analysis of reference epigenomes in 20 rice varieties. Nat Commun 11:2658
Zhu M et al (2022) Single-cell genomics revolutionizes plant development studies across scales. Development 149. https://doi.org/10.1242/dev.200179
Zhu X et al (2023a) Single-cell resolution analysis reveals the preparation for reprogramming the fate of stem cell niche in cotton lateral meristem. Genome Biol 24:194
Zhu J et al (2023b) Single-cell profiling of Arabidopsis leaves to *Pseudomonas syringae* infection. Cell Rep 42:112676
Zong J et al (2022) A rice single-cell transcriptomic atlas defines the developmental trajectories of rice floret and inflorescence meristems. New Phytol 234:226–241

Chapter 12
Trait Discovery in the Post-NGS Era

Mukesh Rathore, Farkhandah Jan, Nikita Aggarwal, Mahpara Bashir, Shri Hari Prasad, Mohd Anwar Khan, Parvaze Ahmad Sofi, Annapurna Chitikineni, Dinesh Kumar Saini, and Reyazul Rouf Mir

Abstract Improving agronomic traits is a key objective in crop improvement programmes, with many of these traits governed by complex quantitative inheritance. The identification of quantitative trait loci (QTLs), followed by their fine mapping and cloning, plays a critical role in trait mapping. Pre-next-generation sequencing (pre-NGS) methods had limited scope and precision in trait discovery. In contrast, the advent of NGS has revolutionised trait discovery, improved our understanding of complex trait genetics and facilitated the efficient use of genomic regions in molecular breeding, QTL/gene cloning, and functional validation studies. This chapter provides an overview of the transformative impact of post-NGS tools and approaches on trait mapping in crops. We explore the role of widely used NGS platforms and advanced genotyping methods in improving genetic resolution. The shift from single-reference genomes to de novo genome assemblies and pan-genomics is also explored, highlighting their role in capturing genetic diversity across species. Advancements in QTL mapping, meta-QTL analysis, association mapping, fine mapping, and cloning are highlighted, focusing on their role in understanding the genetic control of complex traits in the post-NGS era. Further, how the integration of genomics and phenomics can accelerate trait discovery is

M. Rathore · F. Jan · N. Aggarwal · M. Bashir · M. A. Khan · P. A. Sofi
Division of Genetics & Plant Breeding, Faculty of Agriculture (FoA), SKUAST-Kashmir, Sopore, Jammu and Kashmir, India

S. H. Prasad · A. Chitikineni
WA State Agricultural Biotechnology Centre, Centre for Crop and Food Innovation, Food Futures Institute, Murdoch University, Murdoch, WA, Australia

D. K. Saini
Department of Plant and Soil Science, Texas Tech University, Lubbock, TX, USA

R. R. Mir (✉)
Division of Genetics & Plant Breeding, Faculty of Agriculture (FoA), SKUAST-Kashmir, Sopore, Jammu and Kashmir, India

WA State Agricultural Biotechnology Centre, Centre for Crop and Food Innovation, Food Futures Institute, Murdoch University, Murdoch, WA, Australia
e-mail: rrmir@skuastkashmir.ac.in

V. Garg et al. (eds.), *DNA of Sustainability*, Sustainability Sciences in Asia and Africa,
https://doi.org/10.1007/978-981-95-5109-5_12

discussed. Finally, we outline future research areas, specifically the potential of emerging tools such as single-cell genomics, synthetic biology, systems biology, and AI-driven approaches, in trait mapping for crop improvement.

Keywords Mapping · QTL · Sequencing · Molecular breeding · Genomic regions

12.1 Introduction

Plant functional traits refer to the inherent characteristics or attributes of plants that influence their ecological roles and interactions within the environment (Violle et al. 2007). These traits are pivotal for understanding how plants adapt to diverse ecological niches, compete with other species, and respond to environmental fluctuations. In agricultural contexts, functional traits hold immense significance as they directly contribute to crop improvement. Identifying traits that confer tolerance to biotic and abiotic stresses is a key objective for scientists and breeders striving to develop resilient crop varieties (Roychowdhury et al. 2020). Traits related to stress tolerance, yield, disease resistance, and nutritional content form the foundation for breeding robust and high-performing crop varieties and hybrids.

However, many functional traits in plants are controlled by complex genetic networks, involving numerous genes and intricate interactions. This complexity makes trait discovery a challenging yet essential endeavour. In the early 1990s, DNA-based markers such as restriction fragment length polymorphism (RFLP), amplified fragment length polymorphism (AFLP), and simple sequence repeat (SSR) markers were widely used for dissecting traits. While groundbreaking at the time, these methods faced limitations in mapping complex traits that involved multiple loci across the genome (Deepa et al. 2020). Conventional QTL mapping methods also had drawbacks, including incomplete gene identification, labour-intensive processes, and lengthy workflows (Bohra et al. 2020).

The advent of NGS technologies have revolutionised trait discovery in plant genomics. Platforms such as Illumina, Pacific Biosciences, and Oxford Nanopore have made it possible to sequence entire genomes or specific regions rapidly and cost-effectively (Abdi et al. 2024). These technologies facilitate the identification of genetic variations and mutations associated with specific traits (Satam et al. 2023). NGS has also enabled parallel sequencing, generating millions of DNA sequences in a single run while reducing costs by over 1000-fold since its inception (Park and Kim 2016). This unprecedented accessibility to genetic data has accelerated genetic mapping studies, particularly genome-wide association studies (GWAS) and the discovery of novel loci linked to critical traits in various crops, including rice (Gu et al. 2022), wheat (Sandhu et al. 2021), rapeseed (Pal et al. 2021), potato (Sharma et al. 2024), maize (Bocianowski 2024), and *Cannabis sativa* (de Ronne et al. 2024). Moreover, NGS has facilitated the development of reference genomes and pan-genomes, which capture the full spectrum of genetic diversity within species (Gong et al. 2023).

This chapter explores the transformative role of NGS platforms in plant genomics and their contribution to advanced genotyping techniques, such as genotyping-by-sequencing (GBS-derived SNPs), SNP arrays, whole-genome sequencing, and RADSeq. These methods have significantly enhanced the precision of trait discovery. Additionally, we examine the transition from reference genomes to comprehensive pan-genomes, which provide deeper insights into genetic diversity within plant species. In the post-NGS era, advanced frameworks for trait discovery, including improvements in QTL mapping, GWAS, and fine-mapping techniques, have emerged. Finally, we highlight the integration of phenomics with genomics as a critical approach to bridge the gap between genotype and phenotype, paving the way for more accurate and efficient trait discovery.

12.2 Revolutionising Trait Discovery: Genomic Technologies and Tools

12.2.1 Next-Generation Sequencing Platforms

Understanding the genome of any living organism begins with determining the precise order of nucleotides in its complete DNA sequence, a process known as sequencing (Bansal et al. 2018). Since its inception, sequencing has been a cornerstone of molecular biology, driving significant efforts to develop advanced tools and methodologies for unravelling genetic information. The first revolution in DNA sequencing emerged in the 1970s with two landmark techniques: Sanger's chain-termination method and Maxam–Gilbert's chemical degradation method (Low and Tammi 2023). Among these, the Sanger sequencing method, with its modifications, became the predominant technology, ultimately overshadowing the Maxam–Gilbert approach due to its reliability and scalability.

As a 'first-generation sequencing' technology, automated Sanger sequencing played a pivotal role in achieving groundbreaking milestones such as the completion of the human genome project and sequencing various animal and plant genomes (Barba et al. 2014).

The BAC system (for bacterial artificial chromosome) is based on *Escherichia coli* and its single-copy plasmid F factor. It is capable of maintaining human genomic DNA fragments of >300 kilobase pairs (Kb). Because of high cloning efficiency, easy manipulation of the cloned DNA, and stable maintenance of inserted DNA, the BAC system may facilitate the construction of DNA libraries of complex genomes with fuller representation and subsequent rapid analysis of complex genomic structure (Shizuya et al. 1992).

Expressed sequence tags (ESTs) are fragments of mRNA sequences derived through single sequencing reactions performed on randomly selected clones from cDNA libraries. However, with improvements in accuracy and coverage, they are beginning to find application in fields such as phylogenetics, transcript profiling, and proteomics (Parkinson and Blaxter 2009).

The development of novel high-throughput DNA sequencing methods has provided a new method for both mapping and quantifying transcriptomes. This method, termed RNA-Seq (RNA sequencing), has clear advantages over existing approaches and is expected to revolutionise the manner in which eukaryotic transcriptomes are analysed. It analyses the complete set of RNA transcripts in a cell or tissue (the transcriptome) and provides quantitative information about gene expression levels and alternative splicing (Wang et al. 2009).

Targeted sequencing or resequencing is a method for only sequencing part of a whole genome or regions of interest without sequencing the entire genome of a sample. In order to only focus on specific or clinically relevant regions of a genome or DNA sample, it requires a pre-sequencing DNA preparation step called Target Enrichment where target DNA sequences are either directly amplified (amplicon or multiplex PCR-based) or captured (hybrid capture-based) and then subsequently sequenced using DNA sequencers (Mamanova et al. 2010). Despite its success, Sanger sequencing's high cost and time-consuming nature limited its application in large-scale genome projects (Annapureddy et al. 2024).

To overcome these limitations, researchers in the late twentieth and early twenty-first centuries focused on developing faster, more cost-effective, and highly efficient sequencing methodologies. These efforts culminated in the advent of NGS, a transformative technology that simplified the sequencing process by eliminating the need for DNA cloning (Kyriakidou et al. 2018). NGS introduced massively parallel sequencing, enabling millions, or billions of DNA molecules to be sequenced simultaneously (Kumar et al. 2024). The rapid advancement of NGS technologies, complemented by progress in bioinformatics, has revolutionised genetics and biology (Lelieveld et al. 2016). These developments have allowed researchers to generate draft genome sequences of diverse organisms with unprecedented efficiency, reducing sequencing costs by more than 1000-fold since NGS's introduction (Deepa et al. 2020). NGS is now categorised into second-generation and third-generation sequencing technologies, differentiated by factors such as sequencing chemistries, read lengths, and polymerase chain reaction (PCR) requirements (Browne and Freeman 2024). Figure 12.1 illustrates the technological milestones in sequencing, from the pioneering first-generation methods to the advancements in traits improvement during the post-NGS era.

12.2.1.1 Second-Generation Sequencing Platforms

Second-generation sequencing technologies revolutionised genomic research by offering high parallelism, greater sequencing yield, simpler operations, and significantly reduced cost per read compared to first-generation methods. These platforms primarily perform short-read sequencing, producing DNA fragments of 75–400 base pairs through techniques such as sequencing by hybridisation (SBH), sequencing by synthesis (SBS), and pyrosequencing (Browne and Freeman 2024). Unlike first-generation methods, all second-generation technologies amplify DNA prior to sequencing to enhance signal strength, enabling the parallel sequencing of millions of DNA fragments. The sequencing process generally involves three main steps:

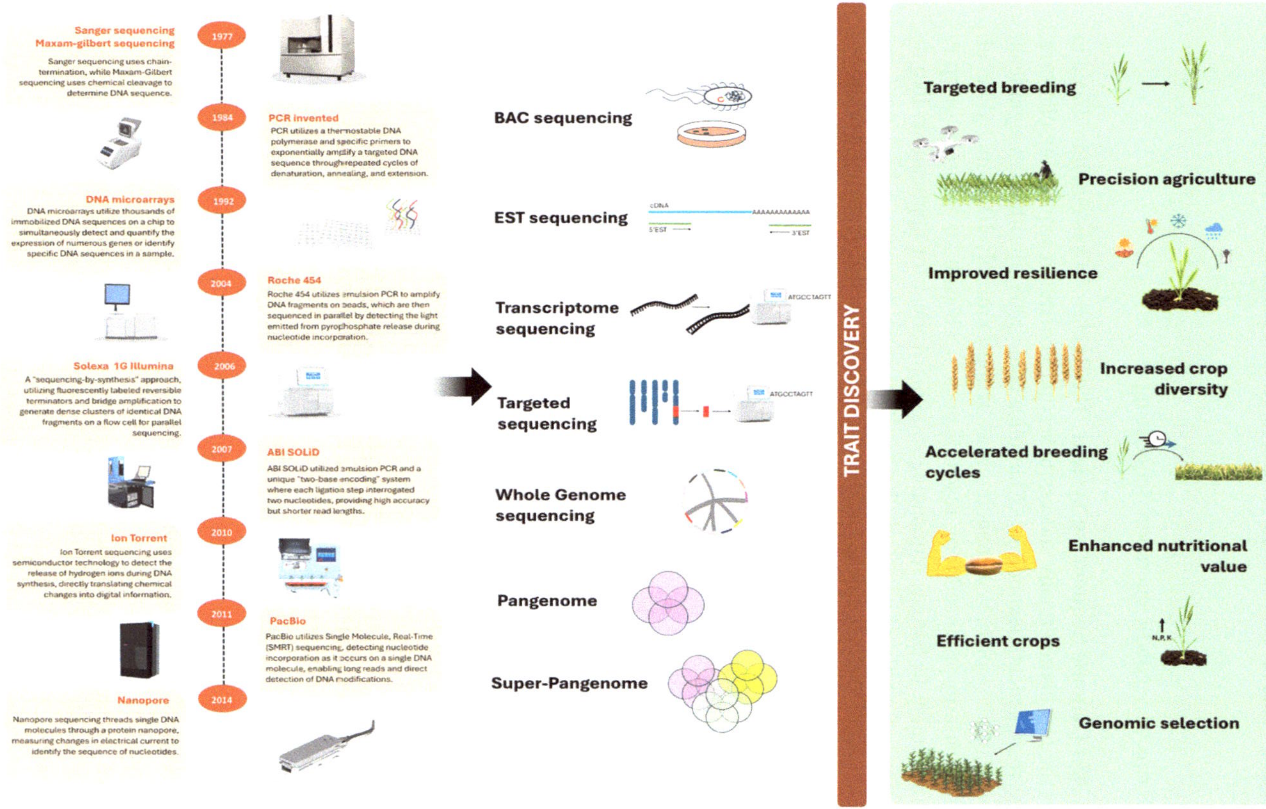

Fig. 12.1 Technological advancements in SNP discovery, genotyping, and trait discovery in the post-NGS era. The illustration highlights the technological advancements in SNP discovery and genotyping, emphasising their vital role in trait discovery during the post-NGS era. The panel on the left illustrates the evolution of sequencing technologies, the middle panel highlights the diverse resources employed for trait discovery, and the panel on the right depicts the impact of trait discovery on crop improvement in the post-NGS landscape

library preparation, amplification, and imaging/sequencing (Bansal et al. 2018; Shariat and Neysi 2022). Prominent platforms in this category include Roche 454, Illumina Solexa, Ion Torrent/Proton, and ABI SOLiD systems (Soni and Singh 2022).

Key Platforms of Second-Generation Sequencing

A. Roche 454 Genome Sequencing System The Roche 454 system, introduced in 2005, was the first commercially successful second-generation sequencing platform, utilising pyrosequencing technology (Margulies et al. 2005). It offered a significant speed advantage, completing sequencing runs in approximately 10 hours (Aly et al. 2015). The platform produced long reads of up to 1 kb with an accuracy of up to 99.9%, comparable to Sanger sequencing, making it ideal for applications like de novo genome assembly (Liao et al. 2019). Its unique approach, which does not require fluorescent labels or nucleic acid probes, streamlined sequencing workflows. However, limitations such as insertion and deletion errors during sequencing have reduced its utility in certain applications (Liang et al. 2017).

B. Illumina Sequencing Illumina, Inc., is the most widely used NGS platform, dominating the industry with multiple sequencing instruments (Kanzi et al. 2020). This technology is based on sequencing by synthesis (SBS) with reversible dye terminators, initially developed by Solexa before being acquired by Illumina in 2007 (Wang 2021). Illumina's ability to detect nucleotide incorporation during DNA strand synthesis allows highly accurate paired-end sequencing with an average error rate of 0.1% per nucleotide (Hu et al. 2021). Despite its exceptional accuracy, shorter read lengths and substitution errors pose challenges for de novo genome assembly (Tyagi et al. 2022).

C. ABI/SOLiD Sequencing Applied Biosystems introduced the SOLiD platform in 2007, employing supported oligonucleotide ligation and detection (SOLiD) technology (Wang 2021). This system relies on the ligation and synthesis of fluorescently labelled oligonucleotides for high-throughput sequencing (Li and Xu 2019). The SOLiD platform boasts a high accuracy rate, with an error rate as low as 0.06% (Peterson et al. 2010). However, its limitations include short read lengths (maximum 75 bp), long run times, and susceptibility to decoding errors, making it less suitable for applications like de novo genome assembly (Liao et al. 2019).

D. Ion Torrent Sequencing Developed by the creators of the 454 system, Ion Torrent technology introduced a novel sequencing method that detects hydrogen ions released during nucleotide incorporation, bypassing the need for fluorescence-based detection (Rothberg et al. 2011). This approach leverages ion semiconductor chips to identify chemical changes, enabling fast sequencing runs (2–4 hours) with read lengths between 200 and 600 bp (Merriman et al. 2012). Although Ion Torrent sequencing offers rapid results, it struggles with homopolymer stretches and repeat regions, leading to insertion and deletion errors despite minimal substitution errors (Brozynska et al. 2014).

These platforms enabled whole-genome sequencing, transcriptome analysis, and targeted sequencing. These advancements have accelerated the identification of genetic variations, including SNPs and structural variants (SVs), which are essential for plant adaptation, stress tolerance, and yield improvement. They also support marker-assisted selection, genomic prediction, and the study of gene regulation, epigenetic modifications, and plant–microbe interactions under stress, advancing crop improvement. While the limitations of individual platforms, such as shorter read lengths or sequencing biases, remain, the cumulative impact of second-generation sequencing has revolutionised plant research. These advancements have laid a robust foundation for the integration of third-generation sequencing technologies, which promise to address current challenges by offering longer reads, improved resolution, and more comprehensive genome assemblies.

12.2.1.2 Third-Generation Sequencing Platforms

Second-generation sequencing platforms, though revolutionary, rely on PCR amplification of DNA templates, which can introduce errors and prolong run times (Pervez et al. 2022). The advent of third-generation sequencing technologies has addressed these limitations by providing longer read lengths, faster sequencing, and improved accuracy while reducing costs and operational complexity (Hu et al. 2021). These platforms enable direct sequencing of nucleotides from biological samples without requiring PCR amplification, thereby streamlining workflows and minimising errors associated with amplification steps (Athanasopoulou et al. 2021). Third-generation sequencing has found extensive applications in genome sequencing, structural variation detection, transcriptome profiling, and methylation studies (Levy and Myers 2016). The prominent platforms are PacBio single-molecule real-time (SMRT) sequencing and Oxford Nanopore sequencing, each with unique strengths and limitations.

A. PacBio SMRT Sequencing Developed by Pacific Biosciences, PacBio SMRT Technology provides significantly longer read lengths (up to 20 kb or more) than second-generation methods, making it highly effective for resolving complex genomic regions, transcriptome analysis, and epigenetic research (Rhoads and Au 2015). While the system offers an average read length of approximately 1300 bp and a runtime of less than 1 day (Wong et al. 2019), its lower throughput compared to second-generation platforms is a notable limitation (Hu et al. 2021).

B. Oxford Nanopore Sequencing Oxford Nanopore Technologies introduced the MinION in 2014, a portable sequencing device capable of generating ultra-long reads exceeding 1 Mb, with typical read lengths ranging from 10–100 kb (Akacin et al. 2022). This platform sequences DNA by measuring disruptions in an electric current as DNA strands pass through nanopores, enabling rapid sequencing with minimal sample preparation (Lee et al. 2016). Early iterations of the technology exhibited raw read accuracies of ~85%, but recent advancements have raised accuracy to over 95% (https://nanoporetech.com/news/news-new-research-algorithms-yield-accuracy-gainsnanopore-sequencing). Despite these improvements, challenges such as high single-base error rates and random errors persist, impacting the reliability of individual reads.

In summary, third-generation sequencing technologies have significantly advanced the field of genomics by overcoming critical limitations of second-generation platforms. Their ability to produce long reads and sequence nucleotides directly positions them as invaluable tools for addressing complex biological questions and accelerating genomic research.

12.2.1.3 Applications of NGS Platforms in Plant Genomics

NGS platforms have been pivotal in sequencing complex plant genomes, starting with the sequencing of the common wheat (*Triticum aestivum*) genome for the landrace 'Chinese Spring' using Roche 454 pyrosequencing platforms (GS FLX Titanium and GS FLX1), achieving ~5× genome coverage. Progenitors such as *Triticum monococcum* (A-genome donor of bread wheat) were sequenced using Illumina methods, while *Aegilops tauschii* (D-genome donor) utilised Roche 454, Illumina Solexa, and ABI SOLiD platforms (N'Diaye et al. 2023). Advancements in NGS have enabled the assembly of polyploid genomes using PacBio long reads, as demonstrated in allotetraploid species like *Chenopodium quinoa* (quinoa) (Jarvis et al. 2017), *Eleusine coracana* (finger millet) (Hatakeyama et al. 2018), and *Coffea arabica* (Arabica coffee) (Cheng et al. 2017). Similarly, nanopore sequencing has been applied to crops such as *Gossypium* spp. (cotton) (Wang et al. 2022), *Solanum tuberosum* (potato) (Hoopes et al. 2022), *Citrus* spp. (citrus) (Huang et al. 2023), and *Arabidopsis thaliana* (Lian et al. 2024).

The utility of NGS for comparative genomics is exemplified by Zhang et al. (2021), who contrasted conventional DNA markers, such as DNA barcodes, with rice-specific nuclear and chloroplast genome barcodes. Analysing Illumina whole-genome sequence reads of Dongxiang wild rice (DXWR) and Nipponbare revealed 2536 SVs in Nipponbare, primarily deletions, linked to 731 QTLs associated with yield, anatomy, and quality traits. Domestication of Nipponbare also introduced genes linked to traits like plant height, spikelet number, panicle length, and biomass yield (Nautiyal et al. 2021). In soybean, WGS of parental lines Guizao 1 and B13 using the Illumina HiSeq X Ten platform provided 8× coverage, enabling the identification of SVs associated with agronomic traits, including lodging resistance (Chen et al. 2024).

12.2.2 *High-Resolution Genotyping*

High-resolution genotyping technologies have transformed trait discovery in crops, offering unprecedented precision and depth in genomic analysis. The widespread integration of NGS-based assays into population genotyping has facilitated high-density linkage mapping across diverse crop species, enabling more efficient identification of genetic markers associated with important traits (Varshney et al. 2019). By leveraging these advanced tools, researchers can dissect complex genetic architectures with greater accuracy, paving the way for refined trait discovery and improved crop breeding outcomes. Each method within high-resolution genotyping brings unique strengths to this endeavour, collectively enhancing our ability to understand and manipulate crop genomes.

12.2.2.1 Genotyping-by-Sequencing (GBS)

Genotyping-by-Sequencing (GBS) is a widely adopted, high-throughput sequencing method that has proven invaluable in identifying SNPs and small InDel variations, particularly in crops with complex genomes, such as wheat and barley. This technology has become integral to GWAS and MAS, offering a powerful approach for precision breeding (Bajgain et al. 2016). For instance, a recent GWAS in sorghum utilising GBS-derived SNPs uncovered resistance genes linked to grain mould fungal disease resistance (Nida et al. 2019). In addition, GBS-based SNPs have been associated with chilling tolerance QTLs, contributing valuable insights to sorghum breeding efforts (Hao et al. 2021). GBS has also been instrumental in mapping resistance genes, such as the Hessian fly resistance genes *h4* and *H7*, to chromosomes 1A and 6A in wheat, explaining significant portions of the phenotypic variation (Liu et al. 2020).

Furthermore, a GWAS involving 102,147 SNP markers from 439 elite spring wheat breeding lines identified a QTL for Fusarium head blight (FHB) resistance on chromosome arm 1AL, which accounted for 5.3% of the phenotypic variation (Liu et al. 2019). In another study, 14,063 polymorphic GBS-SNP markers led to the identification of eight QTLs associated with spot blotch disease resistance (Tomar et al. 2021). Similarly, GBS has been employed in rice breeding, where 73,147 novel SNP markers were discovered, providing a valuable resource for identifying QTLs for grain yield and flowering time (Spindel et al. 2015). The diverse applications and approaches within GBS have been extensively reviewed, further emphasising its pivotal role in advancing crop genomics and improving breeding programmes (Krishna et al. 2023).

12.2.2.2 SNP Arrays

NGS technologies generate vast amounts of sequencing data, which have paved the way for the development of chip-based marker platforms, such as SNP arrays, for high-throughput genotyping. Unlike NGS and PCR-based markers, SNP arrays offer enhanced flexibility in customising sample sizes and selecting data points. This flexibility, combined with high-density scanning capabilities and robust call rates, makes SNP arrays a powerful tool for precise and efficient genotyping (Song et al. 2023).

In wheat, several SNP arrays have been developed with varying densities, such as the 15K (Chen et al. 2021), 35K (Mu et al. 2019), 55K (Xiong et al. 2021), 90K (Jia et al. 2018), 660K (Fan et al. 2018), and 820K (Winfield et al. 2016) arrays, each tailored for specific applications. These arrays have become integral to wheat improvement programmes, particularly in the breeding of rust disease-resistant varieties (Geethanjali et al. 2024). Similarly, in rice, SNP arrays like RiceSNP6K (Yu et al. 2014), RiceSNP50K (Chen et al. 2014), and HDRA700K (McCouch et al. 2016) have been developed on the Illumina Infinium platform, significantly advancing rice genomics and facilitating trait improvement.

In cotton, the Cotton SNP 63K SNP array was used in a GWAS involving 719 diverse upland cotton accessions. The study, which included multiple environmental tests for fibre quality traits, identified 10,511 polymorphic SNPs and 46 significant SNPs associated with five key fibre quality traits (Sun et al. 2017). Similarly, in chickpea, the Axiom® CicerSNP Array was employed in two recombinant inbred line (RIL) populations to identify numerous QTLs related to drought tolerance. These QTLs were linked to five groups of traits, including root traits, morphological traits, phenological traits, yield, and yield-related traits (Roorkiwal et al. 2018). Overall, SNP arrays have proven to be an invaluable tool in crop genomics, enabling precise trait identification and aiding in the development of improved crop varieties.

12.2.2.3 Single Primer Enrichment Technology

Single primer enrichment technology (SPET) is a targeted genotyping method that differs from GBS by simplifying panel design. Unlike GBS, which requires primer pairs for amplification, SPET uses a single primer to sequence regions flanking the target loci, which streamlines the process and enhances multiplexing capabilities compared to PCR-based genotyping (Barchi et al. 2019). This simplification makes SPET an efficient tool for large-scale genotyping studies. As shown by Scaglione et al. (2019), SPET is highly effective for genotyping numerous loci, making it particularly valuable for high-throughput genotyping across diverse species. In plants, SPET has been successfully applied to a variety of species, including maize (*Zea mays* L.), black poplar (*Populus nigra* L.) (Scaglione et al. 2019), oil palm (*Elaeis guineensis* Jacq.) (Herrero et al. 2020), cultivated and wild species of tomato and eggplant (*Solanum* spp.) (Barchi et al. 2019), and peach (*Prunus persica* L.) (Baccichet et al. 2022). These studies demonstrate the versatility of SPET in genotyping germplasm collections and breeding populations. Its applications are diverse, ranging from population structure analysis and phylogenetic studies to high-density linkage map construction and association mapping, further highlighting its potential as a powerful tool in modern plant genomics.

12.2.2.4 Restriction-Site Associated DNA Sequencing (RADSeq)

Restriction-site associated DNA sequencing (RADSeq) is considered one of the most significant breakthroughs in genomics over the past decade (Andrews et al.

2016). This method, along with its various adaptations, has become a key tool in genomics-assisted breeding (GAB), particularly in GWAS and GS (Yang et al. 2012; Glaubitz et al. 2014). RADSeq has revolutionised the identification of large numbers of SNPs, enabling the discovery of thousands of variants with applications in crop improvement. For instance, Hsu et al. (2022) used RADSeq to sequence 31 accessions of Taiwanese peanut germplasm, identifying 17,610 SNPs. Similarly, Wang et al. (2017) applied RADSeq to develop a high-density genetic map for foxtail millet, detecting over 2.6 million SNP loci from the reference genome and genotyping 124 F_2 progenies with 9968 SNP markers. These examples illustrate how RADSeq enhances the precision and efficiency of genomic studies, facilitating the discovery of important genetic traits and boosting breeding programmes across diverse species.

12.2.2.5 Whole-Genome Sequencing (WGS)

Advancements in NGS technologies and the reduction in sequencing costs have made WGS a practical tool for population-scale genomic research (Park and Kim 2016). WGS allows for the identification of genetic variants and the creation of comprehensive genome assemblies, providing valuable insights into various biological processes. These high-throughput sequencing data can be utilised in genomics-assisted breeding, leveraging diverse variant types, including SNPs, InDels, and larger SVs. The combination of these variants into variant graphs, or pan-genomes, enhances the understanding of genetic diversity (Jensen et al. 2020). The increased density of variants obtained through WGS also improves the ability to identify causal variants, distinguishing them from variants that are merely in linkage disequilibrium (LD) with causal variants.

WGS enhances the power of large association panels, similar to how GBS improved SSR markers, greatly advancing GWAS and genomic selection in various crops. For instance, Boatwright et al. (2022) conducted WGS on 400 sorghum accessions from the Sorghum association panel (SAP), achieving an average coverage of 38×. This generated a high-density marker set consisting of 43,983,694 variants, including ~38 million SNPs, ~five million InDels, and ~ 170,000 CNVs. Notably, this marker set revealed novel genomic associations for plant height and tannin content, which were not detectable using lower-density markers. Furthermore, WGS uncovered variants in 5-kb bins that were absent in GBS data, covering half of the genome. In a similar study, WGS of 355 accessions yielded 799.58 billion 100-bp paired-end reads, with an average coverage of 14.52× per accession, identifying 76,874,471 high-quality SNPs and 5,208,800 Indels. These SNPs were validated using the wheat 660K SNP array, showing 99.74% concordance, demonstrating that the resequencing data are not only abundant but also of high quality (Niu et al. 2023).

12.2.3 From Reference Genomes to Pangenomes

The availability of reference genomes for most modern crops has significantly advanced our ability to conduct genome-wide surveys of SNPs and marker–trait association studies, linking genetic variations to phenotypic traits (Tao et al. 2019). However, our understanding of genetic variation has been limited by the reliance on resequencing approaches based on a single reference genome for a given species. These methods often fail to capture highly polymorphic regions or regions absent from the reference genome, thereby hindering the identification of SVs such as copy number variants (CNVs) and presence/absence variants (PAVs) (Varshney et al. 2017). Mounting evidence suggests that SVs play crucial roles in determining agronomic traits and are abundant in crops (Zhao et al. 2018).

To address these limitations, a more comprehensive approach is needed to capture the full spectrum of genetic variation within a species. One promising approach is pan genomics, which aims to represent the complete genetic diversity of a species (Tettelin et al. 2005). The concept of the pangenome, originally introduced in bacterial genomics, encompasses the entire genetic makeup of a species, including the core genome, genes shared across nearly all individuals, and the accessory genome, which consists of genes variably present among different individuals or strains (Schreiber et al. 2024). Recent advancements in sequencing technologies, including short-read Illumina sequencing, long-read PacBio sequencing, and Nanopore sequencing, as well as improved bioinformatics tools, have enabled the creation of high-quality genome assemblies for multiple accessions or varieties of a single species (Naithani et al. 2023). These developments have ushered crop genomics into the pan-genomic era, providing a more comprehensive understanding of genetic diversity.

The first step in developing a pangenome infrastructure involves selecting a diverse panel of genotypes that maximally represent genetic variation with minimal overlap. This process requires careful consideration of both cultivated varieties and their wild relatives, which harbour significant genetic diversity and are often genetically distant from domesticated crops (Yan et al. 2023). Traditional resequencing approaches often fail to capture the full genomic landscape of wild crop relatives (CWRs). However, pan-genomic studies systematically characterise the genomic content of CWRs, uncovering unique SVs and SNPs that can be exploited for crop improvement (Supriya et al. 2024). In the past decade, several studies using de novo assembly and resequencing approaches have highlighted the genomic diversity of CWRs. For instance, resequencing 14 cultivated and 17 wild soybean accessions revealed greater allelic diversity in wild accessions and identified regions of high linkage disequilibrium, shedding light on domestication and evolution (Lam et al. 2010). Similarly, resequencing 75 maize lines, including wild relatives, landraces, and improved varieties, provided valuable insights into the evolution of modern maize (Hufford et al. 2012).

The variations identified through pan-genomic analyses can serve as markers for marker-assisted selection, facilitating the incorporation of desirable traits from

CWRs into domesticated cultivars. A growing body of pan-genomic studies in major crops and related species, including *Solanum tuberosum* (Hoopes et al. 2022), *Gossypium* spp. (Wang et al. 2022), *Solanum* spp. (Li et al. 2023), *Capsicum* spp. (Liu et al. 2023), *Arabidopsis thaliana* (Lian et al. 2024), *Avena sativa* (Avni et al. 2024), and *Cicer arietinum* L. (Khan et al. 2024), have revealed new insights into crop diversity and the improvement of agronomically important traits.

12.3 Trait Discovery Frameworks in the Post-NGS Era

12.3.1 QTL Mapping and Meta-QTL Analysis in the Post-NGS Era

The post-NGS era has ushered in significant advancements in the mapping of QTLs, particularly through the use of biparental and multi-parental populations (Pascual et al. 2016). These innovations have enhanced our understanding of the genetic architecture underlying complex traits, such as yield and stress tolerance in crops, facilitating the identification of key genetic factors that govern these traits. QTL mapping is a statistical approach used to link molecular markers to quantitative traits by analysing populations derived from crosses between genetically diverse parents. This process allows researchers to identify associations between phenotypic traits and genotypic data (Wu et al. 2007). The use of high-density molecular markers, particularly SNPs, has increased the resolution of QTL mapping, enabling more precise localisation of QTLs associated with specific traits (Zou et al. 2012). In biparental populations, researchers typically rely on linkage maps constructed from markers segregating in the progeny. This approach has proven effective for identifying major QTLs that account for significant phenotypic variations (Pascual et al. 2016). However, biparental mapping has limitations, particularly its ability to capture the full breadth of genetic diversity within a species. To overcome this, multi-parental populations, such as those created through nested association mapping (NAM) or multiparental advanced generation intercross (MAGIC) designs, offer notable advantages. By incorporating genetic variation from multiple parents, these populations increase allelic diversity, thus enhancing the power of QTL detection and providing deeper insights into allele-environment interactions (Scott et al. 2020).

Meta-QTL analysis represents an advanced method for synthesising QTL data from multiple studies to refine QTL locations and estimate their effects across different environments and genetic backgrounds. This approach aims to identify stable, major, and reliable MQTLs with reduced confidence intervals, improving the accuracy of genetic mapping (Kumar et al. 2023). Meta-QTL analysis has been successfully applied to a variety of traits in several crops such as rice (Shashiprabha et al. 2022), wheat (Zheng et al. 2021; Pal et al. 2022; Saini et al. 2022; Kumar et al. 2023), and maize (Rossi et al. 2019). By consolidating QTL data from various studies, this method helps identify consensus regions that remain stable across different genetic backgrounds, providing valuable insights into the genetic basis of complex traits. For instance, a comprehensive meta-QTL analysis by Kumar et al. (2023) examining stripe rust resistance in bread wheat identified 505 QTLs spread across all chromosomes. These QTLs were grouped into 67 MQTLs, which were further refined to 29 high-confidence MQTLs, significantly reducing the CIs of the original QTLs. The CIs of the identified MQTLs ranged from 0 to 11.68 cM, with an average CI of 1.97 cM, demonstrating the power of meta-analysis in refining QTL locations and enhancing the precision of MAS strategies (Soriano et al. 2021). Moreover, MQTL analysis allows for the identification of pleiotropic effects, where a single genomic region influences multiple traits. This feature is particularly valuable in breeding programmes aiming to simultaneously improve complex traits, such as yield and stress tolerance.

12.3.1.1 Pan-Genome-Based QTL Mapping

The advent of NGS technologies has revolutionised the field of genetics, particularly in QTL mapping. One of the most notable advancements is the shift from reference-centric approaches to pan-genome-based methodologies (Nautiyal et al. 2021). This shift enables researchers to capitalise on the genetic diversity found in multiple genomes, thus enhancing the resolution and accuracy of QTL mapping. Recent studies have demonstrated the significant advantages of pan-genome-based QTL mapping. For instance, a study developed a robust workflow for integrating QTL data into pan-genomes, enabling the extraction and analysis of homologous regions across various genomes. This method not only aided in identifying causal genes linked to specific traits but also provided a deeper understanding of SVs that may influence these traits (Li et al. 2022). By leveraging pan-genomes, researchers can gain a more holistic view of genetic variation, considering multiple genomes simultaneously rather than relying on a single reference genome. This approach is particularly important as it allows for the identification of QTLs that may be present in some genotypes but absent in others, thereby capturing a broader spectrum of genetic diversity (Della Coletta et al. 2021). For instance, in rice, SVs were found to be widespread even among closely related genomes, highlighting the importance of accounting for these variations in QTL mapping (Qin et al. 2021).

The integration of advanced computational tools has been instrumental in advancing pan-genome-based QTL mapping. Tools such as PanTools facilitate the annotation of pan-genomes with known QTLs, allowing for the detection of homologous regions across different genomes (Li et al. 2022). The workflow typically involves annotating QTL data within the pan-genome, identifying homologous genes, and constructing an overview of these regions and their relationships. This comprehensive approach enhances the ability to identify and validate QTLs linked to specific traits. Additionally, algorithms like PanGenie have been developed to efficiently genotype variants using haplotype-resolved pan-genomes, significantly improving the speed and accuracy of QTL analyses (Kaur et al. 2024). By bypassing traditional read-mapping techniques, PanGenie can rapidly process short-read sequencing data to identify a broad range of genetic variations, including those that are often overlooked in conventional methods (Li et al. 2022).

The application of pan-genome-based QTL mapping holds immense promise for trait discovery in crops. By harnessing diverse genetic resources and advanced computational tools, researchers can uncover novel alleles associated with key agronomic traits. This approach not only enhances our understanding of complex traits but also supports breeding programmes aimed at developing resilient crop varieties that can thrive under changing environmental conditions (Li et al. 2022). These methodologies demonstrate how pan-genomic approaches can streamline the identification of beneficial traits and accelerate the breeding of improved crop varieties.

12.3.2 Genome-Wide Association Studies in the Post-NGS Era

Genome-wide association studies (GWAS) have advanced significantly in the post-NGS era, particularly in methods for analysing SNPs and haplotypes. These improvements have enhanced our understanding of genetic associations with complex traits, facilitating the identification of genetic loci underlying traits and strengthening the relationship between genetic markers and phenotypic characteristics across diverse crop varieties (Saini et al. 2022). In plant breeding, missing heritability is less of an issue due to the identification of genetic variants that clearly contribute to trait variation. The effectiveness of GWAS depends on the proportion of phenotypic variance explained by each SNP, which is influenced by the allelic effect size and its prevalence (Korte and Farlow 2013). However, rare variants and small effect sizes still pose challenges (Asimit and Zeggini 2010).

Recent genotyping methods like GBS, RADseq, WGS, and SNP arrays have revolutionised GWAS, enabling large-scale genotyping of germplasm collections (Zargar et al. 2015; Elshire et al. 2011; Poland et al. 2012). Despite these advances, challenges remain, including the biallelic nature of SNPs, rare alleles, and linkage drag (Voss-Fels and Snowdon 2016). Often, GWAS identifies genomic loci associated with traits through linkage disequilibrium (LD) with nearby genes or regulatory elements, rather than directly pinpointing causative loci. To improve precision, incorporating haplotypes into GWAS is a promising approach.

Haplotypes, sets of alleles from different polymorphisms inherited together, offer a more comprehensive approach than individual SNPs in GWAS (Garg 2021). Haplotype-based GWAS analyses combinations of alleles across multiple loci, capturing interactions between variants that may have a stronger impact on phenotypes. Haplotype blocks are typically multi-allelic, providing more information than biallelic SNPs, and are more frequent, reflecting ongoing recombination and mutations within associated genes. This approach reduces false positives and offers deeper insights into causal haplotypes, with studies showing it enhances allelic effects, increasing phenotypic variation explained by 34–50% compared to single-SNP approaches (Hamblin and Jannink 2011).

12.3.3 Fine Mapping and QTL Cloning

Fine mapping and QTL cloning are crucial for identifying and introgressing beneficial alleles into crop species. DNA markers linked to QTL regions enable rapid and accurate selections, but the QTL regions identified through traditional mapping often span large genetic intervals, typically several centimorgans (cM) or megabases (Mb) on physical maps. These broad regions may contain multiple genes, making it difficult to pinpoint the specific locus responsible for a trait. Moreover, introgressing large QTL regions using flanking markers may inadvertently carry undesirable genes, a phenomenon known as linkage drag, which can harm the performance of improved cultivars (Varshney et al. 2014).

To overcome these challenges, the resolution of mapping techniques must be enhanced to precisely place QTLs within the smallest possible genomic region. This process, known as fine mapping, has been greatly accelerated by recent advancements in NGS technologies, such as reduced representation libraries (RRLs), restriction-site-associated DNA sequencing (RAD), GBS, whole-genome resequencing (WGRS), and skim-GBS. These technologies allow researchers to identify and map a vast number of SNPs across large sample sets in a single analysis (Varshney et al. 2019). Fine mapping has become more efficient due to several factors, including (i) the availability of reference genomes for most crop species, (ii) the development of high-resolution mapping populations like MAGIC and NAM, (iii) the ability to generate higher-density genetic maps, (iv) the use of QTL-seq and RenSeq techniques for direct gene identification, (v) gene annotation data that help shortlist candidate genes, and (vi) the application of mutant populations and fast gene editing tools for functional validation. These advances have reduced the time required for fine mapping, with some projects now completed in as little as 10–20 months, compared to the 10–15 years previously needed (Deepa et al. 2020).

High-density SNP mapping is a major driver of the rapid progress in fine mapping. Techniques such as SNP arrays and high-density genotyping via resequencing provide comprehensive marker profiles across large populations, enhancing the resolution of QTLs and facilitating the discovery of causal variants in associated regions (Roorkiwal et al. 2018; Saxena et al. 2018). This has significantly improved the efficiency and precision of QTL cloning, particularly with the advent of NGS technologies, which have revolutionised trait mapping. Despite these advancements, challenges remain in the post-NGS era, particularly with sequencing noise, which can obscure true genetic signals, and the complexities of plant genomes, especially in non-model species. To address these issues, improved statistical models and sophisticated computational approaches are being developed to minimise errors and enhance QTL detection accuracy.

12.4 Bridging Phenomics with Post-NGS Genomics for Trait Discovery

The integration of advanced phenotyping techniques with post-NGS genomics is essential for improving the precision and efficiency of trait mapping. This synergy is particularly crucial for developing crops capable of adapting to changing environmental conditions and meeting the growing global demand for food (Xiao et al. 2022). High-throughput phenotyping (HTP) employs cutting-edge technologies such as drones and automated imaging systems to collect large-scale phenotypic data quickly. This enables the assessment of various traits, including growth rate, biomass, and stress responses, across extensive plant populations (Gill et al. 2022). When combined with genomic data, HTP provides a wealth of valuable insights that facilitate dynamic, real-time phenotypic data, which can be directly linked to genotypic information, enhancing the understanding of genotype–phenotype relationships and enabling more precise identification of trait associations (Montesinos-López et al. 2024). Furthermore, by merging HTP data with genomic information, researchers can more accurately identify QTLs linked to specific traits, improving trait mapping and leading to more focused and effective breeding strategies.

Genomic selection (GS) leverages genomic data to predict the performance of breeding lines based on their genetic makeup. Integrating phenotypic information particularly high-throughput phenomics and environmental data in to GS models further enhance prediction accuracy. This significantly enhances the prediction accuracy of plant performance across diverse environments, making breeding decisions more reliable (Montesinos-López et al. 2024). Additionally, by predicting traits early in the breeding cycle, phenotypic data enables breeders to select individuals with desirable characteristics before maturity, accelerating the breeding process and shortening the time required to develop superior cultivars (Srivastava et al. 2024). Understanding the environmental factors that influence crop growth is vital for successful trait discovery, and the integration of environmental data, or enviromics, provides critical insights into genotype performance under various conditions. The inclusion of environmental data addresses genotype-by-environment interactions, improving the reliability and adaptability of trait prediction models as environmental conditions change (Crossa et al. 2021). Moreover, by integrating genomic, phenomics, and environmental information, along with historical weather data and crop growth models, researchers can develop predictive models that estimate how different cultivars will perform under future climate scenarios, providing breeders with actionable insights for long-term crop improvement (Crossa et al. 2021).

Machine learning (ML) techniques are increasingly applied to analyse the vast datasets generated from integrated genomic and phenomics studies, offering powerful tools for enhancing trait discovery. ML algorithms, such as support vector machines and random forests, are used to classify accessions based on phenomics traits, aiding in the identification of hybrids with desirable characteristics. These methods allow breeders to make data-driven decisions from complex datasets

(Rodriguez et al. 2022). Additionally, ML facilitates the identification of functional traits linked to specific genetic markers, deepening the understanding of how these traits contribute to overall plant performance and providing valuable information for the development of high-performing cultivars (Volk et al. 2021). In summary, the integration of phenomics and post-NGS genomics, along with machine learning and environmental data, is transforming the landscape of trait discovery. This comprehensive approach not only improves the efficiency of breeding programmes but also ensures the development of crops that are resilient to future environmental challenges.

12.5 Conclusion and Future Prospects

The post-NGS era has significantly transformed the landscape of trait discovery, greatly enhancing the precision of gene mapping and cloning for important agronomic traits. Key factors that have streamlined trait mapping across various crops include (i) the availability of reference genomes for most crops, (ii) the development of high-resolution mapping populations, such as nested-association mapping (NAM) and Multi-parental Advanced Generation Inter-Cross (MAGIC) populations, (iii) high-density and ultra-high-density linkage maps, and (iv) gene annotation, in silico expression databases, and gene-regulatory network databases to prioritise high-confidence candidate genes. These advancements have contributed to a better understanding of trait architectures, which is crucial for advancing crop improvement initiatives aimed at addressing global food security challenges, particularly under changing climate conditions.

Looking ahead, further advancements in various approaches, such as single-cell genomics, synthetic biology, systems biology, and AI-driven tools, have the potential to significantly enhance trait discovery. Single-cell genomics allows for high-resolution analysis of cellular diversity and responses within plant tissues (Depuydt et al. 2023). By integrating multi-omics approaches at the single-cell level, key genes and their associated molecular mechanisms underlying complex traits can be identified, leading to more precise trait mapping. Synthetic biology, which enables the construction of new biological systems and the introduction of multiple genes in a single event (Joshi and Hanson 2024), offers an accelerated pathway for discovering traits that contribute to higher grain yield, stress tolerance, and improved nutritional quality. Systems biology, by combining large-scale data from genomics, transcriptomics, proteomics, and metabolomics, facilitates a comprehensive understanding of how genes, RNA transcripts, proteins, and metabolites interact within a plant system. This approach is essential for accurately identifying genetic factors that control specific traits in crops, thereby driving improved trait discovery in crop improvement programmes (Saini et al. 2024).

Furthermore, the application of AI and machine learning technologies to analyse large datasets, including genomic, phenotypic, and environmental information, holds promise for identifying correlations and developing predictive models for desirable traits. These innovations will not only improve trait discovery but also accelerate breeding decisions, making crop improvement more efficient and timelier (Shaikh et al. 2022). In conclusion, the integration of cutting-edge technologies such as single-cell genomics, synthetic biology, systems biology, and AI will further propel the field of trait discovery, contributing to the development of crops that are more resilient, higher-yielding, and better suited to meet the challenges of future food security.

References

Abdi G, Tarighat MA, Jain M, Tendulkar R, Tendulkar M, Barwant M (2024) Revolutionizing genomics: exploring the potential of next-generation sequencing. In: Advances in bioinformatics. Springer Nature Singapore, pp 1–33. https://doi.org/10.1007/978-981-99-8401-5_1

Akacin I, Ersoy Ş, Doluca O, Güngörmüşler M (2022) Comparing the significance of the utilization of next generation and third generation sequencing technologies in microbial metagenomics. Microbiol Res 264:127154. https://doi.org/10.1016/j.micres.2022.127154

Aly S, Mostafa E, Sabri D (2015) Overviews of "next-generation sequencing". Res Rep Forensic Med Sci 5:1–5. https://doi.org/10.2147/RRFMS.S57998

Andrews KR, Good JM, Miller MR, Luikart G, Hohenlohe PA (2016) Harnessing the power of RADseq for ecological and evolutionary genomics. Nat Rev Genet 17(2):81–92. https://doi.org/10.1038/nrg.2015.28

Annapureddy P, Gupta V, Kumar G (2024) The journey of genome sequencing. Preprints. https://doi.org/10.20944/preprints202408.1719.v1

Asimit J, Zeggini E (2010) Rare variant association analysis methods for complex traits. Annu Rev Genet 44(1):293–308. https://doi.org/10.1146/annurev-genet-102209-163421

Athanasopoulou K, Boti MA, Adamopoulos PG, Skourou PC, Scorilas A (2021) Third-generation sequencing: the spearhead towards the radical transformation of modern genomics. Life 12(1):30. https://doi.org/10.3390/life12010030

Avni R, Kamal N, Bitz L, Jellen E, Bekele W, Angessa T et al (2024) A pangenome and pantranscriptome of hexaploid oat. bioRxiv. https://doi.org/10.1101/2024.10.23.619697

Baccichet I, Chiozzotto R, Scaglione D, Bassi D, Rossini L, Cirilli M (2022) Genetic dissection of fruit maturity date in apricot (P. Armeniaca L.) through a single primer enrichment technology (SPET) approach. BMC Genomics 23(1):712. https://doi.org/10.1186/s12864-022-08901-1

Bajgain P, Rouse MN, Anderson JA (2016) Comparing genotyping-by-sequencing and single nucleotide polymorphism chip genotyping for quantitative trait loci mapping in wheat. Crop Sci 56(1):232–248. https://doi.org/10.2135/cropsci2015.06.0389

Bansal G, Narta K, Teltumbade MR (2018) Next-generation sequencing: technology, advancements, and applications. In: Bioinformatics: sequences, structures, phylogeny, pp 15–46. https://doi.org/10.1007/978-981-13-1562-6_2

Barba M, Czosnek H, Hadidi A (2014) Historical perspective, development and applications of next-generation sequencing in plant virology. Viruses 6(1):106–136. https://doi.org/10.3390/v6010106

Barchi L, Acquadro A, Alonso D, Aprea G, Bassolino L, Demurtas O et al (2019) Single primer enrichment technology (SPET) for high-throughput genotyping in tomato and eggplant germplasm. Front Plant Sci 10:1005. https://doi.org/10.3389/fpls.2019.01005

Boatwright JL, Sapkota S, Jin H, Schnable JC, Brenton Z, Boyles R, Kresovich S (2022) Sorghum association panel whole-genome sequencing establishes cornerstone resource for dissecting genomic diversity. Plant J 111(3):888–904. https://doi.org/10.1111/tpj.15853

Bocianowski J (2024) Using NGS technology and association mapping to identify candidate genes associated with Fusarium stalk rot resistance. Genes 15(1):106. https://doi.org/10.3390/genes15010106

Bohra A, Chand Jha U, Godwin ID, Kumar Varshney R (2020) Genomic interventions for sustainable agriculture. Plant Biotechnol J 18(12):2388–2405. https://doi.org/10.1111/pbi.13472

Browne TN, Freeman M (2024) Next generation sequencing: forensic applications and policy considerations. Wiley Interdiscip Rev Forensic Sci:e1531. https://doi.org/10.1002/wfs2.1531

Brozynska M, Furtado A, Henry RJ (2014) Direct chloroplast sequencing: comparison of sequencing platforms and analysis tools for whole chloroplast barcoding. PLoS One 9(10):e110387. https://doi.org/10.1371/journal.pone.0110387

Chen H, Xie W, He H, Yu H, Chen W, Li J et al (2014) A high-density SNP genotyping array for rice biology and molecular breeding. Mol Plant 7(3):541–553. https://doi.org/10.1093/mp/sst135

Chen S, Zhang Z, Sun Y, Li D, Gao D, Zhan K, Cheng S (2021) Identification of quantitative trait loci for Fusarium head blight (FHB) resistance in the cross between wheat landrace N553 and elite cultivar Yangmai 13. Mol Breed 41(3):24. https://doi.org/10.1007/s11032-021-01220-5

Chen B, Chai C, Duan M, Yang X, Cai Z, Jia J et al (2024) Identification of quantitative trait loci for lodging and related agronomic traits in soybean (Glycine max [L.] Merr.). BMC Genomics 25(1):900. https://doi.org/10.1186/s12864-024-10794-1

Cheng B, Furtado A, Henry RJ (2017) Long-read sequencing of the coffee bean transcriptome reveals the diversity of full-length transcripts. Gigascience 6(11). https://doi.org/10.1093/gigascience/gix086

Crossa J, Fritsche-Neto R, Montesinos-Lopez OA, Costa-Neto G, Dreisigacker S, Montesinos-Lopez A, Bentley AR (2021) The modern plant breeding triangle: optimizing the use of genomics, phenomics, and enviromics data. Front Plant Sci 12:651480. https://doi.org/10.3389/fpls.2021.651480

de Ronne M, Lapierre É, Torkamaneh D (2024) Genetic insights into agronomic and morphological traits of drug-type cannabis revealed by genome-wide association studies. Sci Rep 14(1):9162. https://doi.org/10.1038/s41598-024-58931-w

Deepa J, Abhishek B, Mahendar T, Rajeev KV (2020) Fine mapping and gene cloning in the postNGS era: advances and prospects. Theor Appl Genet 133:1791–1810. https://doi.org/10.1007/s00122-020-03560-w

Della Coletta R, Qiu Y, Ou S, Hufford MB, Hirsch CN (2021) How the pan-genome is changing crop genomics and improvement. Genome Biol 22:1–19. https://doi.org/10.1186/s13059-020-02224-8

Depuydt T, De Rybel B, Vandepoele K (2023) Charting plant gene functions in the multi-omics and single-cell era. Trends Plant Sci 28(3):283–296. https://doi.org/10.1016/j.tplants.2022.09.008

Elshire RJ, Glaubitz JC, Sun Q, Poland JA, Kawamoto K, Buckler ES, Mitchell SE (2011) A robust, simple genotyping-by-sequencing (GBS) approach for high diversity species. PLoS One 6(5):e19379. https://doi.org/10.1371/journal.pone.0019379

Fan X, Zhang W, Zhang N, Chen M, Zheng S, Zhao C et al (2018) Identification of QTL regions for seedling root traits and their effect on nitrogen use efficiency in wheat (Triticum aestivum L.). Theor Appl Genet 131:2677–2698. https://doi.org/10.1007/s00122-018-3183-6

Garg S (2021) Computational methods for chromosome-scale haplotype reconstruction. Genome Biol 22(1):101. https://doi.org/10.1186/s13059-021-02328-9

Geethanjali S, Kadirvel P, Periyannan S (2024) Wheat improvement through advances in single nucleotide polymorphism (SNP) detection and genotyping with a special emphasis on rust resistance. Theor Appl Genet 137(10):224. https://doi.org/10.1007/s00122-024-04730-w

Gill T, Gill SK, Saini DK, Chopra Y, de Koff JP, Sandhu KS (2022) A comprehensive review of high throughput phenotyping and machine learning for plant stress phenotyping. Phenomics 2:156–183. https://doi.org/10.1007/s43657-022-00048-z

Glaubitz JC, Casstevens TM, Lu F, Harriman J, Elshire RJ, Sun Q, Buckler ES (2014) TASSEL-GBS: a high capacity genotyping by sequencing analysis pipeline. PLoS One 9(2):e90346. https://doi.org/10.1371/journal.pone.0090346

Gong Y, Li Y, Liu X, Ma Y, Jiang L (2023) A review of the pangenome: how it affects our understanding of genomic variation, selection and breeding in domestic animals? J Anim Sci Biotechnol 14(1):73. https://doi.org/10.1186/s40104-023-00860-1

Gu H, Liang S, Zhao J (2022) Novel sequencing and genomic technologies revolutionized rice genomic study and breeding. Agronomy 12(1):218. https://doi.org/10.3390/agronomy12010218

Hamblin MT, Jannink JL (2011) Factors affecting the power of haplotype markers in association studies. Plant Genome 4(2). https://doi.org/10.3835/plantgenome2011.03.0008

Hao H, Li Z, Leng C, Lu C, Luo H, Liu Y et al (2021) Sorghum breeding in the genomic era: opportunities and challenges. Theor Appl Genet 134:1899–1924. https://doi.org/10.1007/s00122-021-03789-z

Hatakeyama M, Aluri S, Balachadran MT, Sivarajan SR, Patrignani A, Grüter S et al (2018) Multiple hybrid de novo genome assembly of finger millet, an orphan allotetraploid crop. DNA Res 25(1):39–47. https://doi.org/10.1093/dnares/dsx036

Herrero J, Santika B, Herrán A, Erika P, Sarimana U, Wendra F et al (2020) Construction of a high-density linkage map in oil palm using SPET markers. Sci Rep 10(1):9998. https://doi.org/10.1038/s41598-020-67118-y

Hoopes G, Meng X, Hamilton JP, Achakkagari SR, Guesdes FDAF, Bolger ME et al (2022) Phased, chromosome-scale genome assemblies of tetraploid potato reveal a complex genome, transcriptome, and predicted proteome landscape underpinning genetic diversity. Mol Plant 15(3):520–536. https://doi.org/10.1016/j.molp.2022.01.003

Hsu YM, Wang SS, Tseng YC, Lee SR, Fang H, Hung WC et al (2022) Assessment of genetic diversity and SNP marker development within peanut germplasm in Taiwan by RADseq. Sci Rep 12(1):14495. https://doi.org/10.1038/s41598-022-18737-0, https://doi.org/10.3390/plants10112260

Hu T, Chitnis N, Monos D, Dinh A (2021) Next-generation sequencing technologies: an overview. Hum Immunol 82(11):801–811. https://doi.org/10.1016/j.humimm.2021.02.012

Huang Y, He J, Xu Y, Zheng W, Wang S, Chen P et al (2023) Pangenome analysis provides insight into the evolution of the orange subfamily and a key gene for citric acid accumulation in citrus fruits. Nat Genet 55(11):1964–1975. https://doi.org/10.1038/s41588-023-01516-6

Hufford MB, Xu X, Van Heerwaarden J, Pyhäjärvi T, Chia JM, Cartwright RA et al (2012) Comparative population genomics of maize domestication and improvement. Nat Genet 44(7):808–811. https://doi.org/10.1038/ng.2309

Jarvis DE, Ho YS, Lightfoot DJ, Schmöckel SM, Li B, Borm TJ et al (2017) The genome of Chenopodium quinoa. Nature 542(7641):307–312. https://doi.org/10.1038/nature21370

Jensen SE, Charles JR, Muleta K, Bradbury PJ, Casstevens T, Deshpande SP et al (2020) A sorghum practical haplotype graph facilitates genome-wide imputation and cost-effective genomic prediction. Plant Genome 13(1):e20009. https://doi.org/10.1002/tpg2.20009

Jia A, Ren Y, Gao F, Yin G, Liu J, Guo L et al (2018) Mapping and validation of a new QTL for adult-plant resistance to powdery mildew in Chinese elite bread wheat line Zhou8425B. Theor Appl Genet 131:1063–1071. https://doi.org/10.1007/s00122-018-3058-x

Joshi J, Hanson AD (2024) A pilot oral history of plant synthetic biology. Plant Physiol 195(1):36–47. https://doi.org/10.1093/plphys/kiad585

Kanzi AM, San JE, Chimukangara B, Wilkinson E, Fish M, Ramsuran V, De Oliveira T (2020) Next generation sequencing and bioinformatics analysis of family genetic inheritance. Front Genet 11:544162. https://doi.org/10.3389/fgene.2020.544162

Kaur H, Shannon LM, Samac DA (2024) A stepwise guide for pangenome development in crop plants: an alfalfa (Medicago sativa) case study. BMC Genomics 25(1):1022. https://doi.org/10.1186/s12864-024-10931-w

Khan AW, Garg V, Sun S, Gupta S, Dudchenko O, Roorkiwal M et al (2024) Cicer super-pangenome provides insights into species evolution and agronomic trait loci for crop improvement in chickpea. Nat Genet:1–10. https://doi.org/10.1038/s41588-024-01760-4

Korte A, Farlow A (2013) The advantages and limitations of trait analysis with GWAS: a review. Plant Methods 9:1–9. https://doi.org/10.1186/1746-4811-9-29

Krishna TPA, Veeramuthu D, Maharajan T, Soosaimanickam M (2023) The era of plant breeding: conventional breeding to genomics-assisted breeding for crop improvement. Curr Genomics 24(1):24–35. https://doi.org/10.2174/1389202924666230517115912

Kumar S, Saini DK, Jan F, Jan S, Tahir M, Djalovic I et al (2023) Comprehensive meta-QTL analysis for dissecting the genetic architecture of stripe rust resistance in bread wheat. BMC Genomics 24(1):259. https://doi.org/10.1186/s12864-023-09336-y

Kumar KR, Cowley MJ, Davis RL (2024) Next-generation sequencing and emerging technologies. In: Semin thromb hemost. Thieme Medical Publishers. https://doi.org/10.1055/s-0044-1786397

Kyriakidou M, Tai HH, Anglin NL, Ellis D, Strömvik MV (2018) Current strategies of polyploid plant genome sequence assembly. Front Plant Sci 9:1660. https://doi.org/10.3389/fpls.2018.01660

Lam HM, Xu X, Liu X, Chen W, Yang G, Wong FL et al (2010) Resequencing of 31 wild and cultivated soybean genomes identifies patterns of genetic diversity and selection. Nat Genet 42(12):1053–1059. https://doi.org/10.1038/ng.715

Lee H, Gurtowski J, Yoo S, Nattestad M, Marcus S, Goodwin S et al (2016) Third-generation sequencing and the future of genomics. bioRxiv:048603. https://doi.org/10.1101/048603

Lelieveld SH, Veltman JA, Gilissen C (2016) Novel bioinformatic developments for exome sequencing. Hum Genet 135:603–614. https://doi.org/10.1007/s00439-016-1658-6

Levy SE, Myers RM (2016) Advancements in next-generation sequencing. Annu Rev Genomics Hum Genet 17(1):95–115. https://doi.org/10.1146/annurev-genom-083115-022413

Li Y, Xu XY (2019) Research progress of high-throughput sequencing technology. China Med Eng 27:32–37. https://doi.org/10.19338/j.issn.1672-2019.2019.03.009

Li W, Liu J, Zhang H, Liu Z, Wang Y, Xing L et al (2022) Plant pan-genomics: recent advances, new challenges, and roads ahead. J Genet Genomics 49(9):833–846. https://doi.org/10.1016/j.jgg.2022.06.004

Li N, He Q, Wang J, Wang B, Zhao J, Huang S et al (2023) Super-pangenome analyses highlight genomic diversity and structural variation across wild and cultivated tomato species. Nat Genet 55(5):852–860. https://doi.org/10.1038/s41588-023-01340-y

Lian Q, Huettel B, Walkemeier B, Mayjonade B, Lopez-Roques C, Gil L et al (2024) A pan-genome of 69 Arabidopsis thaliana accessions reveals a conserved genome structure throughout the global species range. Nat Genet:1–10. https://doi.org/10.1038/s41588-024-01715-9

Liang SB, Liu JY, Liu J, Yang JT, Li JL, Zhang YM (2017) Next-generation sequencing applications for crop genomes. China Biotechnol 37(2):111–120. https://doi.org/10.13523/j.cb.20170216

Liao X, Li M, Zou Y, Wu FX, Wang J (2019) Current challenges and solutions of de novo assembly. Quant Biol 7(2):90–109. https://doi.org/10.1007/s40484-019-0166-9

Liu Y, Salsman E, Fiedler JD, Hegstad JB, Green A, Mergoum M et al (2019) Genetic mapping and prediction analysis of FHB resistance in a hard red spring wheat breeding population. Front Plant Sci 10:1007. https://doi.org/10.3389/fpls.2019.01007

Liu G, Liu X, Xu Y, Bernardo A, Chen M, Li Y et al (2020) Reassigning Hessian fly resistance genes H7 and H8 to chromosomes 6A and 2B of the wheat cultivar 'Seneca' using genotyping-by-sequencing. Crop Sci 60(3):1488–1498. https://doi.org/10.1002/csc2.20148

Liu F, Zhao J, Sun H, Xiong C, Sun X, Wang X et al (2023) Genomes of cultivated and wild Capsicum species provide insights into pepper domestication and population differentiation. Nat Commun 14(1):5487. https://doi.org/10.1038/s41467-023-41251-4

Low L, Tammi MT (2023) Introduction to next generation sequencing technologies. In: Practical bioinformatics for beginners: from raw sequence analysis to machine learning applications, pp 1–22. https://doi.org/10.1142/12908
Mamanova L, Coffey AJ, Scott CE, Kozarewa I, Turner EH, Kumar A et al (2010) Target-enrichment strategies for next-generation sequencing. Nat Methods 7(2):111–118. https://doi.org/10.1038/nmeth.1419
Margulies M, Egholm M, Altman WE, Attiya S, Bader JS, Bemben LA et al (2005) Genome sequencing in microfabricated high-density picolitre reactors. Nature 437(7057):376–380. https://doi.org/10.1038/nature03959
McCouch SR, Wright MH, Tung CW, Maron LG, McNally KL, Fitzgerald M et al (2016) Open access resources for genome-wide association mapping in rice. Nat Commun 7(1):10532. https://doi.org/10.1038/ncomms10532
Merriman B, R&DTeam IT, Rothberg JM (2012) Progress in ion torrent semiconductor chip-based sequencing. Electrophoresis 33(23):3397–3417. https://doi.org/10.1002/elps.201200424
Montesinos-López OA, Herr AW, Crossa J, Montesinos-López A, Carter AH (2024) Enhancing winter wheat prediction with genomics, phenomics and environmental data. BMC Genomics 25(1):544. https://doi.org/10.1186/s12864-024-10438-4
Mu J, Huang S, Liu S, Zeng Q, Dai M, Wang Q et al (2019) Genetic architecture of wheat stripe rust resistance revealed by combining QTL mapping using SNP-based genetic maps and bulked segregant analysis. Theor Appl Genet 132:443–455. https://doi.org/10.1007/s00122-018-3231-2
N'Diaye A, Walkowiak S, Pozniak C (2023) Wheat sequencing: the pan-genome and opportunities for accelerating breeding. In: The wheat genome. Springer International Publishing, Cham, pp 273–288
Naithani S, Deng CH, Sahu SK, Jaiswal P (2023) Exploring pan-genomes: an overview of resources and tools for unraveling structure, function, and evolution of crop genes and genomes. Biomolecules 13(9):1403. https://doi.org/10.3390/biom13091403
Nautiyal AK, Ahuja V, Kshirsagar S, Dasgupta D (2021) Recent advancement in NGS technologies. In: Bioinformatics in rice research: theories and techniques, pp 585–609. https://doi.org/10.1007/978-981-16-3993-7_26
Nida H, Girma G, Mekonen M, Lee S, Seyoum A, Dessalegn K et al (2019) Identification of sorghum grain mold resistance loci through genome wide association mapping. J Cereal Sci 85: 295–304. https://doi.org/10.1016/j.jcs.2018.12.016
Niu J, Ma S, Zheng S, Zhang C, Lu Y, Si Y et al (2023) Whole-genome sequencing of diverse wheat accessions uncovers genetic changes during modern breeding in China and the United States. Plant Cell 35(12):4199–4216. https://doi.org/10.1093/plcell/koad229
Pal L, Sandhu SK, Bhatia D, Sethi S (2021) Genome-wide association study for candidate genes controlling seed yield and its components in rapeseed (Brassica napus subsp. napus). Physiol Mol Biol Plants 27:1933–1951. https://doi.org/10.1007/s12298-021-01060-9
Pal N, Jan I, Saini DK, Kumar K, Kumar A, Sharma PK et al (2022) Meta-QTLs for multiple disease resistance involving three rusts in common wheat (Triticum aestivum L.). Theor Appl Genet 135(7):2385–2405. https://doi.org/10.1007/s00122-022-04119-7
Park ST, Kim J (2016) Trends in next-generation sequencing and a new era for whole genome sequencing. Int Neurourol J 20(Suppl 2):S76. https://doi.org/10.5213/inj.1632742.371
Parkinson J, Blaxter M (2009) Expressed sequence tags: an overview. In: Expressed sequence tags (ESTs) generation and analysis, pp 1–12
Pascual L, Albert E, Sauvage C, Duangjit J, Bouchet JP, Bitton F et al (2016) Dissecting quantitative trait variation in the resequencing era: complementarity of bi-parental, multi-parental and association panels. Plant Sci 242:120–130. https://doi.org/10.1016/j.plantsci.2015.06.017
Pervez MT, Hasnain MJU, Abbas SH, Moustafa MF, Aslam N, Shah SSM (2022) A comprehensive review of performance of next-generation sequencing platforms. Biomed Res Int 2022(1): 3457806. https://doi.org/10.1155/2022/3457806

Peterson TW, Nam SJ, Darby A (2010) Next gen sequencing survey. North America equity research. https://doi.org/10.1007/978-1-60327-136-3_1
Poland JA, Brown PJ, Sorrells ME, Jannink JL (2012) Development of high-density genetic maps for barley and wheat using a novel two-enzyme genotyping-by-sequencing approach. PLoS One 7(2):e32253. https://doi.org/10.1371/journal.pone.0032253
Qin P, Lu H, Du H, Wang H, Chen W, Chen Z et al (2021) Pan-genome analysis of 33 genetically diverse rice accessions reveals hidden genomic variations. Cell 184(13):3542–3558. https://doi.org/10.1016/j.cell.2021.04.046
Rhoads A, Au KF (2015) Pac Bio sequencing and its applications. Genomics Proteomics Bioinformatics 13(5):278–289. https://doi.org/10.1016/j.gpb.2015.08.002
Rodriguez DFC, Urban MO, Santaella M, Gereda JM, Contreras AD, Wenzl P (2022) Using phenomics to identify and integrate traits of interest for better-performing common beans: a validation study on an interspecific hybrid and its Acutifolii parents. Front Plant Sci 13: 1008666. https://doi.org/10.3389/fpls.2022.1008666
Roorkiwal M, Jain A, Kale SM, Doddamani D, Chitikineni A, Thudi M, Varshney RK (2018) Development and evaluation of high-density Axiom® Cicer SNP Array for high-resolution genetic mapping and breeding applications in chickpea. Plant Biotechnol J 16(4):890–901. https://doi.org/10.1111/pbi.12836
Rossi EA, Ruiz M, Rueda Calderón MA, Bruno CI, Bonamico NC, Balzarini MG (2019) Meta-analysis of QTL studies for resistance to fungi and viruses in maize. Crop Sci 59(1):125–139. https://doi.org/10.2135/cropsci2018.05.0330
Rothberg JM, Hinz W, Rearick TM, Schultz J, Mileski W, Davey M et al (2011) An integrated semiconductor device enabling non-optical genome sequencing. Nature 475(7356):348–352. https://doi.org/10.1038/nature10242
Roychowdhury R, Choudhury S, Hasanuzzaman, M, Srivastava S (eds) (2020) Sustainable agriculture in the era of climate change https://doi.org/10.1007/978-3-030-45669-6
Saini DK, Chopra Y, Singh J, Sandhu KS, Kumar A, Bazzer S, Srivastava P (2022) Comprehensive evaluation of mapping complex traits in wheat using genome-wide association studies. Mol Breed 42:1–52. https://doi.org/10.1007/s11032-021-01272-7
Saini DK, Kumar S, Saini K, Goyal S (2024) Omics and system biology for developing climate-ready cereals: an overview. In Omics and system biology approaches for delivering better cereals. CRC Press, pp 1–25
Sandhu KS, Mihalyov PD, Lewien MJ, Pumphrey MO, Carter AH (2021) Genome-wide association studies and genomic selection for grain protein content stability in a nested association mapping population of spring wheat. bioRxiv. https://doi.org/10.3390/agronomy11122528
Satam H, Joshi K, Mangrolia U, Waghoo S, Zaidi G, Rawool S et al (2023) Next-generation sequencing technology: current trends and advancements. Biology 12(7):997. https://doi.org/10.3390/biology12070997
Saxena RK, Rathore A, Bohra A, Yadav P, Das RR, Khan AW et al (2018) Development and application of high-density Axiom Cajanus SNP array with 56K SNPs to understand the genome architecture of released cultivars and founder genotypes. Plant Genome 11(3):180005. https://doi.org/10.3835/plantgenome2018.01.0005
Scaglione D, Pinosio S, Marroni F, Di Centa E, Fornasiero A, Magris G et al (2019) Single primer enrichment technology as a tool for massive genotyping: a benchmark on black poplar and maize. Ann Bot 124(4):543–551. https://doi.org/10.1093/aob/mcz054
Scott MF, Ladejobi O, Amer S, Bentley AR, Biernaskie J, Boden SA et al (2020) Multi-parent populations in crops: a toolbox integrating genomics and genetic mapping with breeding. Heredity 125(6):396–416. http://creativecommons.org/licenses/by/4.0/
Schreiber M, Jayakodi M, Stein N, Mascher M (2024) Plant pangenomes for crop improvement, biodiversity and evolution. Nat Rev Genet 1–15. https://doi.org/10.1038/s41576-024-00691-4
Shaikh TA, Rasool T, Lone FR (2022) Towards leveraging the role of machine learning and artificial intelligence in precision agriculture and smart farming. Comput Electron Agric 198: 107119. https://doi.org/10.1016/j.compag.2022.107119

Shariat M, Neysi S (2022) Introduction to NGS platforms and their role in plant genetic modification. J Agric Sci Eng 4(1):20–34

Sharma SK, McLean K, Hedley PE, Dale F, Daniels S, Bryan GJ (2024) Genotyping-by-sequencing targets genic regions and improves resolution of genome-wide association studies in autotetraploid potato. Theor Appl Genet 137(8):180. https://doi.org/10.1007/s00122-024-04651-8

Shashiprabha HED, Abeysundara SP, Ariyarathna HACK (2022) Meta-QTL analysis identified stable quantitative trait loci (QTLs) and associated resistance gene analogues in rice. J Natl Sci Found Sri Lanka 50(1):35. https://www.researchgate.net/publication/360003410

Shizuya H, Birren B, Kim UJ, Mancino V, Slepak T, Tachiiri Y, Simon M (1992) Cloning and stable maintenance of 300-kilobase-pair fragments of human DNA in Escherichia coli using an F-factor-based vector. Proc Natl Acad Sci 89(18):8794–8797. https://doi.org/10.1073/pnas.89.18.8794

Song L, Wang R, Yang X, Zhang A, Liu D (2023) Molecular markers and their applications in marker-assisted selection (MAS) in bread wheat (Triticum aestivum L.). Agriculture 13(3):642. https://doi.org/10.3390/agriculture13030642

Soni HK, Singh A (2022) Application of next-generation sequencing in plant molecular ecology. In: Plant ecogenomics. Apple Academic Press, pp 47–81. eBook ISBN-9781003282006

Soriano JM, Colasuonno P, Marcotuli I, Gadaleta A (2021) Meta-QTL analysis and identification of candidate genes for quality, abiotic and biotic stress in durum wheat. Sci Rep 11(1):11877. https://doi.org/10.1038/s41598-021-91446-2

Spindel J, Begum H, Akdemir D, Virk P, Collard B, Redona E et al (2015) Genomic selection and association mapping in rice (Oryza sativa): effect of trait genetic architecture, training population composition, marker number and statistical model on accuracy of rice genomic selection in elite, tropical rice breeding lines. PLoS Genet 11(2):e1004982. https://doi.org/10.1371/journal.pgen.1004982

Srivastava RP, Yadav S, Singh MK (2024) Integrating genomics and phenomics in agricultural breeding: a comprehensive review. AJRRA 17(4):116–125

Sun Z, Wang X, Liu Z, Gu Q, Zhang Y, Li Z et al (2017) Genome-wide association study discovered genetic variation and candidate genes of fibre quality traits in Gossypium hirsutum L. Plant Biotechnol J 15(8):982–996. https://doi.org/10.1111/pbi.12693

Supriya P, Naveen A, Vasanthrao T, Mallesh G, Danakumara T (2024) The role of pan-genomics in understanding genetic diversity and structural variations for developing climate-resilient crops. Int J Adv Biochem Res SP-8(7):568–575. https://doi.org/10.33545/26174693.2024.v8.i7Sg.2045

Tao Y, Zhao X, Mace E, Henry R, Jordan D (2019) Exploring and exploiting pan-genomics for crop improvement. Mol Plant 12(2):156–169. https://doi.org/10.1016/j.molp.2018.12.016

Tettelin H, Masignani V, Cieslewicz MJ, Donati C, Medini D, Ward NL et al (2005) Genome analysis of multiple pathogenic isolates of Streptococcus agalactiae: implications for the microbial "pan-genome". Proc Natl Acad Sci 102(39):13950–13955. https://doi.org/10.1073/pnas.0506758102

Tomar V, Singh D, Dhillon GS, Singh RP, Poland J, Joshi AK et al (2021) New QTLs for spot blotch disease resistance in wheat (Triticum aestivum L.) using genome-wide association mapping. Front Genet 11:613217. https://doi.org/10.3389/fgene.2020.613217

Tyagi P, Singh D, Mathur S, Singh A, Ranjan R (2022) Upcoming progress of transcriptomics studies on plants: an overview. Front Plant Sci 13:1030890. https://doi.org/10.3389/fpls.2022.1030890

Varshney RK, Thudi M, Nayak SN, Gaur PM, Kashiwagi J, Krishnamurthy L et al (2014) Genetic dissection of drought tolerance in chickpea (Cicer arietinum L.). Theor Appl Genet 127:445–462. https://doi.org/10.1007/s00122-013-2230-6

Varshney RK, Shi C, Thudi M, Mariac C, Wallace J, Qi P et al (2017) Pearl millet genome sequence provides a resource to improve agronomic traits in arid environments. Nat Biotechnol 35(10): 969–976. https://doi.org/10.1038/nbt.3943

Varshney RK, Pandey MK, Bohra A, Singh VK, Thudi M, Saxena RK (2019) Toward the sequence-based breeding in legumes in the post-genome sequencing era. Theor Appl Genet 132(3):797–816. https://doi.org/10.1007/s00122-018-3252-x

Violle C, Navas ML, Vile D, Kazakou E, Fortunel C, Hummel I, Garnier E (2007) Let the concept of trait be functional! Oikos 116(5):882–892. https://doi.org/10.1111/j.0030-1299.2007.15559.x

Volk GM, Byrne PF, Coyne CJ, Flint-Garcia S, Reeves PA, Richards C (2021) Integrating genomic and phenomic approaches to support plant genetic resources conservation and use. Plants 10(11):2260

Voss-Fels K, Snowdon RJ (2016) Understanding and utilizing crop genome diversity via high-resolution genotyping. Plant Biotechnol J 14(4):1086–1094. https://doi.org/10.1111/pbi.12456

Wang M (2021) Next-generation sequencing (NGS). In: Clinical molecular diagnostics, pp 305–327. https://doi.org/10.1007/978-981-16-1037-0_23

Wang Z, Gerstein M, Snyder M (2009) RNA-Seq: a revolutionary tool for transcriptomics. Nat Rev Genet 10(1):57–63. https://doi.org/10.1038/nrg2484

Wang J, Wang Z, Du X, Yang H, Han F, Han Y et al (2017) A high-density genetic map and QTL analysis of agronomic traits in foxtail millet [Setariaitalica (L.) P. Beauv.] using RAD-seq. PLoS One 12(6):e0179717. https://doi.org/10.1371/journal.pone.0179717

Wang M, Li J, Qi Z, Long Y, Pei L, Huang X et al (2022) Genomic innovation and regulatory rewiring during evolution of the cotton genus Gossypium. Nat Genet 54(12):1959–1971. https://doi.org/10.1038/s41588-022-01237-2

Winfield MO, Allen AM, Burridge AJ, Barker GL, Benbow HR, Wilkinson PA et al (2016) High-density SNP genotyping array for hexaploid wheat and its secondary and tertiary gene pool. Plant Biotechnol J 14(5):1195–1206. https://doi.org/10.1111/pbi.12485

Wong KC, Zhang J, Yan S, Li X, Lin Q, Kwong S, Liang C (2019) DNA sequencing technologies: sequencing data protocols and bioinformatics tools. ACM Comput Surv (CSUR) 52(5):1–30. https://doi.org/10.1145/3340286

Wu R, Ma C, Casella G (2007) Statistical genetics of quantitative traits: linkage, maps and QTL. Springer Science & Business Media

Xiao Q, Bai X, Zhang C, He Y (2022) Advanced high-throughput plant phenotyping techniques for genome-wide association studies: a review. J Adv Res 35:215–230. https://doi.org/10.1016/j.jare.2021.05.002

Xiong H, Li Y, Guo H, Xie Y, Zhao L, Gu J et al (2021) Genetic mapping by integration of 55K SNP array and KASP markers reveals candidate genes for important agronomic traits in hexaploid wheat. Front Plant Sci 12:628478. https://doi.org/10.3389/fpls.2021.628478

Yan H, Sun M, Zhang Z, Jin Y, Zhang A, Lin C et al (2023) Pangenomic analysis identifies structural variation associated with heat tolerance in pearl millet. Nat Genet 55(3):507–518. https://doi.org/10.1038/s41588-023-01302-4

Yang H, Tao Y, Zheng Z, Li C, Sweetingham MW, Howieson JG (2012) Application of next-generation sequencing for rapid marker development in molecular plant breeding: a case study on anthracnose disease resistance in Lupinusangustifolius L. BMC Genomics 13:1–12. https://doi.org/10.1186/1471-2164-13-318

Yu H, Xie W, Li J, Zhou F, Zhang Q (2014) A whole-genome SNP array (RICE 6 K) for genomic breeding in rice. Plant Biotechnol J 12(1):28–37. https://doi.org/10.1111/pbi.12113

Zargar SM, Raatz B, Sonah H, Nazir M, Bhat JA, Dar ZA et al (2015) Recent advances in molecular marker techniques: insight into QTL mapping, GWAS and genomic selection in plants. J Crop Sci Biotechnol 18:293–308. https://doi.org/10.1007/s12892-015-0037-5

Zhang W, Sun Y, Liu J, Xu C, Zou X, Chen X et al (2021) DNA barcoding of Oryza: conventional, specific, and super barcodes. Plant Mol Biol 105:215–228. https://doi.org/10.1007/s11103-020-01054-3

Zhao Q, Feng Q, Lu H, Li Y, Wang A, Tian Q et al (2018) Pan-genome analysis highlights the extent of genomic variation in cultivated and wild rice. Nat Genet 50(2):278–284. https://doi.org/10.1038/s41588-018-0041-z

Zheng T, Hua C, Li L, Sun Z, Yuan M, Bai G et al (2021) Integration of meta-QTL discovery with omics: towards a molecular breeding platform for improving wheat resistance to Fusarium head blight. Crop J 9(4):739–749. https://doi.org/10.1016/j.cj.2020.10.006

Zou G, Zhai G, Feng Q, Yan S, Wang A, Zhao Q et al (2012) Identification of QTLs for eight agronomically important traits using an ultra-high-density map based on SNPs generated from high-throughput sequencing in sorghum under contrasting photoperiods. J Exp Bot 63(15): 5451–5462. https://doi.org/10.1093/jxb/ers205

Chapter 13
Tailoring Future Crops with Genomics-Informed Gene Editing

Pooja Shukla, Muthiah Naveenkumar, Rishitha Magunta, Laha Supriya, and Mehanathan Muthamilarasan

Abstract The rapid increase in global population and climate change pose significant challenges to food security, demanding the development of resilient and high-yielding crops. Advances in genomics and gene editing technologies offer precise and efficient solutions for crop improvement. Given this, the chapter discusses cutting-edge gene editing tools, including CRISPR/Cas, TALENs, ZFNs, and meganucleases, with a focus on their applications in improving traits. Integrating genomic selection and genome-wide association studies with high-throughput phenotyping enables the identification of candidate genes associated with the target traits. Multi-omics approaches further facilitate deciphering the gene function and regulation, enabling targeted interventions for crop enhancement using genomics approaches. The combination of speed breeding with gene editing accelerates the development of superior cultivars within a short time frame. Additionally, innovations such as tissue culture-free genome editing and nanoparticle-mediated delivery systems offer practical solutions for diverse crops. Despite the transformative potential of gene editing, challenges related to off-target effects and gene editing accuracy remain, necessitating robust bioinformatics tools and precision-targeting strategies to circumvent these shortcomings. Altogether, the convergence of genomics and gene editing holds the potential for sustainable agriculture, facilitating the development of climate-resilient and nutritionally enhanced crop varieties that will help secure food and nutritional security for the ever-growing population.

Keywords Gene editing · GWAS · Genomic selection · Multi-omics · Crop improvement · Climate resilience · Food security · High-throughput phenotyping

P. Shukla · M. Naveenkumar · R. Magunta · L. Supriya · M. Muthamilarasan (✉)
The Millet Lab, Department of Plant Sciences, School of Life Sciences, University of Hyderabad, Hyderabad, Telangana, India
e-mail: muthu@uohyd.ac.in

V. Garg et al. (eds.), *DNA of Sustainability*, Sustainability Sciences in Asia and Africa,
https://doi.org/10.1007/978-981-95-5109-5_13

13.1 Crop-Trait Enhancement: Why Is It Necessary for the Current Scenario

Crops cultivated in the current global and climatic scenario face undue pressure to perform well under predicted but inevitable challenges. With the worldwide population expected to continue growing by 2030 (FAO 2019), a steep rise in food demand poses a significant threat to global food security, and an increase in food, feed, and fuel demand is anticipated. According to the UN Population Report, around 9.7 billion people will inhabit Earth by 2050 (https://news.un.org/en/story/2019/06/1040621), with India expected to become the most populous country, surpassing China and sub-Saharan Africa. Additionally, there is expected to be a high increase in the proportion of the older population, while the life expectancy is expected to decrease in poorer countries (https://news.un.org/en/story/2019/06/1040621). Beyond quantity, nutritional security remains a critical concern, with rising cases of malnutrition and hidden hunger emphasising the need for nutritional improvement to restore global nutritional security. Additionally, the frequent weather changes, such as increasing temperatures, uneven rainfall, and erratic weather patterns, further demand the development of climate-resilient crops capable of withstanding environmental fluctuations while maintaining stable yields and superior postharvest quality.

Hence, crop trait enhancement becomes imperative to achieve the goals of improving agricultural productivity, nutritional biofortification, climate change resilience, and postharvest quality. The current trend towards crop genetic improvement necessitates the availability of genome assembly, characterisation of the available germplasm, identification of gene function, genomic breeding, and functional validation using gene editing approaches (Varshney et al. 2020). A reverse domestication approach can be adopted by targeting domestication-related traits in wild species of crops, which can lead to the development of custom-designed lines that are superior in terms of abiotic and biotic stress tolerance. Furthermore, an 'ExpressEdit' approach that harnesses speed through a combination of gene editing approaches and speed-breeding methods can provide expedited line fixation (FAO 2019; Barmukh et al. 2024). By integrating advanced molecular tools with precision breeding, we can develop resilient, high-yielding, and nutritionally superior crops that sustain future agricultural demands while mitigating the impacts of climate change and global food insecurity.

13.2 From Conventional to Modern Breeding: Achievements and Challenges

Conventional breeding aligns with nature's law of inheritance and relies primarily on the observations, experiences, and judgment skills of breeders (Allard 1961). This method continues to be a fundamental tool in developing superior crop varieties with stable traits through selection, hybridisation (crossing it with other parents for creating variability and trait improvement/enhancement), and mutation breeding (inducing mutations for desired changes in the trait). Despite its usefulness, conventional breeding is a slow and prolonged process, requiring more than 10 years to release a new cultivar (Bharti and Chimata 2019). Additionally, being overly dependent on the expression of phenotypic characters, selected 'superior' cultivars tend to yield non-reliable and inconsistent results due to the influence that genotype-environment interactions have on the phenotype (Lee et al. 2015). Additionally, conventional breeding only selects dominant genes.

The integration of breeding with other branches of science and modern technologies is known as modern breeding. It integrates high-throughput phenotyping with genomic selection to explore and characterise the performance of a variety, which is superior and stable across distant locations (Ewing et al. 2019). Determination of a pure gene line becomes easier with modern breeding methods. Here, recessive alleles can also be identified using markers and specific gene sites. Plant breeders have played a crucial role in identifying and selecting diverse germplasm for breeding, as well as harnessing natural and artificial mutations that benefit trait enhancement. For instance, increasing yield per area would require targeting increased yield (in terms of the number of grains) per spike, an increased number of branches per plant, and a larger size and weight of each grain. Similarly, improving nutritional content would aim to target germplasm with higher vitamin and mineral levels. Similarly, improved climate resilience would require targeting higher tolerance to heat, cold, drought, salinity, heavy metals, and other environmental stressors. While improved tolerance to biotic stress involves identifying resistance-conferring loci, introducing target loci into elite varieties while balancing the energy requirements for resistance and minimising yield penalties remains a challenge. With the rapidly changing global weather and climatic patterns, crops are affected by multiple stressors simultaneously. Hence, the development of multi-stress-resistant varieties through modern breeding methods, especially without a yield penalty, remains a challenge (Kumar et al. 2011). Another important limitation of breeding is genetic erosion. A rapid decline in genetic diversity has been reported in Canadian crop gene pools due to breeding efforts (Fu and Dong 2015). Molecular markers are useful in analysing genetic diversity and phylogeny. Advanced molecular plant breeding deploys the omics technologies. Due to the cost-effective next-generation sequencing (NGS) technologies, molecular marker development has accelerated. Genomics-assisted breeding largely depends on the identification of molecular markers tightly linked to traits of interest, using marker-assisted methods such as marker-assisted backcrossing (MABC), marker-assisted recurrent selection (MARS), and genomic selection (GS), in addition to gene linkage and QTL mapping.

Genomic selection utilises the assistance of genome-wide markers to speculate the effect of quantitative gene loci and predict the estimated breeding value (Wang et al. 2018a, b). Successful mapping of small-effect genes, which govern most key agronomic and economic crop traits, has also been achieved through genomic selection (Robertsen et al. 2019). Conclusively, genomic selection is preferred to work with traits that are otherwise difficult to recognise through phenotypic expression (Sweeney et al. 2019). Deployment of marker-assisted selection (MAS) allows for the identification and selection of superior alleles at the seedling stage itself, especially for traits that express only in the late reproductive stage, are less heritable, or are tedious. Markers are a reliable method for determining the genomic regions underlying the desired trait expression, as they are not affected by the growing environment or inheritance, and neither do they alter the phenotypic expression of the plant under study (Ashraf et al. 2012). Modern breeding, empowered by genomic advancements and molecular tools, holds the potential to develop high-yielding, climate-resilient crops while addressing the challenges of genetic erosion and multi-stress tolerance.

13.3 Genomics

'Omics' is derived from the Greek word 'ome', meaning 'whole', which essentially refers to scientific disciplines that encompass complete biological systems and their processes. Omics approaches refer to various technologies that aid in the analysis of different biological molecules and processes, including genomics, transcriptomics, proteomics, metabolomics, phenomics, and ionomics. The advances in omics technologies have introduced and made available tools and methodologies that have revolutionised crop improvement interventions, approaches, strategies, and outcomes. Integrated multi-omics techniques and an interdisciplinary outlook deploy a systems biology approach, aimed at developing meaningful interpretations, bridging the shortcomings and understanding complex networks in plant science, such as the link between genotype and phenotype, and responses to multiple stresses. Deciphering a particular trait requires a comprehensive, integrated multi-omics approach that aids in the deep analysis of involved genetic, structural, and functional components, such as the detection of potential candidate genes, proteins, metabolic pathway networks, and interaction players linked to critical agronomic traits or stress responses.

Genomics, in essence, is the study of gene and their function in a cell and an organism. Genomic tools facilitate the analysis of genes and their encoded products. While initially genes were studied 'one gene at a time,' the genomics approach encompasses genes and the genome as a whole. With the aim of understanding and characterising both the structural and functional aspects of the genome, genomics can be further divided into structural genomics and functional genomics as approaches. Structural genomics focuses on the physical organisation and mapping of DNA sequences to understand genetic architecture. Functional genomics investigates gene expression, interactions, and their influence on biological functions using key molecules such as RNA, proteins, and metabolites.

Furthermore, comparative genomics provides new opportunities for gaining insights into related evolution, selection, divergence, and other phenomena by simply aligning the sequences under analysis. The fundamental principle of comparative genomics relies on the fact that DNA sequences responsible for conserved function must also be conserved among individuals, while sequences imparting different functions between individuals must be divergent. Comparison of genomes also serves as a powerful tool to gain insights into the types of genes present between species, as well as their degree of conservation or nonconservation, and whether they are functional or nonfunctional. Additionally, it provides insights into the coding, noncoding, and regulatory regions present (Hardison 2003). Regions of noncoding sequences with high similarity among species are potential candidates for functional regions (Pennacchio and Rubin 2001), which have been reported to be regulatory sequences (Loots et al. 2000). However, predicting the exact function of noncoding sequences remains a challenge. Building a training model by aligning all these regulatory sequences with known functions can help develop a statistical model for likelihood estimation (Elnitski et al. 2003). The multidisciplinary approach, arising from the combination of biologists, physicists, chemists, engineers, and computer scientists, leads to the exploration of the complex communication network of genes and proteins. Furthermore, analysis of raw data generated from sequencing and re-sequencing efforts reveals the need for bioinformatics intervention to facilitate further analysis and information generation through integrative biology. Integrative biology combines biological data with a mathematical formulation. Altogether, the power and potential of the genomics approach are indeed promising and impressive, leading to big changes in the paradigms of biology and multidisciplinary science.

13.4 Role of High-Throughput Phenotyping in Accelerating Genomics-Assisted Trait Discovery

High-throughput phenotyping plays a critical role in studying the phenome and genome of a crop of interest. As phenotypic information is often influenced by environmental factors, rapid multi-locational and multi-seasonal large-scale screening of various traits that contribute to yield, quality, tolerance, and resistance becomes a necessary step towards selecting the best phenotype, thereby paving the way for breeding. Phenotyping is a labour-intensive and time-consuming activity. High-throughput phenotyping relies on automated sensing, computation, data acquisition, and data analysis to generate large-scale phenotypic data. As the data recording is robust, fewer replications also become reliable. Various non-destructive imaging techniques for effective plant phenotyping include visible light imaging, thermal imaging, laser imaging, fluorescence imaging, hyperspectral imaging, 3D imaging, magnetic resonance imaging, positron emission tomography, and computed tomography (Jangra et al. 2021). The vast amount of data thus generated is analysed using machine learning approaches, which employ a range of statistical and probabilistic tools. Therefore, precise phenotyping of the traits, when combined with genomic information, facilitates the exploration of the essential genetic structures underlying the trait of interest.

A plethora of success stories integrates phenotyping with GWAS to dissect complex traits, such as trait characterisation and the identification of QTLs and candidate genes (Xiao et al. 2022). This hassle-free approach, irrespective of the crop ploidy, has eased the process and has redirected the candidate gene identification approach of complex traits in crop species (Yang et al. 2014; Rasheed et al. 2014; Crowell et al. 2016; Rebolledo et al. 2016; Xavier et al. 2017; Wu et al. 2019; Wang et al. 2019a, b), biomass (Neumann et al. 2017; Chen et al. 2019), crop lodging (Singh et al. 2019), abiotic stress tolerance and response (Al-Tamimi et al. 2016; Guo et al. 2018; Pham et al. 2019; Rapacz et al. 2019; Campbell et al. 2020; Li et al. 2020), biotic stress resistance (Kuska et al. 2015; Fordyce et al. 2018; Yates et al. 2019), and nutrition (Sun et al. 2019) in major staple cereal crops. Although advancements in high-throughput phenotyping technology are being leveraged to identify the genetic players responsible, it is recognised and understood that automated phenotyping can still not replace manual phenotyping and observational experience. The significant shortcomings include the need for low-cost, high-throughput phenotyping technology, accurate characterisation, a lack of skilled manpower that is both technically and subjectively sound, and the establishment and protection of phenotypic databases available in the public domain.

13.5 Role of Sequencing, Re-sequencing and Annotation Efforts Towards Identification of Targets

As genotyping efforts have accelerated the generation of molecular data, the integration of multi-omics, systems biology, and bioinformatics can help build efficient predictive models for large molecular datasets through machine learning and data algorithms. The data generated can be used to strategise gene targeting in a time- and cost-effective manner. Crop improvement requires genetic dissection of complex traits for which significant variation and genetic diversity within the species population remain imperative. The genetic variations are due to SNPs, InDels, and structural variations (SVs) caused by large insertions/deletions, large-scale duplications, presence/absence variants (PAVs), and copy number variants (CNVs). Dedicated whole-genome sequencing and resequencing efforts, in addition to high-quality assemblies, have been instrumental in detecting these genetic variants, thereby resolving their significance in desired agronomic target traits and their evolution. SV detection with base-level resolution can be accelerated by pairwise sequence alignment of multiple high-quality genome sequences. To facilitate an efficient analysis and characterisation of genetic diversity, the concept of the 'pangenome' was proposed. The pangenome of phylogenetically linked, closely related accessions facilitates inter-genomic comparisons and the identification of genes that constitute the core genome and the dispensable genome (Pronozin et al.

2021). As a single reference genome can never represent the complete set of genomic elements and genetic variation repertoire that a crop species and its variants collectively hold, pan-genome assembly proves to be a comprehensive approach to studying the genetic variation available for its exploitation in tailoring and genomic design of the crop for the future.

As more variants are sequenced, more unique genes become available for exploration of their function and traits. Additionally, pan genomes provide access to and representation of a new genome that can be used as a reference for alignment, as aligning with the already available reference genome introduces bias regarding the uniqueness of the genome assembly under question (Bayer et al. 2020). Pan genomes and super-pan genomes for many important crop such as mustard (Golicz et al. 2016), bread wheat (Montenegro et al. 2017), rice (Sun et al. 2017a, b; Zhao et al. 2018a, b; Qin et al. 2021; Shang et al. 2022), chickpea (*Cicer arietinum* L.) (Varshney et al. 2021; Khan et al. 2024), soybean (Liu et al. 2020), capsicum (Ou et al. 2018), foxtail millet (He et al. 2023), sorghum (Tao et al. 2021), broomcorn millet (Chen et al. 2023), and cotton (He et al. 2024) have been developed that possess unlimited potential to accelerate safer crop improvement programmes. Along the same lines, pan-transcriptomes consider the mRNA (transcript nucleotide sequences), their expression level data, and isoforms available to predict the presence of a gene, provided it is expressed. Genetic diversity in protein-coding sequences, in addition to variations and mutations, has been reported in important crops such as maize (Hirsch et al. 2014; Jin et al. 2016), barley (Ma et al. 2019a, b; Guo et al. 2025), sugarcane (Chen et al. 2025), potato (Petek et al. 2020), and alfalfa (Medina et al. 2021). Therefore, advanced phenotyping and sequencing technologies should be best harnessed to get a better understanding of the gene pools and genetic diversity.

13.6 Role of the In Silico Approach for Genome-Wide Identification of the Putative Targets

The enormous amount of genomic data generated through high-throughput sequencing gets stored in public comprehensive genomic databases such as Ensembl (https://www.ensembl.org/; Hubbard et al. 2002), Plant GDB (http://www.plantgdb.org/; Dong et al. 2004), Phytozome (http://www.phytozome.net/; Goodstein et al. 2012), Gramene (http://www.gramene.org/; Ware et al. 2002), to mention a few, in which, due to easy public access, the rate of submission of genome sequencing, transcriptome, and genotypic data is increasing exponentially. Exploration of these datasets aids in the easy genome-wide identification (GWI) of gene families present in a crop species, elucidating their gene structure patterns, functions, expression patterns, possible roles in developmental and hormonal pathways, and evolutionary relationships with gene families of other crop species of interest. The *in silico* analysis provides further information on the evolution of different plant species and their variation resulting from evolutionary events, such as whole-genome duplication (WGD). The computational approach employs two methods: a BlastP search-based method and an HMM (Hidden Markov Model)-based method. The BlastP method scans the genome of interest for sequences with conserved protein domains specific to the gene family and its function under question. The criteria for screening are to have a lower *e*-value, such as an *e*-value $\leq 10^{-5}$ (mostly preferred) and an *e*-value $\leq 1e^{-10}$. However, HMM-based analysis requires building an HMM profile that is used to search for similar genes through a query against the genome of interest. The genes are selected for further study based on their sensitive *e*-value (i.e. *e*-value $\leq 1e^{-10}$). The domain integrity of the selected genes is confirmed using various databases such as SMART (http://SMART.embl-heidelberg.de; Schultz et al. 2000), Pfam database (http://pfam.sanger.ac.uk/; http://pfam.janelia.org/; Finn et al. 2003), NCBI's conserved domain database (CDD) (http://www.ncbi.nlm.nih.gov/Structure/cdd/cdd.shtml; Marchler-Bauer et al. 2002), InterProScan (http://www.ebi.ac.uk/InterProScan/; Jones et al. 2014), and PROSITE (http://www.expasy.org/prosite/; Hulo et al. 2006) to avoid any false-positive results.

Furthermore, the gene-structure pattern and intron–exon organisation in genes can be analysed using a gene structure display server (http://gsds.cbi.pku.edu.cn/; Guo et al. 2007), which provides insights into gene family evolution, intron–exon divergence, and alternative splicing. Syntenic analysis can be essential for gaining insights into segmental duplication, tandem duplication, and transposition. Databases such as the multiple collinearity scan toolkit (MCScanX) (http://chibba.pgml.uga.edu/mcscan2/; Wang et al. 2012) and plant genome duplication database (http://chibba.agtec.uga.edu/duplication/ Lee et al. 2012) can help study the synteny and collinearity relationships, mainly constructed through the Circos software. Phylogeny analysis is performed using the full-length amino acid sequences of the genes, which are aligned using Clustal (http://www.clustal.org/) or MUSCLE (http://www.drive5.com/muscle; Edgar 2004). The aligned sequences are then analysed using different algorithms by MEGA (http://www.megasoftware.net/; Kumar et al. 2016), such as the neighbour-joining method or the maximum-likelihood method with a reliable bootstrap test replication of 1000. The use of both methods is recommended, as it enhances the consistency of the data, and the same output confirms the agreement between the two methods (Safder et al. 2022).

Additionally, the expression patterns of gene families and their members can be analysed using RNA-seq data available in public databases. Digital expression analysis provides genome-wide, tissue-specific, and stage-specific expression data. The expression values are represented in RPKM (reads per kilobase per million mapped reads) and FPKM (fragments per kilobase per million mapped reads), which are estimated by dividing the mean of all gene expression values and then transforming the obtained ratios to $\log_2$. The values are then used to generate a heat map, representing the expression levels in multiple tissues. Therefore, based on the contrasting expressions of the genes, they can be investigated further. Critical targets involved in protein–protein interactions can also be identified using an in silico approach, indicating the nature of proteins involved and the degree of interactions that occur. Node-link and matrix-based diagrams are types that display the relationships between immediate and non-immediate members, where proximity in the diagram is indicative of the strength of the interaction. Bio-analytic resource (BAR) is a computational tool that provides access to information related to protein–protein interactions, DNA–protein interactions, and transcriptional and promoter data for various plant species. Similarly, STRING (http://string-db.org/; Szklarczyk et al. 2015) also provides predicted protein–protein association data based on three-dimensional structural information. Genome-wide transcription factor binding sites can be detected using approaches such as ChIP-chip, ChIP-seq, and ChIP-Exo, which detect protein–DNA interactions (Ma et al. 2019a, b). Identification of cis-regulatory elements can be performed using a position weight matrix (PWM) model, which scans the sequence to identify potential binding sites of transcriptional factors based on their binding specificity (Stormo 2000; Stormo and Zhao 2010).

Open databases such as JASPAR (http://jaspar.cgb.ki.se/; Sandelin et al. 2004a, b) and UniPROBE (http://thebrain.bwh.harvard.edu/uniprobe/; Newburger and Bulyk 2009) provide curated and experimentally determined PWMs. The sequence conservation also aids in the detection of a single binding site. ConSite (http://www.phylofoot.org/consite; Sandelin et al. 2004a, b) and rVista (http://rvista.dcode.org/; Loots and Ovcharenko 2004) are two tools that perform PWM scanning. Transcription Factor binding site LOCater (TFLOC) employs a similar approach to identify PWM matches. Detection of cis-regulatory modules can also be done using motif clustering and comparative genomics (Aerts 2012). There is a need for high-level software that can integrate heterogeneous datasets and algorithms from multiple data sources for the identification and interpretation of candidate targets in regulatory biology (Aerts 2012).

Conclusively, different *in silico* analyses enable the identification of gene families and subfamilies with similar functions to have similar gene structures, exon–intron patterns, and common conserved domains, forming a monophyletic group. Therefore, GWI acts as a reliable primer to narrow down the target genes and understand their putative gene functions, which can be further validated using genome editing tools. Thus, it saves a significant amount of time to identify a potential target for editing and enhancing approaches.

13.7 GWAS for High-Confidence Candidate Gene Identification

Genome-wide association is an approach that deploys sequencing efforts and genome variation to find the link between genotype and phenotype. The approach becomes a preliminary necessity to decipher the key genetic players underpinning the target complex agronomic trait. Genome-wide association studies (GWAS) have become the method of choice and one of the key reliable approaches for identifying genetic contributors that regulate complex traits. The concept of association mapping was first applied in maize to associate Dwarf8 polymorphisms with the flowering trait, after evaluating sequence variations observed in 92 maize inbred lines (Thornsberry et al. 2001). The method exploits wide agronomic, molecular, or biochemical phenotypic trait variation in extremely diverse panels of genetic backgrounds to identify and examine genome-wide genetic variants. The true power of GWAS to identify a true SNP-trait association for complex traits lies in the availability of high phenotypic variance in the population under investigation. This can also be explained by the contrastingly different phenotypes of the allelic variants and their high frequency in the screening population (Asimit and Zeggini 2010; Gibson 2011). False-negative results can arise due to low-frequency causal alleles and issues in population structure correlation (Korte and Farlow 2013). Therefore, rare variants, when present in strong association with other non-causative rare variants, tend to drag with them many other synthetic associations (Dickson et al. 2010).

The confidence in meaningful associations can be increased by considering factors such as sample size and genetic heterogeneity. Heterogeneity can be overcome by including 'competing' variants as cofactors within a mixed model setting, or increasing the sample size with a highly phenotypically diverse population (Segura et al. 2012; Feng and Zhu 2012). Additionally, the use of the linear mixed model (LMM) approach significantly reduces the false-positive associations (Zhang et al. 2010). Although statistical inference provides effective data, the validation of deploying laboratory molecular experimentation techniques, such as gene knockouts, gene expression, or gene complementation, remains nonnegotiable as high throughput phenotyping, high-density genotyping, genotyping by sequencing (GBS), and re-sequencing efforts reveal genome-wide SNPs that, when further coupled with mixed model framework and rigorous statistical analysis, lead to mapping of the target QTL and precise identification of the candidate gene for the trait of interest. GWAS is suggested to be a complementary approach to QTL mapping, where they can mitigate each other's limitations when conducted together (Korte and Farlow 2013). However, with limitations that can be addressed to improve the method, GWAS remains a reliable tool for genotype–phenotype association.

GWAS has been successfully providing a bird's-eye view in deciphering the genetic structure and identifying novel trait associations for important crop traits, including agronomy, yield, biomass, lodging, nutrition, climate resilience, and infection resistance. Achievements of important crop-trait-gene identification through GWAS are reported for rice (Wang et al. 2020a, b), maize (Shikha et al. 2021; Sahito et al. 2024), wheat (Saini et al. 2021; Zia et al. 2022), pearl millet (Srivastava et al. 2020; Pujar et al. 2020), foxtail millet (Jaiswal et al. 2019; Wei et al. 2021; Liu et al. 2022; Liu et al. 2024a, b), finger millet (Sood et al. 2023; Kannababu et al. 2025), proso millet (Vetriventhan et al. 2024), and cotton (Yasir et al. 2022), to mention a few. Therefore, continuous collaborative efforts will continue contributing to the transformation of future crops.

13.8 The Need to Incorporate Genomics to Enhance Target Gene Editing Precision

Since the gene editing methods can also lead to unwanted cleavages, careful target site selection becomes paramount. Using bioinformatics and analytics, repeated and homologous regions can be easily identified and avoided. Therefore, there is a nonnegotiable need to incorporate genomics and genomics-assisted tools when finalising the target gene for the trait of interest. This is because the precision in gene editing technologies stems from the informed, advanced knowledge gained through genomics. Because most commercial crop traits are polygenic and the crops of interest are polyploid, the preliminary steps require assistance from genomics tools to narrow down and confirm the target gene or SNP while minimising the probability of a false target. Genomics approaches such as GWI and GWAS aid in the error-proof identification of candidate genes to proceed further with genome editing methods. GWAS has revolutionised the discovery of agronomic traits and their associated genes in important crop plants that are otherwise difficult to identify through traditional genetic and breeding techniques, owing to their complex polygenic nature and high ploidy. Therefore, towards tailoring future crops for improved tolerance and enhanced yield-related traits using genetic engineering and genome editing technology, GWAS indeed provides an increased accuracy in target gene identification (Fig. 13.1).

13.9 Targeted Crop-Trait Improvement Through Gene Editing

Combining recent advances in science and technology, gene editing methods enable the precise modification and alteration of a few nucleotides or specific genome regions, leading to the insertion or deletion of nucleotides (Voytas 2013; Voytas and Gao 2014). Gene editing techniques are most promising for the functional characterisation of genes that contribute to the improvement of crop traits. While forward genetics aims at identifying the genetic basis of a phenotype of interest, through approaches such as mutagenesis, phenotype-selection, mapping-by-sequencing, map-based cloning, and instant positional cloning (Peters et al. 2003); reverse genetics approach deploying genetic manipulation methods such as loss and gain of gene function, is used to study the impact of the changed sequence on the resulting phenotype (Gilchrist and Haughn 2010). Forward genetics provides unbiased results that elucidate genetic regulatory networks and the relationship between genotype and phenotype, thereby maintaining its high relevance in the post-genomic era (Khan et al. 2021; Kumar et al. 2022). Given the current climate scenario and urgency, the accuracy and precision delivered by genome-edited crops undoubtedly improve the agronomic performance and quality of the crop while simultaneously saving years of developmental time and the cost of conventional breeding. Gene

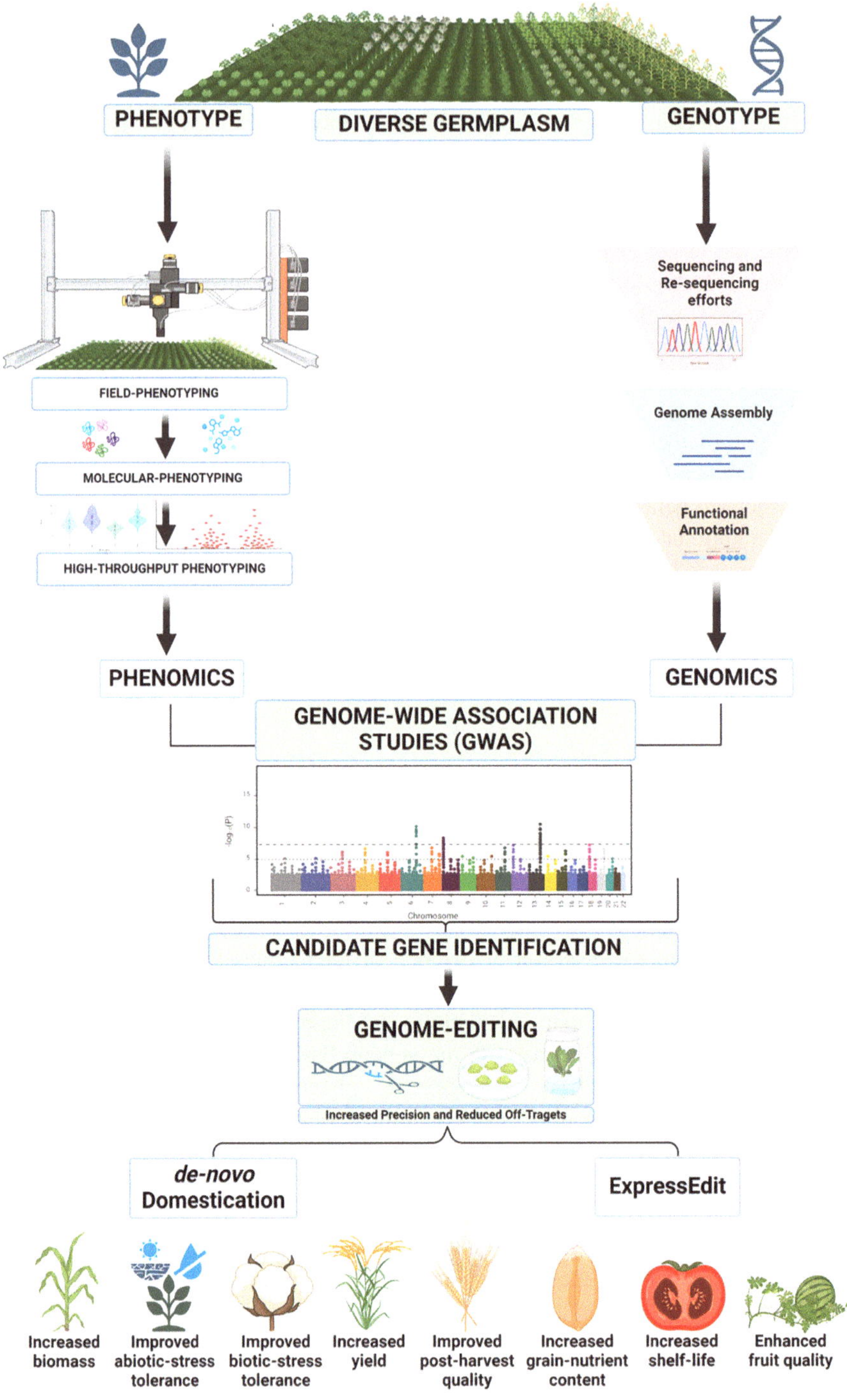

Fig. 13.1 Integrating genomics and phenotyping for precision crop improvement. The figure illustrates a comprehensive framework for crop trait improvement, integrating genomics,

editing is the method of choice that allows precise modification of sequences of interest and rapid introduction of genetic diversity, thus accelerating crop variety improvement, especially the polyploid crops, resulting in crops with enhanced traits such as higher yield with improved climate resilience, biotic stress tolerance, nutritional quality, and shelf-life (Fig. 13.2).

13.10 Programmable Sequence-Specific Nucleases

13.10.1 Meganucleases

Meganucleases, also known as homing endonucleases, are distinct enzymes that identify and cleave specific DNA sequences, typically ranging from 14 to 40 base pairs in length (Stoddard 2005). These enzymes were initially discovered in mobile genetic elements, such as inteins and introns, which help their movement by causing double-strand breaks (DSBs) at specific genomic locations (Belfort and Roberts 1997). Owing to their elevated specificity, they serve as a valuable tool for accurate gene editing across diverse organisms (Stoddard 2005). Their structure comprises a characteristic conserved DNA-binding domain that can interact with their long recognition sequences (Stoddard 2005). The LAGLIDADG family features a unique αββαβββα motif that enables accurate DNA binding and cleavage (Stoddard 2005). Meganucleases possess a characteristic capacity to identify and bind to specific DNA sequences, ultimately leading to the induction of a double-strand break (DSB) (Stoddard 2005). The intrinsic repair mechanisms of the cell, such as non-homologous end joining (NHEJ) and homologous recombination (HR), later facilitate the rectification of this break, thereby resulting in targeted genetic alterations (Pâques and Duchateau 2007). Although the engineering of meganucleases to recognise novel DNA sequences has remained a challenge due to their complex structure-function relationships (Stoddard 2011), advancements in protein engineering have enabled the development of customised meganucleases with altered specificities (Stoddard 2011). Techniques such as biolistic particle delivery and Agrobacterium-mediated transformation are primarily used for delivering genetic material into plant cells (Voytas and Gao 2014).

Fig. 13.1 (continued) phenomics, and genome editing. Diverse germplasm and phenotype data are obtained through field phenotyping, molecular phenotyping, and high-throughput phenotyping approaches. Concurrently, genomics approaches involving sequencing, re-sequencing, genome assembly, and functional annotation facilitate gene discovery. Genome-wide association studies (GWAS) integrate both phenotypic and genotypic data to identify candidate genes for targeted crop improvement. Candidate gene identification is followed by precision genome editing using advanced molecular tools to enhance traits such as biomass production, abiotic and biotic stress tolerance, yield improvement, postharvest quality, nutritional content, shelf life, and fruit quality. Innovative strategies like de novo domestication and 'ExpressEdit' are employed to accelerate breeding and genetic improvement

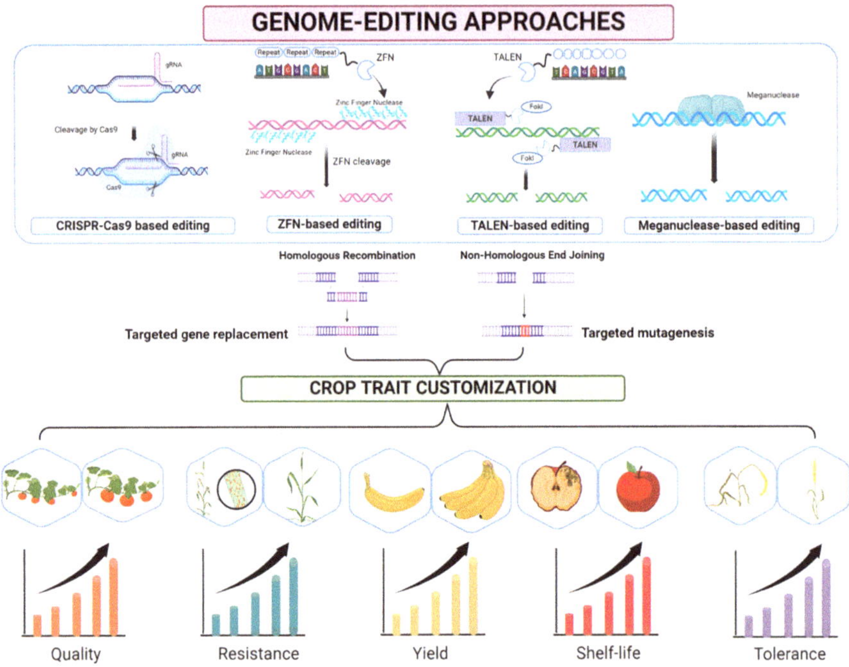

Fig. 13.2 Genome editing approaches for tailoring the traits. The figure depicts various genome editing techniques used for crop trait improvement, including CRISPR-Cas9-based editing, ZFN-based editing, TALEN-based editing, and meganuclease-based editing. CRISPR-Cas9 utilises guide RNA (gRNA) to direct site-specific cleavage by Cas9. Zinc Finger Nucleases (ZFNs) employ zinc finger motifs to bind DNA and the FokI nuclease domain for cutting. Transcription activator-like effector nucleases (TALENs) use repeat-variable di-residues (RVDs) to recognise DNA, linked with FokI for cleavage. Meganucleases recognise long DNA sequences and induce double-strand breaks. Editing outcomes involve homologous recombination for gene replacement or non-homologous end joining for targeted mutagenesis. Crop trait customisation through genome editing targets improvements in quality, resistance, yield, shelf life, and stress tolerance

Meganucleases were used in wheat and cotton for accurate genome modifications. In wheat (*Triticum aestivum*), a customised meganuclease facilitated a double-cut, resulting in the excision of integrated DNA cassettes that contain the DsRed reporter gene while simultaneously maintaining the integrity of the target locus for subsequent gene stacking applications (Youssef et al. 2018). In cotton (*Gossypium hirsutum*), a reengineered meganuclease targeted a sequence near an insect-resistance transgene, precisely inserting *hppd* and *epsps* genes through homologous recombination (D'Halluin et al. 2013). Although modifying meganucleases to target new sequences is complex and labour-intensive, and their efficient delivery into plant cells remains a significant hurdle, the precision provided by meganuclease editing not only reduces off-target effects but also minimises unintended genomic alterations, resulting in lowered cytotoxicity (Stoddard 2005).

13.10.2 Zinc Finger Nuclease

Zinc finger nucleases (ZFNs) were among the earliest site-specific gene editing tools developed and extensively employed in crop improvement (Kim et al. 1996). The structure consists of two components: the DNA-binding zinc finger domain and the FokI endonuclease domain. The DNA-binding specificity of zinc finger proteins (ZFPs) and the DNA-cleaving activity of the FokI nuclease domain are combined to form ZFNs (Kim et al. 1996). In eukaryotic transcription factors, the Cys2His2 zinc fingers are commonly found in DNA-binding motifs, with multiple fingers making tandem contacts along the DNA (Miller et al. 1985). A zinc ion stabilises the conserved ββα structure, which is composed of approximately 30 amino acids. The α-helix interacts with DNA bases to guarantee sequence-specific recognition. The α-helix contacts the bases in the major groove of DNA in a sequence-specific manner (Wolfe et al. 2000). Multiple Cys2His2 zinc finger motifs comprise a ZFN's DNA-binding domain, as each motif can recognise and bind to a distinct three-nucleotide sequence located in the major groove of DNA (Pavletich and Pabo 1991). The modular nature of these motifs enables the assembly of customised ZFNs that bind to extended DNA sequences, thereby increasing target specificity (Urnov et al. 2010).

During the early development of ZFNs, individual nucleases used three zinc fingers per monomer, recognising a 9-bp target that facilitates ZFN dimers to specify an 18-bp-long recognition site (Porteus and Carroll 2005). Further research led to an increase in the number of fingers per ZFN (up to six fingers), which enhanced specificity towards longer and rarer cleavage sites within genomes (Urnov et al. 2010). The specificity of ZFNs is determined by specific residues in the α-helix, which can dictate trinucleotide recognition through amino acid-nucleotide interactions (Segal et al. 2003). However, context-dependent effects can be observed in binding affinity, which means that the interactions between adjacent zinc fingers can modify the overall binding affinity and specificity, sometimes leading to the recognition of unintended sequences and off-target effects (Carroll 2011). Notably, various approaches to zinc finger engineering have been developed to overcome these challenges. The modular assembly approach involves selecting individual zinc fingers for each trinucleotide and linking them into a multifunctional ZFN that targets a composite DNA sequence (Kim et al. 2009). Another commonly used approach is the OPEN (Oligomerized Pool Engineering) system, which utilises bacterial selection techniques to identify zinc finger combinations with high specificity and reduced cross-reactivity (Maeder et al. 2008).

The FokI endonuclease domain is crucial for the function of zinc finger nucleases (ZFNs), which introduce double-strand breaks in DNA. Derived originally from *Flavobacterium okeanokoites*, FokI is a Type-IIS restriction enzyme characterised by specific DNA-binding and catalytic domains (Bitinaite et al. 1998). In ZFNs, FokI has a non-specific DNA cleavage domain that is combined with a custom-designed zinc finger protein (ZFP), making it a valuable tool for targeted gene editing (Kim et al. 1996). A key feature of the FokI nuclease domain is its requirement for dimerization to achieve catalytic activity. The two ZFN monomers dimerize through their FokI domains upon binding to the corresponding DNA target sites, forming an active nuclease complex that causes double-strand breaks (Smith 2001). To further improve specificity, researchers have engineered obligate heterodimeric variants of the FokI nuclease domain, designed to preferentially form heterodimers over homodimers, thereby reducing unwanted cleavage at non-specific sites (Miller et al. 2007). Mutations in the dimerization interface have optimised cleavage efficiency, and additionally, engineering the linker connecting the ZFP to the FokI nuclease has allowed adjustments to the spacing requirements between the two monomers for further refining target specificity (Szczepek et al. 2007). Multiple zinc fingers can be assembled to target long stretches of DNA sequences with high specificity (Pabo et al. 2001). The two ZFN monomers must dimerize after binding to specific DNA sequences to activate the FokI nuclease and ensure site-specific DNA cleavage (Miller et al. 2007). Therefore, the use of heterodimeric FokI variants can minimise unwanted homodimerization and reduce off-target cleavages. The optimal spacing between two ZFN binding sites is generally 5–7 base pairs, allowing FokI to introduce precise double-strand cleavage at specific loci (Doyon et al. 2011).

ZFN's cleavage efficiency relies on several factors, including the number and specificity of zinc fingers, chromatin accessibility, and the cellular environment (Ramalingam et al. 2011). Regarding the risk of creating genomic instability, the specificity of the method can be enhanced by modifying linker regions and refining the zinc finger design further (Ramalingam et al. 2011). ZFNs can be introduced as plasmid DNA, mRNA, or protein, each with different efficiencies and risks. The most common methods for delivering ZFNs in plant systems are Agrobacterium-mediated transformation, biolistic particle delivery, and protoplast transfection (Voytas and Gao 2014).

After a DSB is introduced, the cell repairs the break either by non-homologous end joining (NHEJ) or homology-directed repair (HDR). While non-homologous end joining (NHEJ) is the predominant repair pathway in eukaryotic cells particularly in plants (Gaj et al. 2013) that directly ligates the ends of damaged DNA without the use of a homologous template, often leading to small insertions or deletions (InDels) that impair gene functions, in turn serving as gene knockout applications (Lieber 2010), in contrast, homology-directed repair (HDR) repairs the break using a homologous donor template that helps with precise gene modifications (Urnov et al. 2010) taking place during the S and G2 phases of the cell cycle (when a sister chromatid is accessible as the repair template (Heyer et al. 2010)). While NHEJ is seen as error-prone, HDR ensures accurate genome editing, comparatively (Puchta 2005). However, efficient strategies such as supplying synthetic donor templates and modifying repair pathway regulators can enhance HDR efficiency (Ceccaldi et al. 2016). Conclusively, the preference for NHEJ or HDR depends on the cell cycle stage. The development of efficient methods to synchronise the cell cycle can be utilised to enhance the specificity of gene editing during the S and G2 phases when homology-directed repair (HDR) is active (Chu et al. 2015).

A breakthrough in ZFN-based crop improvement was the development of herbicide-resistant maize (*Zea mays*). The target gene *IPK1* produces the enzyme inositol-1,3,4,5,6-pentakisphosphate 2-kinase, which catalyses the last step of phytic acid biosynthesis in maize seeds. This targeted mutation reduced phytate levels and has also acquired herbicide resistance (Shukla et al. 2009). ZFNs are used to target nine endogenous genes in soybean (*Glycine max*), including DICER-LIKE (DCL) and RNA-silencing genes (Curtin et al. 2011). ZFNs facilitate the creation of highly specific genome modifications by reducing unintended mutations (Urnov et al. 2010) while simultaneously targeting multiple genes (Curtin et al. 2011). ZFN-edited crops without foreign DNA insertions are often classified as non-GMOs, providing relief from regulatory concerns (Voytas and Gao 2014). However, the emergence of alternative gene editing techniques, such as TALENs and CRISPR, has shifted attention away from ZFNs, even though it remains a valuable tool for targeted modifications.

13.10.3 Transcription Activator-Like Effector Nucleases

Transcription activator-like effector nucleases (TALENs) are an effective gene editing tool that utilises the DNA-binding specificity of transcription activator-like effectors (TALEs), which are fused to the catalytic domain of the FokI endonuclease (Boch et al. 2009). TALENs provide an effective way for targeted modifications, including gene knockouts, insertions, and replacements, and also offer a powerful approach to functional genomics and crop improvement (Cermak et al. 2011). Cells encode DNA-binding proteins to regulate gene expression, replication, and inheritance. Nature has evolved conserved motifs, such as zinc fingers, helix-turn-helix, and leucine zippers, for DNA recognition. However, studies by Boch et al. (2009) and Moscou and Bogdanove (2009) report a unique DNA-binding mechanism in *Xanthomonas* pathogens-mediated transcription activator-like (TAL) effectors, which are delivered through the type III secretion system that in turn activates plant susceptibility genes and leads to bacterial infection (Yang et al. 2006). Transcription activator-like effectors (TALEs) are key virulence proteins released by *Xanthomonas* species, which are responsible for major plant diseases (Gu et al. 2005; White and Yang 2009). These effectors belong to the AvrBs3/PthA family and play a significant role in bacterial pathogenesis by directly manipulating host gene expression (Bonas et al. 1993; Swarup et al. 1992).

Bacteria deliver TALEs into plant cells through the type III secretion system (T3SS). TALEs enter the nucleus and function as transcriptional activators (Schornack et al. 2013; Kay and Bonas 2009). Their capacity to recognise and bind to specific DNA sequences induces susceptibility genes, promoting bacterial infection, and disease progression (Yang et al. 2006). The conserved domain architecture of TALEs regulates nuclear localisation, DNA binding, and the activation of host gene expression, playing a crucial role in host–pathogen interactions (Boch and Bonas 2010). Repeat-variable residues (RVDs) at the 12th and 13th positions determine the DNA specificity of the central repeat domain of TAL effectors, which consists of 34 amino acid units. Each repeat corresponds to a single nucleotide in the target sequence that regulates precise DNA recognition (Moscou and Bogdanove 2009). Computational and genetic analyses have confirmed this code, which helps us design artificial TAL effectors with custom DNA-binding properties (Boch et al. 2009).

TALENs are composed of three key domains that define their function, namely the TALE DNA-binding domain, nuclear localisation signals (NLS), and FokI nuclease domain. The TALE domain is made of tandem repeats, also known as modular repeats. Each repeat is approximately 33-35 amino acids in length and selectively binds to a single nucleotide in the target DNA (Boch et al. 2009). Each repeat has repeat variable Di-residues (RVDs), which are amino acid residues at the 12th and 13th positions that determine the DNA-binding specificity. The most common RVDs in TALEs are NI, NG, HD, NN, and HG, which can recognise adenine (A), thymine (T), cytosine (C), guanine/adenine (G/A), and thymine (T), respectively (Boch et al. 2009; Moscou and Bogdanove 2009). Over 20 distinct RVDs have been found in TALEs. Structural analyses reveal that TALEs bind to DNA in a right-handed superhelical conformation, where each repeat unit adopts a left-handed, two-helix bundle conformation, exposing the RVD loop to the DNA major groove (Deng et al. 2012; Mak et al. 2013). Histidine (His) or asparagine (Asn) is reported as the first residue of each RVD that won't directly interact with DNA but stabilises the RVD loop by forming a hydrogen bond with Ala8 in the repeat structure (Sun and Zhao 2013). The second residue (position 13) determines sequence specificity through base-specific interactions, for example, in HD RVDs that are specific to cytosine. Here, Asp 13, with its carboxylate group, forms a hydrogen bond with the cytosine amine group, providing specificity through the electrostatic exclusion of other bases. Similarly, NG and HG RVDs recognise thymine through van der Waals interactions between Gly13 and the methyl group of thymine. In contrast, NI RVDs recognise adenine through nonpolar interactions between Ile13 and the C8 and N7 atoms of adenine. The nuclear localisation signals are short peptide sequence that has the role of directing the TALEN protein to the cell nucleus (Li et al. 2011). Multiple NLS sequences can be incorporated, or optimising NLS composition can improve the nuclear import efficiency of TALENs (Zhang et al. 2013). The length of the spacer region is a significant factor that affects TALEN activity (Cermak et al. 2011). Optimal spacer lengths can typically range from 12 to 24 base pairs, depending on the specific TALEN architecture (Miller et al. 2011). Research has shown that spacers of 12–15 base pairs are optimal for high TALEN activity (Doyon et al. 2011). If the spacer is too short or too long, the efficiency of FokI dimerization and DNA cleavage may be compromised (Tesson et al. 2011).

TALENs function depends on the DNA recognition capability of TALEs, which originate from *Xanthomonas* species, and the FokI endonuclease that induces double-stranded breaks at specific loci. The process and mechanism of TALEN-mediated genome editing can be divided into several key steps, such as DNA recognition by TALEs, where TALEs bind to DNA in a 1:1 manner, meaning each repeat binds to a single nucleotide (Deng et al. 2012). This modularity allows for precise targeting of genomic loci. TALENs have a pair of FokI nuclease domains, with each monomer binding to opposite sides of the target sequence, thereby introducing a double-stranded cut only at the intended site (Miller et al. 2011). Following TALEN-induced double-strand breaks (DSBs), DNA repair occurs via two main pathways: non-homologous end joining (NHEJ) and homology-directed repair (HDR).

One of the earliest methods for TALEN assembly is the restriction enzyme and ligation (REAL) method, which relies on a set of predefined restriction sites flanking different repeat modules. This approach is beneficial for the sequential ligation of repeat units, but it is time-consuming and limited in scalability (Sander et al. 2011). Golden gate cloning is the most efficient and commonly used method for TALEN assembly, utilising type IIS restriction enzymes such as BsaI and Esp3I. These enzymes will cut outside their recognition sequences, providing more specific and precise ligation of multiple TALE repeats in a single reaction. This method significantly reduces the time required for TALE assembly and can facilitate the high-throughput production of customised TALENs (Engler et al. 2009; Cermak et al. 2011). High-throughput approaches, such as fast ligation-based automatable solid-phase high-throughput (FLASH), have been further developed to enhance the efficiency of the TALEN construction process. FLASH utilises solid-phase oligonucleotide synthesis, combined with ligation-based assembly, for the rapid and automated construction of TALEN repeat arrays in a single reaction (Reyon et al. 2012). Solid-phase ligation is another method that immobilises TALE repeats on a solid support, making it easier to assemble repeat modules step-by-step with minimal errors. This technique improves efficiency and reproducibility while decreasing the need for labour-intensive cloning steps (Briggs et al. 2012). Newer variants, such as unit assembly and iterative capillary electrophoresis (UACE), have also been developed to reduce recombination errors and increase accuracy during repeat ligation. In addition to these cloning techniques, TALEN scaffolds have been improved for better performance and modifications to the C-terminal region of TALEN scaffolds have made it possible to target sequences that do not have the 5′-T at position zero that traditional TALEs require for effective DNA binding (Mak et al. 2013; Sun et al. 2012). The most common methods used for delivering TALENs in plant systems are Agrobacterium-mediated transformation, biolistic particle delivery, and protoplast transfection (Voytas and Gao 2014).

Despite their modularity, TAL effectors have limitations, including their inability to recognise guanine efficiently. Structural studies may clarify additional specificity determinants and potential dimerization effects. The modular DNA-binding mechanism of TAL effectors, similar to zinc finger proteins, has enabled applications in synthetic biology and genome engineering, including the development of TALENs for targeted gene modifications (Boch et al. 2009; Christian et al. 2010; Cermak et al. 2011).

TALENs have been utilised for precise gene modifications in various crops, improving disease resistance, stress tolerance, and other agronomic traits. *OsSWEET14* (*Os11N3*) of rice, which encodes a sucrose-efflux transporter (SWEET), is a susceptibility gene that is exploited by *Xanthomonas oryzae* pv. Oryzae (Xoo) uses its TAL effectors AvrXa7 and PthXo3. Since no natural *OsSWEET14* polymorphisms prevent *Xoo* induction, achieving resistance through traditional breeding remained a challenge. However, TALEN-based genome editing disrupted EBEs without affecting *the function of OsSWEET14, thereby* conferring resistance (Li et al. 2012). In *Wheat (Triticum aestivum),* TALEN-mediated knockouts of *MLO* genes have conferred resistance to powdery mildew caused by *Blumeria graminis f. sp. tritici* (Wang et al. 2014). Targeted mutagenesis of the GA signalling repressor *PROCERA* (PRO) in tomato (*Solanum lycopersicum*) using TALENs under an estrogen-inducible promoter resulted in the generation of pro alleles with heritable mutations, from which homozygous segregants exhibited phenotypes characterised by an increased GA response (Lor et al. 2014). Mutations introduced into the FAD2-1A and FAD2-1B genes of soybeans (Glycine max) resulted in a significant reduction in polyunsaturated fat content. In homozygous mutants, oleic acid levels increased from 20% to 80%, while linoleic acid levels declined from 50% to below 4%, thereby rapidly improving oil stability and shelf life in a single generation (Haun et al. 2014). Similarly, induced heritable mutations in the *OsBADH2* gene of rice, which encodes betaine aldehyde dehydrogenase, lead to the production of fragrance compound 2-acetyl-1-pyrroline (2AP), resulting in aromatic rice varieties with improved fragrance (Shan et al. 2015). Knocking out the VInv gene (encoding the enzyme vacuolar invertase, responsible for breaking down sucrose into glucose and fructose) in *Ranger Russet* potatoes prevented the accumulation of reducing sugars and reduced acrylamide formation during high-temperature processing. This resulted in lighter-coloured, less bitter chips with improved food safety (Clasen et al. 2016).

13.11 CRISPR-Cas Gene Editing System

The CRISPR-Cas9 gene editing system has indeed revolutionised the approach to crop trait improvement. The method not only allows precision towards the target but also significantly reduces the time from process initiation to the final product. The approach has been used to improve agronomic, yield, nutritional, abiotic stress tolerance, and biotic stress resistance traits in most commercial crops, including rice, wheat, maize, tomato, potato, cotton, soybean, sorghum, banana, and peanut (Table 13.1). The discovery of the CRISPR-Cas system began with the identification of unusual repetitive DNA sequences in prokaryotic genomes, first reported in 1987 by Ishino's research group. This discovery was based on their observation of the unexplained presence of an array of short palindromic repeats interspersed with unique sequences in the *Escherichia coli* genome. Further studies identified similar sequences in various bacterial and archaeal genomes as well, indicating that it is a conserved genetic feature (Mojica et al. 2000). Later, in the early 2000s, several independent research groups analysed these repeat sequences and subsequently named them clustered regularly interspaced short palindromic repeats (CRISPR) (Jansen et al. 2002). It was known that the spacer sequences within CRISPR arrays were derived from plasmid and viral DNA, which are involved in bacterial adaptive immunity. This finding has led to the hypothesis that CRISPR is a defence mechanism in bacteria that functions by storing genetic information from past infections to

Table 13.1 List of important cereal and commercial crops where the CRISPR-Cas9 based gene editing method was deployed towards crop trait improvement

Crop	Target gene	Associated trait	References
Banana	*MaACO1*	Modifies ethylene biosynthesis, delaying ripening and extending shelf life	Hu et al. (2020)
	LcyE	Enhances β-carotene accumulation by modulating carotenoid biosynthesis	Kaur et al. (2020)
	MaACO1	Prolongs fruit shelf life by modulating ethylene biosynthesis	Hu et al. (2021)
Cotton	*GhFAD2-1A/D*, *GhPEPC1*, and *GhPEPEC2*	Enhances cottonseed oil quality by modifying fatty acid composition	Chen et al. (2020)
Grapefruit	*CsLOB1* promoter	Improves resistance to citrus canker disease via transcriptional modulation	Jia et al. (2017)
Maize	*Wx1*	Encodes granule-bound starch synthase; mutations result in waxy (amylose-free) starch	Nelson and Rines (1962)
	PPR and *RPL*	Pentatricopeptide repeat and ribosomal protein L genes; their modification reduces zein protein content	Tang et al. (2016)
	Psy1	Phytoene synthase gene; mutations alter carotenoid biosynthesis and kernel colour	Zhu et al. (2016)
	ALS	Confers herbicide resistance by altering acetolactate synthase activity	Svitashev et al. (2015)
	ARGOS8	Enhances grain yield under drought stress by regulating ethylene responses	Shi et al. (2017)
Peach	*PpAAT1*	Regulates ester biosynthesis, influencing fruit aroma and flavour	Peng et al. (2020)
Peanut	*FAD2A* and *FAD2B*	Modifies linoleic acid content, improving oil quality	Yuan et al. (2019)
	Arahy.4E7qku and *Arahy*	Reduces palmitic acid content to enhance nutritional value	Tang et al. (2022)
Physalis sp.	*ClV1*	Controls fruit size by regulating CLAVATA signalling pathways	Lemmon et al. (2018)
Potato	*elF4E* and *elF (iso)4E*	Provides virus resistance and prevents cold-induced sweetening via translation initiation factor modifications	Hameed et al. (2020)

(continued)

Table 13.1 (continued)

Crop	Target gene	Associated trait	References
Rape	*BnaFAD2*	Increases oleic acid content by altering fatty acid desaturation	Huang et al. (2020)
	BnTT8	Regulates seed oil biosynthesis and grain protein content (GPC)	Zhai et al. (2020)
	BnITPK	Reduces phytic acid content, improving phosphorus bioavailability	Sashidhar et al. (2020)
Rice	*GDEP1, S3*, and *Gn1a*	Regulate grain length and panicle architecture in rice	Shen et al. (2018)
	SPO11-1, *REC8*, *OSD1*, and *MATL*	Key regulators of apomictic reproduction	Xie et al. (2019)
	OsGBSSI	Encodes granule-bound starch synthase, determining amylose content	Xu et al. (2021b)
	OsNramp5	Regulates cadmium uptake, reducing Cd accumulation in rice	Yu et al. (2017)
	REC8, *PAIR1*, *OSD1*, and *BBM1*	Enables clonal seed formation via apomixis	Khanday et al. (2019)
	OsPLDα1	Phospholipase D alpha 1 gene; its suppression lowers phytic acid content	Khan et al. (2019)
	OsGAD3	Encodes glutamate decarboxylase, increasing GABA content in rice grains	Akama et al. (2020)
	OsAAP6 and *OsAAP10*	Amino acid transporters affecting grain protein content	Wang et al. (2020b)
	PAIR1, *REC8*, *OSD1*, and *BBM4*	Essential for synthetic apomixis and clonal reproduction	Wei et al. (2023)
	bHLH024	Regulates Na+/K+ homeostasis, improving salt tolerance	Alam et al. (2022)
	OsSBEI and *OsSBEIIb*	Control starch branching, impacting amylose–amylopectin ratio	Sun et al. (2017a, b)
	OsF3′H, *OsDFR*, *OsLDOX*, and *OsTTG1*	Regulates anthocyanin biosynthesis for pigmentation	Jung et al. (2019) and Yang et al. (2021)
	REC8, *PAIR1*, *OSD1*, and *BBM1*	Induces synthetic apomixis in F1 hybrids	Vernet et al. (2022)
	Waxy	Encodes granule-bound starch synthase; mutations produce waxy (glutinous) starch	Zhang et al. (2018)

(continued)

Table 13.1 (continued)

Crop	Target gene	Associated trait	References
	OsOr	Enhances β-carotene accumulation in rice endosperm	Endo et al. (2019)
	OsGRF3, *GL2*, and *OsGRF4*	Regulate grain size by modulating cell division and expansion	Hao et al. (2019)
	CrtI, *PSY*, and *Osor*	Involved in β-carotene biosynthesis for provitamin A rice	Al-Babili et al. (2006) and Endo et al. (2019)
	OsBADH2	Suppresses fragrance biosynthesis by encoding betaine aldehyde dehydrogenase	Ashokkumar et al. (2020)
	B-type response regulator (*RR22*)	Enhances salt tolerance via cytokinin signalling	Zhang et al. (2019)
	ONAC127 and *ONAC129*	NAC transcription factors reduce seed set under heat stress	Ren et al. (2021)
	FAD2-1A and *FAD2-1B*	Encode fatty acid desaturases; mutations increase oleic acid content	Abe et al. (2018)
	Drought and salt tolerance (*DST*)	Modulates ROS homeostasis, enhancing salt and drought tolerance	Santosh Kumar et al. (2020)
	GS9	Controls slender grain shape through cell elongation	Zhao et al. (2018a and b)
	TGW6, *GS3*, *OsGn1a*, *OsSWEET11*, *GW2*, and *GW5*	Influence grain weight, size, and carbohydrate allocation	Ma et al. (2017), Zhou et al. (2019), and Xu et al. (2016)
	OsGBSSI	Governs amylose biosynthesis, affecting grain texture	Xu et al. (2021a)
Sorghum	*SbBADH2*	Betaine aldehyde dehydrogenase gene; its alteration imparts a unique aroma	Zhang et al. (2022)
Soybean	*GmIPK1*	Lowers phytic acid accumulation in seeds by modulating myo-inositol phosphate metabolism	Song et al. (2022)
	HaHB4	Enhances drought tolerance through HD-Zip transcription factor regulation	Martignago et al. (2020)
	GmGOLS1A and *GmGOLS1B*	Galactinol synthase genes; their manipulation enhances protein and fat content	Le et al. (2020)
	2AP and *GmBADH2*	Involved in 2-acetyl-1-pyrroline biosynthesis, contributing to aroma	Qian et al. (2022)
	FAD2-1A and *FAD2-1B*	Encode fatty acid desaturases; mutations improve oil quality	Kim et al. (2017)

(continued)

Table 13.1 (continued)

Crop	Target gene	Associated trait	References
Strawberry	*FaPG1*	Encodes polygalacturonase, involved in fruit softening and cell wall degradation	Quesada et al. (2009)
Tea	*CsHB1*	Reduces caffeine biosynthesis via transcriptional regulation	Ma et al. (2021)
Tomato	*SlGAD2* and *SlGAD3*	Enhances y-aminobutyric acid (GABA) accumulation, improving stress tolerance and nutrition	Nonaka et al. (2017)
	CLV3, *fas*, *lc*, *OVATE*, and *ENO*	Controls fruit morphology, influencing shape and size	Yuste-Lisbona et al. (2020), Zsögön et al. (2018), Chu et al. (2019), and Tanksley (2004)
	SlAGL6	Promotes seedless fruit development via MADS-box transcription factor regulation	Gupta et al. (2021)
	CrtR-b2, *SlMYB12*, *CRTISO*, *ANT1*, and *Psy1*	Regulates carotenoid biosynthesis and fruit pigmentation	Vu et al. (2020), D'Ambrosio et al. (2018), Deng et al. (2018), Čermák et al. (2015), and Shlush et al. (2020)
	ALS	Confers resistance to bacterial speck disease through gene editing	Ortigosa et al. (2019)
	GGP1	Regulates ascorbic acid (Vitamin C) biosynthesis	Li et al. (2018b)
	slyPDS	Increases lycopene content by inhibiting phytoene desaturase	Li et al. (2018b)
	GID1a	Enhances drought tolerance and harvest index by modulating gibberellin signalling	Illouz-Eliaz et al. (2020)
	CycB	Alters carotenoid metabolism, increasing β-carotene (provitamin A)	Zsögön et al. (2018)
	SlIAA9	Induces parthenocarpy by disrupting auxin signalling pathways	Ueta et al. (2017) and Nguyen et al. (2024)
	SlGT30	Increases fruit size and weight while improving drought tolerance	Lv et al. (2024)
	ALC	Delays fruit softening and extending shelf life	Yu et al. (2017)
	SPO11-1, *REC8*, *TAM*, and *OSD1*	Facilitates clonal gamete formation for synthetic apomixis	Wang et al. (2024)
Wheat	*EDR1*	Confers resistance to powdery mildew (*Blumeria graminis* f. sp. tritici)	Zhang et al. (2017)

(continued)

Table 13.1 (continued)

Crop	Target gene	Associated trait	References
	α-Gliadin	A gluten component; its reduction leads to low-gluten wheat	Sánchez-León et al. (2018)
	TaGW2	Affects grain size through modulation of cell number	Wang et al. (2018a, b)
	TaGW2 and *TaARR12*	Improves drought resistance and grain yield via growth regulation	Li et al. (2024)
	TaGW7	Influences grain shape by regulating cell division	Wang et al. (2019a, b)

recognise and eliminate invading nucleic acids during future infections (Mojica et al. 2005; Bolotin et al. 2005). The ability of *Streptococcus thermophilus* to acquire new spacers from bacteriophages has provided experimental validation of its functional role and has been shown to confer resistance to future infections (Barrangou et al. 2007). This directly demonstrated that prokaryotes have an adaptive immune system in the form of CRISPR-Cas systems. By studying the molecular mechanism of CRISPR-mediated immunity, it was understood that CRISPR-derived RNAs (crRNAs) can direct Cas proteins to degrade foreign genetic elements (Brouns et al. 2008).

CRISPR-Cas systems are classified into two major classes. Class 1 systems utilise multi-subunit protein complexes for DNA interference, whereas Class 2 systems cleave DNA using a single effector protein, such as Cas9 (Makarova et al. 2011). Among these, the Cas9 protein derived from *Streptococcus pyogenes* was identified as a key enzyme, as it can mediate RNA-guided DNA cleavage and function independently without the need for other multi-protein complexes (Gasiunas et al. 2012). In the same year, similar findings were reported by Charpentier and Doudna (Jinek et al. 2012), who showed that the system could be simplified by combining crRNA and tracrRNA into a single guide RNA (sgRNA). Cong et al. (2013) first adapted CRISPR-Cas9 for eukaryotic gene editing, as they programmed Cas9 from *S. thermophilus* and *S. pyogenes* for precise DNA cleavage in human and mouse cells. They also demonstrated multiplex editing, which led to homology-directed repair (Cong et al. 2013). This laid the foundation for its use as a genome editing tool for targeted gene modification.

The CRISPR-Cas system comprises three main components: the CRISPR array, trans-activating CRISPR RNA (tracrRNA), and Cas proteins, each playing a crucial role in adaptive immunity (Barrangou et al. 2007). The CRISPR array is a genomic locus with repetitive DNA sequences interspersed with unique spacer sequences (Mojica et al. 2005). The direct repeats in CRISPR arrays are typically 24–48 base pairs (bp) long, and they can form hairpin-like secondary structures (Horvath and Barrangou 2010). They also provide recognition sites for Cas protein binding and RNA processing (Jansen et al. 2002). The spacer sequences are usually 21–72 bp long. They are derived initially from previous viral or plasmid infections, which can serve as a memory to identify foreign DNA and provide resistance in future infections (Bolotin et al. 2005). Before the CRISPR array, there is a leader sequence (~100–500 bp long) that contains promoter elements for transcription and regulates the acquisition of new spacers (Pourcel et al. 2005). The transcription of this array gives a long precursor CRISPR RNA (pre-crRNA), which is broken down into smaller crRNAs that guide Cas proteins to target invading DNA (Deltcheva et al. 2011). Cas proteins function in spacer acquisition, crRNA processing, and interference with invading DNA. They can be classified based on the type of CRISPR system to which they belong (Makarova et al. 2015). The most widely used Cas protein is Cas9, a ~160 kDa protein composed of multifunctional, distinct domains that can mediate RNA binding, DNA recognition, and cleavage (Nishimasu et al. 2014). Cas proteins comprise three domains, each playing a critical function. The recognition (REC) lobe contains REC1, REC2, and REC3 domains that interact with the guide RNA (crRNA and tracrRNA or sgRNA), thereby providing sequence specificity (Nishimasu et al. 2014; Jinek et al. 2014). The REC lobe undergoes conformational changes upon binding to RNA, further activating Cas9 for DNA cleavage (Nishimasu et al. 2014). The nuclease (NUC) lobe: HNH domain cleaves the target DNA strand that is complementary to the crRNA (Gasiunas et al. 2012; Jinek et al. 2012). While the RuvC domain makes a double-stranded cut at the noncomplementary DNA strand (Jinek et al. 2012; Nishimasu et al. 2014), the bridge helix, a positively charged α-helix, stabilises guide RNA binding and activates the cleavage function (Nishimasu et al. 2014). The PAM-interacting domain (PID)/C-terminal domain (CTD) recognises the protospacer adjacent motif (PAM), a short sequence (e.g. 5′-NGG for SpCas9) essential for Cas9 to bind and cleave DNA (Anders et al. 2014). The PAM-interacting domain enhances target specificity and prevents Cas9 from cleaving its own DNA (Anders et al. 2014). The different types of Cas proteins have distinct cleavage mechanisms and target preferences. The most common Cas protein variants are Cas12a (Cpf1) that do not require tracrRNA and cuts DNA with staggered overhangs (Zetsche et al. 2015); Cas13a (C2c2), which targets RNA instead of DNA and makes it useful for transcriptome engineering (Abudayyeh et al. 2016), and Cas3 that degrades DNA in a processive manner and can act as a molecular shredder (Makarova et al. 2015).

Based on the composition and function of their interference machinery, CRISPR-Cas systems are classified into two major classes. Class 1 CRISPR-Cas systems use multi-protein complexes for DNA targeting and interference. These systems are further classified into Type I, Type III, and Type IV systems, in which Type I is the well-studied system (Makarova et al. 2015). Type I system utilises a multi-subunit cascade complex to recognise the target and a Cas3 nuclease for DNA degradation (Brouns et al. 2008). Type III system targets both DNA and RNA using the Csm and Cmr effector complexes (Hale et al. 2009). Although the Type IV system is less understood, it is reported to be involved in anti-plasmid defence mechanisms (Makarova et al. 2020). The other class of CRISPR-Cas systems is Class 2 CRISPR-Cas systems that rely on a single effector protein for DNA cleavage. They include Type II, Type V, and Type VI systems (Shmakov et al.

2017), wherein Type II utilises Cas9 for targeted DNA cleavage (Jinek et al. 2012), Type V uses Cas12a (Cpf1), which creates staggered DNA cuts and processes its own crRNA (Zetsche et al. 2015). Type VI utilises Cas13 for RNA interference, which is beneficial for transcriptome engineering (Abudayyeh et al. 2016).

Guide RNAs are essential for guiding Cas proteins to the specific DNA or RNA sequences. CRISPR RNA (crRNA) is a small RNA sequence (~20 nucleotides long) that contains a spacer sequence derived from previous viral infections. It is complementary to the target DNA sequence (Brouns et al. 2008), thus directing Cas9 to cleave matching DNA (Gasiunas et al. 2012). Trans-activating CRISPR RNA (tracrRNA) is required in Type II CRISPR-Cas systems, such as Cas9, to activate crRNA processing and guide Cas9 to the target DNA (Deltcheva et al. 2011). Single-guide RNA (sgRNA) is generated by combining tracrRNA and crRNA into a single molecule, enabling a variety of CRISPR-Cas9 applications (Jinek et al. 2012).

The CRISPR-Cas system functions in three significant steps: adaptation, expression, and interference (Barrangou et al. 2007). Briefly, when a virus infects bacteria, the bacterial Cas proteins recognise these short fragments of the viral DNA and integrate them into the CRISPR array as new spacers. This process is facilitated by Cas1 and Cas2 proteins, which can select protospacers from invading DNA (Nuñez et al. 2014). TracrRNA aids in crRNA maturation in Type II CRISPR-Cas systems (Deltcheva et al. 2011). Pre-crRNA is generated from the CRISPR array, which is subsequently processed to form mature crRNAs (Brouns et al. 2008). The crRNA-tracrRNA or sgRNA then guides Cas proteins to complementary target sequences (Jinek et al. 2012), which subsequently causes a double-strand break (DSB) in DNA upon recognition of the PAM sequence (e.g. NGG in *S. pyogenes*) (Anders et al. 2014). The DNA break caused is later repaired by either non-homologous end joining (NHEJ) or homology-directed repair (HDR) methods (Cong et al. 2013). This mechanism is utilised in the fields of molecular biology and biotechnology to mediate the editing of target genes of interest. CRISPR-Cas-mediated gene editing primarily operates through a sequence-specific recognition and cleavage mechanism. The process involves target recognition, binding, DNA cleavage, and repair of the double-stranded break (Jinek et al. 2012).

The trans-activating crRNA (tracrRNA) and CRISPR RNA (crRNA) combine to form the single-guide RNA (sgRNA), which further directs the Cas9 endonuclease towards the desired DNA sequence (Jinek et al. 2012). The presence of a 20-nucleotide spacer sequence complementary to the target DNA at the 5′ end of the sgRNA facilitates precise base-pairing interactions (Cong et al. 2013). The 3′ end of the sgRNA contains a stem-loop structure crucial for binding to the Cas9 protein, thereby stabilising the ribonucleoprotein complex (Nishimasu et al. 2014). Cas9 first recognises the protospacer adjacent motif (PAM) sequence before DNA cleavage (Anders et al. 2014; Jinek et al. 2012). PAM sequence prevents Cas9 cleavage within the CRISPR array of the host genome as the repeats within the CRISPR locus lack PAM sequences (Mojica et al. 2009). Structural studies have demonstrated that the PAM interacts with the Cas9 C-terminal domain, resulting in conformational changes that enhance DNA binding (Anders et al. 2014). The double-stranded DNA (dsDNA) close to the target site unwinds when Cas9 recognises the PAM sequence, allowing base pairing between the complementary DNA strand and the gRNA spacer region (Jiang et al. 2016). This R-loop formation occurs as the nontarget DNA strand is displaced and the target strand hybridises with the guide RNA (Sternberg et al. 2015). According to Hsu et al. (2013), ambiguities between the gRNA and target DNA sequences significantly lower the cleavage efficiency, which depends on a stable R-loop structure. The Cas9 nuclease causes a double-strand cut at the target site once the R-loop is established through its two nuclease domains, namely HNH (Gasiunas et al. 2012) and RuvC domain (Jinek et al. 2012). The break occurs at 3 bp upstream of the PAM sequence and produces blunt-ended DSBs, mostly (Anders et al. 2014). Cas9 undergoes significant conformational changes upon binding to the target strand, transitioning from an inactive to an active conformation (Nishimasu et al. 2014). Crystal structure studies reveal that Cas9 undergoes a hinge-like movement, facilitating the alignment of the HNH and RuvC domains to simultaneously cleave both DNA strands (Jiang et al. 2016). Engineered variants, such as Cas9 nickase (nCas9), can introduce a single-strand break (SSB or 'nick') instead of a double-strand break (DSB) because they have inactivated the HNH or RuvC domain (Ran et al. 2013). The paired nCas9 variants can be used to enhance specificity by generating double-strand breaks (DSBs) only when two adjacent nicks occur (Mali et al. 2013). DNA repair is a crucial step in CRISPR-Cas gene editing. After cleavage, the cell's endogenous DNA repair machinery takes over and joins the ends using either of the two DNA repair mechanisms: Non-homologous end joining (NHEJ) or homology-directed repair (HDR).

13.12 CRISPR Assembly and Delivery Methods in Plants

The assembly and delivery of CRISPR components are crucial for obtaining efficient and targeted gene editing. The efficiency of CRISPR-based gene editing relies on a proper assembly of CRISPR-Cas components, the selection of an optimal delivery system, and cellular uptake and expression of these components (Mali et al. 2013). CRISPR-based genome editing involves assembling the Cas protein and the guide RNA (gRNA) into a functional ribonucleoprotein (RNP) complex, which can mediate sequence-specific DNA cleavage. The assembly process varies based on the CRISPR system used. *Agrobacterium tumefaciens* is the most commonly used bacterial vector for delivering CRISPR constructs into plant genomes. The CRISPR-Cas9 construct is integrated into the T-DNA region of the binary vector, which is incorporated into the plant genome upon infection (Gelvin 2017). It is used for stable transformation in dicots, such as tomato, soybean, and Arabidopsis (Shan et al. 2013). It can result in low-copy integration of the transgene and reduce gene-silencing issues (Hiei et al. 1994). It is less efficient for monocots, such as wheat and rice, which require additional optimisation (Hwang et al. 2017). The biolistic (particle bombardment) method uses a gene gun to introduce CRISPR DNA constructs into plant cells by coating them onto gold or tungsten particles (Altpeter et al. 2016). In cereal crops such as wheat, rice, and maize, which are challenging to transform using Agrobacterium, biolistics can be employed for both transformation and the transient or stable expression of CRISPR components, eliminating the need for integration (Liu et al. 2019). It can cause random integration and multiple insertions, leading to potential gene disruption (Svitashev et al. 2015). Plant protoplasts, which are cell wall-free plant cells, enable the direct delivery of pre-assembled CRISPR-Cas9 ribonucleoproteins (RNPs) or plasmid DNA (Woo et al. 2015). It exhibits high efficiency and stable transgene integration, thereby reducing other regulatory concerns (Zhang et al. 2016). It facilitates rapid screening of gene- edited events before regenerating whole plants (Andersson et al. 2018). It is restricted to a limited number of plant species, including rice, tobacco, and Arabidopsis. Plant viruses, such as tobacco rattle virus (TRV), barley stripe mosaic virus (BSMV), and cotton leaf crumple virus (CLCrV), are used to deliver the sgRNA and Cas9 into plant cells (Ali et al. 2015). It allows transient gene editing without transgene integration. Its restricted cargo capacity of viral vectors restricts the delivery of large Cas9 variants (Ariga et al. 2020).

Nanoparticles such as lipid nanoparticles, mesoporous silica, and carbon nanotubes are emerging as promising tools for nongenetically modified (non-GM) delivery of CRISPR components (Wang et al. 2019a, b). It facilitates transgene-free gene editing and avoids other regulatory concerns (Demirer et al. 2021). Enhanced specificity and minimised off-target effects make editing more precise and prevent unintended genetic alterations that may cause harmful traits or reduced efficiency. It also helps overcome regulatory approval and maintains genome stability in crop and medical applications (Wiles et al. 2015). Genome-wide homology searches are preferred for selecting gRNAs with minimal off-target sites (Jinek et al. 2012; Fu et al. 2013; Hsu et al. 2013). Additionally, the use of truncated sgRNAs, 17–19 nt in length, significantly reduces off-target activity while retaining high target efficiency (Fu et al. 2014). Cas9 nickase (D10A mutant) with paired sgRNAs creates adjacent single-strand nicks, forming a precise DSB with 50–1500-fold reduced off-target effects (Ran et al. 2013). An integration of dCas9 with the FokI nuclease domain requires paired sgRNAs for dimerization, enhancing specificity similar to ZFN and TALEN (Guilinger et al. 2014; Tsai et al. 2014). Experimental methods such as genotyping and computational predictions help identify and eliminate off-target mutations. In animal models, selective breeding can separate desired edits from off-target alleles (Frock et al. 2015; Tsai et al. 2015; Wang et al. 2015).

CRISPR-Cas9 technology has transformed genetic engineering by facilitating precise genome modifications (Jinek et al. 2012). Advanced methods, such as dead Cas9 (dCas9), base editing, and prime editing, have expanded the CRISPR toolkit for various applications, including transcriptional and epigenetic modifications beyond the creation of double-strand breaks. Dead Cas9 (dCas9) is a catalytically inactive form of the Cas9 protein that can still bind to specific DNA sequences guided by RNA. Still, it cannot induce cleavage as it lacks endonuclease activity (Qi et al. 2013). Due to this property of dCas9, it can be utilised as a platform to recruit various effector domains to specific genomic loci without inducing double-strand breaks. Transcriptional regulation of genes can be done by combining dCas9 with transcriptional activators or repressors. For example, dCas9 fused to repressor domains decreases gene expression, whereas dCas9 fused to activation domains increases target gene expression (Gilbert et al. 2013). Additionally, the linking of epigenetic modifiers, such as histone acetyltransferases and DNA methyltransferases, to dCas9 results in site-specific epigenetic modifications that influence gene expression without altering the primary DNA sequence (Hilton et al. 2015).

Base editing directly converts one nucleotide base pair into another without the use of donor DNA templates or the induction of double-stranded DNA breaks (Komor et al. 2016). This method utilises a deaminase enzyme fused to a catalytically impaired Cas9, enabling accurate single-base modifications. Cytosine base editors (CBEs) convert cytosine (C) to thymine (T) by deaminating cytosine into uracil, which can subsequently be recognised and repaired as thymine by cellular mechanisms. This process allows for the correction of C–G to T–A mutations that are common in various genetic disorders (Komor et al. 2016). Adenine base editors (ABEs) help convert adenine (A) to guanine (G) by deaminating adenine into inosine, which is interpreted as guanine during replication. Consequently, ABEs mediate the correction of A–T to G–C mutations (Gaudelli et al. 2017). Overall, base editing reduces unintended insertions or deletions associated with double-stranded breaks. Base editing is a more efficient and less disruptive alternative to traditional CRISPR-Cas9 editing (Komor et al. 2016).

Prime editing is a flexible genome editing method that can make small insertions and deletions, as well as all 12 possible base-to-base substitutions, without causing double-strand breaks or requiring a donor template DNA (Anzalone et al. 2019). This technique uses a prime editing guide RNA (pegRNA) that identifies the target site and encodes the desired edit in conjunction with a Cas9 nickase fused to a reverse transcriptase enzyme (Anzalone et al. 2019). The prime editing process involves guiding Cas9 nickase to the specific DNA target site by the pegRNA, where it induces a single-strand nick (Anzalone et al. 2019). The reverse transcriptase creates the desired DNA sequence at the nicked site using the RNA template found in the pegRNA. (Anzalone et al. 2019). Cellular repair mechanisms will integrate the new DNA sequence to get the intended edit (Anzalone et al. 2019). In comparison to conventional CRISPR-Cas9 and base editing methods, this approach offers greater accuracy and adaptability, with an increased likelihood of correcting a broad range of genetic mutations (Anzalone et al. 2019).

13.13 Tissue Culture-Free Genome Editing Approaches

Widely recognised as the gene gun technique, biolistic-mediated gene editing facilitates the direct delivery of CRISPR/Cas9 components into shoot apical meristems (SAMs), enabling targeted genetic modifications. This approach has been successfully employed in wheat, where the resulting edits were heritable across generations. However, challenges such as genomic instability, which manifests as significant deletions and chromosomal rearrangements, underscore the need for further optimisation to enhance accuracy and effectiveness (Hamada et al. 2018; Kumagai et al. 2022). Single-guide RNAs (sgRNAs) can be delivered into Cas9-expressing plants via RNA viruses, enabling systemic genome editing without the need for tissue culture-based regeneration. Using viral platforms such as tobacco rattle virus (TRV) and barley stripe mosaic virus (BSMV), this approach has been successfully applied in species including *Nicotiana benthamiana*, Arabidopsis, wheat, and barley. Editing efficiency has been significantly enhanced by advancements such as incorporating mobile sgRNA sequences and exonuclease-driven deletions. However, challenges remain, including limited viral cargo capacity and the inability of some viruses to access meristematic tissues in certain crops (Ali et al. 2015; Ellison et al. 2020; Chen et al. 2022). A novel strategy for overcoming viral exclusion from shoot meristems involves targeting latent axillary meristematic cells in plants expressing Cas9. This strategy, successfully demonstrated in tomato, enabled the regeneration of gene-edited shoots with stable and heritable modifications. Given that axillary bud activation is a fundamental aspect of plant propagation, this method presents a promising avenue for crop improvement. However, its efficiency remains variable across different plant species (Liu et al. 2024). The transient expression of key developmental regulators (DRs), such as WUSCHEL2 (WUS2) and isopentenyl transferase (*IPT*), can induce meristem-like structures from somatic cells, facilitating genome editing independent of conventional regeneration techniques. This strategy has been successfully employed in crops like tomato and grape. However, sustained expression of these regulators often leads to developmental abnormalities, which can be mitigated through temperature-regulated control of IPT expression using a viral vector (Maher et al. 2020; Lian et al. 2022). Certain plants possess a natural capacity for regeneration through root suckering, a trait that can be harnessed for tissue culture-independent genome editing. This approach involves immersing stem cuttings in a solution containing *Agrobacterium rhizogenes* engineered with CRISPR/Cas9 reagents, triggering the formation of gene-edited hairy roots. Subsequent shoot emergence from these roots results in the inheritance of targeted genetic modifications. Successfully demonstrated in *Taraxacum kok-saghyz* and sweet potato, this strategy holds promise for broader application in vegetatively propagated crops (Cao et al. 2023; Lu et al. 2024).

13.14 Social Acceptance and Regulatory Considerations of Genome-Edited Crops

The genetically modified food crops being cultivated since the 1990s have achieved a limited number of consumer preferences. This hesitation is due to a lack of awareness and an unclear distinction between transgenic/genetically modified crops and site-specific genome-edited crops, which are free from foreign DNA. While gene modification implies the addition of a section of DNA (sometimes from other species), gene editing essentially involves snipping or editing out a section of a plant's DNA. The potential of advanced biotechnological tools in transforming agriculture with precision editing and in a time-efficient manner can bring about a new Green Revolution, thus promising a nutritionally secure and sustainable future. This also aligns with two of the United Nations' (UN) sustainable development goals (SDGs) framework, specifically SDG2: zero hunger and SDG13: climate action (https://sdgs.un.org/goals). However, the development and marketability of genome-edited crops in a country are greatly influenced by its regulatory framework. The primary regulatory triggers are highlighted as process-based (e.g. European Union) and product-based (e.g. Canada). The United States follows a hybrid system, and genome-edited plants are reviewed on a case-by-case basis, as the regulatory trigger is process-based, but risk assessment is product-based (Strauss and Sax 2016; McHughen 2016). Argentina does not regulate plants that are null-segregants (i.e. do not retain the transgene in the final product) as GMOs (Eckerstorfer et al. 2019). For instance, the process of transgene delivery mediated through Agrobacterium, where over generations, the transgene is removed by segregation, resulting in transgene-free lines (Shan et al. 2015; Chandrasekaran et al. 2016; Nekrasov et al. 2017). However, on 25 July 2018, the European Court of Justice ruled that any organism developed using genome editing technology must undergo a comprehensive assessment and that its products must fulfil the requirements for detection, identification, and quantification (Eckerstorfer et al. 2019).

Several of the current gene editing techniques are far more precise in introducing single-nucleotide changes without integrating foreign DNA, making it difficult to assess if the mutation occurred naturally or was introduced deliberately, which is a major bottleneck to the acceptance of genome-edited crops. Grohmann et al. (2019) discuss the challenge that might arise for products of gene edited plants, such as seed, food, and feed testing laboratories and enforcement institutions regarding the identification of sequence alterations, the process used, and the genotype created (i.e. the organism with the type of genetic alteration) (Grohmann et al. 2019). However, even after the technical hurdles are tackled, legal obligations such as labelling and traceability will remain.

The initial global legal and ethical discourses surrounding gene edited plants necessitated clarification regarding regulatory mechanisms and a clear distinction between genetically edited crops and genetically modified crops. The regulatory framework regarding edited plants is both stringent, as seen in New Zealand and the European Union (EU), and gene editing friendly, as seen in the United States and Canada (Menz et al. 2020). Drafting of the regulatory framework requires keen consideration of social acceptance, farmers' rights, and intellectual property rights. Additionally, patent innovations are also a way of obtaining 'ethical licensing' related to gene editing. Jennifer Doudna and Emmanuelle Charpentier of the University of California, Berkeley and the University of Vienna, respectively, obtained original CRISPR/Cas9 patents for general use. In contrast, Feng Zhang of the Massachusetts Institute of Technology (MIT), Harvard, and the Broad Institute obtained one patent for use in eukaryotes (Feeney et al. 2018).

However, in light of the current need for tailor-made crops, the Government of India recently released the 'Guidelines for the Safety and Assessment of Genome-

Edited Plants-2022,' which outlines the regulations governing the research and release of genome-edited plants. The guidelines released categorise the site-directed DNA sequence change into three classes based on the expected modification outcome and the repair mechanism directed. The site-directed nuclease 1 (SDN-1) category encompasses any site-directed mutation where the repair is carried out by the cell's repair machinery and does not require a DNA template. The majority of gene editing reports utilise this type of mutagenesis, as sequence similarity facilitates easy target identification and predictable editing. The editing here generally results in small insertions, deletions, or nucleotide residue substitutions (Podevin et al. 2013). Site-directed nuclease 2 (SDN-2) encompasses gene modification that results in altered gene expression patterns. The site-directed mutation requires an exogenous DNA template that acts as a reference and guides DNA repair into the desired change. However, the restriction of homology-directed repair (HDR) to only G2 and late S-phase makes SDN-2 implementation a challenge. Notably, in contrast to the conventional T-DNA delivery method, *Geminiviruses* have been shown to aid an easy HDR repair through S-phase induction (Eini et al. 2022). The site-directed nuclease 3 (SDN-3) involves the site-directed insertions of an extensive DNA sequence, such as the introduction of a new gene (not naturally available in the gene pool). Additionally, an insertion sequence with a flanking sequence similar to that of the host is utilised to direct HDR. Hence, SDN-3-generated plants are not exempt from the regulations implemented for genetically engineered crops related to the manufacture, use, import, export, and storage of hazardous microorganisms/ genetically engineered organisms or cells (Rules 1989), similar to the EFSA GMO panel (EFSA GMO Panel 2020).

13.15 Conclusions and Future Prospects

Sustainability through gene editing is promised, as gene edited crops can not only significantly increase yield per unit land area and confer protection against both biotic and abiotic stressors, but also possess better water-use efficiency (WUE) and nitrogen-use efficiency (NUE), while simultaneously nourishing soil health. Thus, there is an effective reduction in overreliance on intensive irrigation systems and synthetic pesticides. Additionally, gene edited crops have also successfully delivered improved qualitative attributes, such as enhanced nutrition, flavour, and quality, as well as increased shelf life. Deployment of gene editing for crop trait improvement will not only create novel alleles, but deleterious alleles can also be removed. Cumulatively, gene edited crops reduce post-harvest waste, decrease their environmental footprint, and promise sustainability (Shipman et al. 2021). Moral responsibility towards both the planet and its people demands transparency and the incorporation of agricultural and crop gene editing into the current biosafety framework. However, exemption from biosafety regulations should be considered for crops where the editing output does not involve the insertion of foreign DNA or where the editing nature or proportion is similar to that of mutagenesis.

Conclusively, the success stories of gene editing in crop species showcase its remarkable potential for transforming crop improvement programmes. The incorporation of genomics-aided candidate-gene identification approaches into genomics-informed gene editing methods can lead to increased precision in target-gene editing. Collaborative efforts in the same direction can accelerate efforts to improve traits and reshape current agriculture towards sustainability.

Acknowledgements Authors' research in this area is supported by funding from Anusandhan National Research Foundation (erstwhile Science and Engineering Research Board), Department of Science and Technology, Ministry of Science and Technology, Govt. of India [File Nos. CRG/2022/003696 and EEQ/2023/000467], and Department of Biotechnology, Ministry of Science and Technology, Govt. of India [File Nos. HRD-17011/4/2023-HRD-DBT and BT/PR51246/AGIII/103/1476/2023]. The authors acknowledge the 'DST FIST Level C' support to the Department of Plant Sciences, School of Life Sciences, University of Hyderabad, from the Department of Science and Technology, Ministry of Science and Technology, Govt. of India (File No.: SR/FST/LS-III/2023/1155). The authors also acknowledge the DBT BUILDER grant awarded to the School of Life Sciences, University of Hyderabad, from the Department of Biotechnology, Ministry of Science and Technology, Govt. of India (File No.: BT/INF/22/SP41176/2020). The figures are created in BioRender.com.

References

Abe K, Araki E, Suzuki Y, Toki S, Saika H (2018) Production of high oleic/low linoleic rice by genome editing. Plant Physiol Biochem 131:58–62

Abudayyeh OO, Gootenberg JS, Konermann S, Joung J, Slaymaker IM, Cox DB, Shmakov S, Makarova KS, Semenova E, Minakhin L, Severinov K (2016) C2c2 is a single-component programmable RNA-guided RNA-targeting CRISPR effector. Science 353(6299):aaf5573

Aerts S (2012) Computational strategies for the genome-wide identification of cis-regulatory elements and transcriptional targets. Curr Top Dev Biol 98:121–145

Akama K, Akter N, Endo H, Kanesaki M, Endo M, Toki S (2020) An in vivo targeted deletion of the calmodulin-binding domain from rice glutamate decarboxylase 3 (Os GAD3) increases γ-aminobutyric acid content in grains. Rice 13:1–12

Alam MS, Kong J, Tao R, Ahmed T, Alamin M, Alotaibi SS, Abdelsalam NR, Xu JH (2022) CRISPR/Cas9 mediated knockout of the OsbHLH024 transcription factor improves salt stress resistance in rice (Oryza sativa L.). Plants 11(9):1184

Al-Babili S, Hoa TTC, Schaub P (2006) Exploring the potential of the bacterial carotene desaturase CrtI to increase the β-carotene content in Golden Rice. J Exp Bot 57(4):1007–1014

Ali Z, Abul-Faraj A, Li L, Ghosh N, Piatek M, Mahjoub A, Aouida M, Piatek A, Baltes NJ, Voytas DF, Dinesh-Kumar S (2015) Efficient virus-mediated genome editing in plants using the CRISPR/Cas9 system. Mol Plant 8(8):1288–1291

Allard R (1961) Principles of plant breeding. Soil Sci 91:414

Al-Tamimi N, Brien C, Oakey H, Berger B, Saade S, Ho YS, Schmöckel SM, Tester M, Negrão S (2016) Salinity tolerance loci revealed in rice using high-throughput non-invasive phenotyping. Nat Commun 7(1):13342

Altpeter F, Springer NM, Bartley LE, Blechl AE, Brutnell TP, Citovsky V, Conrad LJ, Gelvin SB, Jackson DP, Kausch AP, Lemaux PG (2016) Advancing crop transformation in the era of genome editing. Plant Cell 28(7):1510–1520

Anders C, Niewoehner O, Duerst A, Jinek M (2014) Structural basis of PAM-dependent target DNA recognition by the Cas9 endonuclease. Nature 513(7519):569–573

Andersson M, Turesson H, Olsson N, Fält AS, Ohlsson P, Gonzalez MN, Samuelsson M, Hofvander P (2018) Genome editing in potato via CRISPR-Cas9 ribonucleoprotein delivery. Physiol Plant 164(4):378–384

Anzalone AV, Randolph PB, Davis JR, Sousa AA, Koblan LW, Levy JM, Chen PJ, Wilson C, Newby GA, Raguram A, Liu DR (2019) Search-and-replace genome editing without double-strand breaks or donor DNA. Nature 576(7785):149–157

Ariga H, Toki S, Ishibashi K (2020) Potato virus X vector-mediated DNA-free genome editing in plants. Plant Cell Physiol 61(11):1946–1953

Ashokkumar S, Jaganathan D, Ramanathan V, Rahman H, Palaniswamy R, Kambale R, Muthurajan R (2020) Creation of novel alleles of fragrance gene OsBADH2 in rice through CRISPR/Cas9 mediated gene editing. PLoS One 15(8):e0237018

Ashraf M, Akram NA, Foolad MR (2012) Marker-assisted selection in plant breeding for salinity tolerance. In: Plant salt tolerance: methods and protocols. Humana Press, Totowa, pp 305–333

Asimit J, Zeggini E (2010) Rare variant association analysis methods for complex traits. Annu Rev Genet 44:293–308

Barmukh R, Thakur N, Shah P (2024) Genomic interventions for improving crop yield and resilience. In: Plant molecular breeding in genomics era: concepts and tools. Springer, pp 63–94

Barrangou R, Fremaux C, Deveau H, Richards M, Boyaval P, Moineau S, Romero DA, Horvath P (2007) CRISPR provides acquired resistance against viruses in prokaryotes. Science 315(5819): 1709–1712

Bayer PE, Golicz AA, Scheben A, Batley J, Edwards D (2020) Plant pan-genomes are the new reference. Nat Plants 6(8):914–920

Belfort M, Roberts RJ (1997) Homing endonucleases: keeping the house in order. Nucleic Acids Res 25(17):3379–3388

Bharti G, Chimata MK (2019) Review on new plant breeding techniques. Int J Sci Res 8:723–730

Bitinaite J, Wah DA, Aggarwal AK, Schildkraut I (1998) Fok I dimerization is required for DNA cleavage. Proc Natl Acad Sci USA 95(18):10570–10575

Boch J, Bonas U (2010) Xanthomonas AvrBs3 family-type III effectors: discovery and function. Annu Rev Phytopathol 48(1):419–436

Boch J, Scholze H, Schornack S, Landgraf A, Hahn S, Kay S, Lahaye T, Nickstadt A, Bonas U (2009) Breaking the code of DNA binding specificity of TAL-type III effectors. Science 326(5959):1509–1512

Bolotin A, Quinquis B, Sorokin A, Ehrlich SD (2005) Clustered regularly interspaced short palindrome repeats (CRISPRs) have spacers of extrachromosomal origin. Microbiol 151(8): 2551–2561

Bonas U, Conrads-Strauch J, Balbo I (1993) Resistance in tomato to Xanthomonas campestris pv vesicatoria is determined by alleles of the pepper-specific avirulence gene avrBs3. Mol Gen Genet 238:261–269

Briggs AW, Rios X, Chari R, Yang L, Zhang F, Mali P, Church GM (2012) Iterative capped assembly: rapid and scalable synthesis of repeat-module DNA such as TAL effectors from individual monomers. Nucleic Acids Res 40(15):e117–e117

Brouns SJ, Jore MM, Lundgren M, Westra ER, Slijkhuis RJ, Snijders AP, Dickman MJ, Makarova KS, Koonin EV, Van Der Oost J (2008) Small CRISPR RNAs guide antiviral defense in prokaryotes. Science 321(5891):960–964

Campbell MT, Grondin A, Walia H, Morota G (2020) Leveraging genome-enabled growth models to study shoot growth responses to water deficit in rice. J Exp Bot 71(18):5669–5679

Cao X, Xie H, Song M, Lu J, Ma P, Huang B, Wang M, Tian Y, Chen F, Peng J, Lang Z (2023) Cut–dip–budding delivery system enables genetic modifications in plants without tissue culture. Innovation 4(1):100345

Carroll D (2011) Genome engineering with zinc-finger nucleases. Genetics 188(4):773–782

Ceccaldi R, Rondinelli B, D'Andrea AD (2016) Repair pathway choices and consequences at the double-strand break. Trends Cell Biol 26(1):52–64

Cermak T, Doyle EL, Christian M, Wang L, Zhang Y, Schmidt C, Baller JA, Somia NV, Bogdanove AJ, Voytas DF (2011) Efficient design and assembly of custom TALEN and other TAL effector-based constructs for DNA targeting. Nucleic Acids Res 39(12):e82

Čermák T, Baltes NJ, Čegan R, Zhang Y, Voytas DF (2015) High-frequency, precise modification of the tomato genome. Genome Biol 16:1–15

Chandrasekaran J, Brumin M, Wolf D, Leibman D, Klap C, Pearlsman M, Sherman A, Arazi T, Gal-On A (2016) Development of broad virus resistance in non-transgenic cucumber using CRISPR/Cas9 technology. Mol Plant Pathol 17(7):1140–1153

Chen TW, Cabrera-Bosquet L, Alvarez Prado S, Perez R, Artzet S, Pradal C, Coupel-Ledru A, Fournier C, Tardieu F (2019) Genetic and environmental dissection of biomass accumulation in multi-genotype maize canopies. J Exp Bot 70(9):2523–2534

Chen Y, Fu M, Li H, Wang L, Liu R, Liu Z, Zhang X, Jin S (2020) High-oleic acid content, nontransgenic allotetraploid cotton (Gossypium hirsutum L.) generated by knockout of GhFAD2 genes with CRISPR/Cas9 system. Plant Biotechnol J 19(3):424

Chen H, Su Z, Tian B, Liu Y, Pang Y, Kavetskyi V, Trick HN, Bai G (2022) Development and optimization of a Barley stripe mosaic virus-mediated gene editing system to improve Fusarium head blight resistance in wheat. Plant Biotechnol J 20(6):1018

Chen J, Liu Y, Liu M, Guo W, Wang Y, He Q, Chen W, Liao Y, Zhang W, Gao Y, Dong K (2023) Pangenome analysis reveals genomic variations associated with domestication traits in broomcorn millet. Nat Genet 55(12):2243–2254

Chen M, Liu P, An R, He X, Zhao P, Huang D, Yang X (2025) Sugarcane pan-transcriptome identifying a master gene ScHCT regulating lignin and sugar traits. J Agric Food Chem 73(2): 1739–1755

Christian M, Cermak T, Doyle EL, Schmidt C, Zhang F, Hummel A, Bogdanove AJ, Voytas DF (2010) Targeting DNA double-strand breaks with TAL effector nucleases. Genetics 186(2): 757–761

Chu VT, Weber T, Wefers B, Wurst W, Sander S, Rajewsky K, Kühn R (2015) Increasing the efficiency of homology-directed repair for CRISPR-Cas9-induced precise gene editing in mammalian cells. Nat Biotechnol 33(5):543–548

Chu YH, Jang JC, Huang Z, van der Knaap E (2019) Tomato locule number and fruit size controlled by natural alleles of lc and fas. Plant Direct 3(7):e00142

Clasen BM, Stoddard TJ, Luo S, Demorest ZL, Li J, Cedrone F, Tibebu R, Davison S, Ray EE, Daulhac A, Coffman A (2016) Improving cold storage and processing traits in potato through targeted gene knockout. Plant Biotechnol J 14(1):169–176

Cong L, Ran FA, Cox D, Lin S, Barretto R, Habib N, Hsu PD, Wu X, Jiang W, Marraffini LA, Zhang F (2013) Multiplex genome engineering using CRISPR/Cas systems. Science 339(6121): 819–823

Crowell S, Korniliev P, Falcao A, Ismail A, Gregorio G, Mezey J, McCouch S (2016) Genome-wide association and high-resolution phenotyping link Oryza sativa panicle traits to numerous trait-specific QTL clusters. Nat Commun 7:10527

Curtin SJ, Zhang F, Sander JD, Haun WJ, Starker C, Baltes NJ, Reyon D, Dahlborg EJ, Goodwin MJ, Coffman AP, Dobbs D (2011) Targeted mutagenesis of duplicated genes in soybean with zinc-finger nucleases. Plant Physiol 156(2):466–473

D'Ambrosio C, Stigliani AL, Giorio G (2018) CRISPR/Cas9 editing of carotenoid genes in tomato. Transgen Res 27(4):367–378

Deltcheva E, Chylinski K, Sharma CM, Gonzales K, Chao Y, Pirzada ZA, Eckert MR, Vogel J, Charpentier E (2011) CRISPR RNA maturation by trans-encoded small RNA and host factor RNase III. Nature 471(7340):602–607

Demirer GS, Silva TN, Jackson CT, Thomas JB, Ehrhardt DW, Rhee SY, Mortimer JC, Landry MP (2021) Nanotechnology to advance CRISPR–Cas genetic engineering of plants. Nat Nanotechnol 16(3):243–250

Deng D, Yan C, Pan X, Mahfouz M, Wang J, Zhu JK, Shi Y, Yan N (2012) Structural basis for sequence-specific recognition of DNA by TAL effectors. Science 335(6069):720–723

Deng L, Wang H, Sun C, Li Q, Jiang H, Du M, Li CB, Li C (2018) Efficient generation of pink-fruited tomatoes using CRISPR/Cas9 system. J Genet Genomics 45(1):51–54
D'Halluin K, Vanderstraeten C, Van Hulle J, Rosolowska J, Van Den Brande I, Pennewaert A, D'Hont K, Bossut M, Jantz D, Ruiter R, Broadhvest J (2013) Targeted molecular trait stacking in cotton through targeted double-strand break induction. Plant Biotechnol J 11(8):933–941
Dickson SP, Wang K, Krantz I, Hakonarson H, Goldstein DB (2010) Rare variants create synthetic genome-wide associations. PLoS Biol 8(1):e1000294. https://doi.org/10.1371/journal.pbio.1000294
Dong Q, Schlueter SD, Brendel V (2004) PlantGDB, plant genome database and analysis tools. Nucleic Acids Res 32(suppl_1):D354–D359
Doyon Y, Vo TD, Mendel MC, Greenberg SG, Wang J, Xia DF, Miller JC, Urnov FD, Gregory PD, Holmes MC (2011) Enhancing zinc-finger-nuclease activity with improved obligate heterodimeric architectures. Nat Methods 8(1):74–79
Eckerstorfer MF, Engelhard M, Heissenberger A, Simon S, Teichmann H (2019) Plants developed by new genetic modification techniques—comparison of existing regulatory frameworks in the EU and non-EU countries. Front Bioeng Biotechnol 7:26
Edgar RC (2004) MUSCLE: multiple sequence alignment with high accuracy and high throughput. Nucleic Acids Res 32(5):1792–1797
EFSA Panel on Genetically Modified Organisms (EFSA GMO Panel), Naegeli H, Bresson JL, Dalmay T, Dewhurst IC, Epstein MM, Firbank LG, Guerche P, Hejatko J, Moreno FJ, Mullins E (2020) Applicability of the EFSA Opinion on site-directed nucleases type 3 for the safety assessment of plants developed using site-directed nucleases type 1 and 2 and oligonucleotide-directed mutagenesis. EFSA J 18:6299
Eini O, Schumann N, Niessen M, Varrelmann M (2022) Targeted mutagenesis in plants using Beet curly top virus for efficient delivery of CRISPR/Cas12a components. Nat Biotechnol 67:1–11
Ellison EE, Nagalakshmi U, Gamo ME, Huang PJ, Dinesh-Kumar S, Voytas DF (2020) Multiplexed heritable gene editing using RNA viruses and mobile single guide RNAs. Nat Plants 6(6):620–624
Elnitski L, Hardison RC, Li J, Yang S, Kolbe D, Eswara P, O'Connor MJ, Schwartz S, Miller W, Chiaromonte F (2003) Distinguishing regulatory DNA from neutral sites. Genom Res 13(1): 64–72
Endo A, Saika H, Takemura M, Misawa N, Toki S (2019) A novel approach to carotenoid accumulation in rice callus by mimicking the cauliflower Orange mutation via genome editing. Rice 12:1–5
Engler C, Gruetzner R, Kandzia R, Marillonnet S (2009) Golden gate shuffling: a one-pot DNA shuffling method based on type IIs restriction enzymes. PLoS One 4(5):e5553
Ewing PM, Runck BC, Kono TYJ, Kantar MB (2019) The home field advantage of modern plant breeding. PLoS One 14:e0227079
FAO (2019) Agriculture and climate change: challenges and opportunities at the global and local level-collaboration on climate-smart agriculture. UN
Feeney O, Cockbain J, Morrison M, Diependaele L, Van Assche K, Sterckx S (2018) Patenting foundational technologies: lessons from CRISPR and other core biotechnologies. Am J Bioeth 18:36–48
Feng T, Zhu X (2012) Detecting rare variants. Methods Mol Biol 850:453–464
Finn R, Griffiths-Jones S, Bateman A (2003) Identifying protein domains with the Pfam database. Curr Protoc Bioinformatics 1(1):2–5
Fordyce RF, Soltis NE, Caseys C, Gwinner R, Corwin JA, Atwell S, Copeland D, Feusier J, Subedy A, Eshbaugh R, Kliebenstein DJ (2018) Digital imaging combined with genome-wide association mapping links loci to plant-pathogen interaction traits. Plant Physiol 178(3): 1406–1422
Frock RL, Hu J, Meyers RM, Ho YJ, Kii E, Alt FW (2015) Genome-wide detection of DNA double-stranded breaks induced by engineered nucleases. Nat Biotechnol 33(2):179–186

Fu YB, Dong YB (2015) Genetic erosion under modern plant breeding: case studies in Canadian crop gene pools. In: Genetic diversity and erosion in plants: indicators and prevention, Cham, Springer International Publishing, pp 89–104

Fu Y, Foden JA, Khayter C, Maeder ML, Reyon D, Joung JK, Sander JD (2013) High-frequency off-target mutagenesis induced by CRISPR-Cas nucleases in human cells. Nat Biotechnol 31(9): 822–826

Fu Y, Sander JD, Reyon D, Cascio VM, Joung JK (2014) Improving CRISPR-Cas nuclease specificity using truncated guide RNAs. Nat Biotechnol 32(3):279–284

Gaj T, Gersbach CA, Barbas CF (2013) ZFN, TALEN, and CRISPR/Cas-based methods for genome engineering. Trends Biotechnol 31(7):397–405

Gasiunas G, Barrangou R, Horvath P, Siksnys V (2012) Cas9–crRNA ribonucleoprotein complex mediates specific DNA cleavage for adaptive immunity in bacteria. Proc Natl Acad Sci USA 109(39):E2579–E2586

Gaudelli NM, Komor AC, Rees HA, Packer MS, Badran AH, Bryson DI, Liu DR (2017) Programmable base editing of A• T to G• C in genomic DNA without DNA cleavage. Nature 551(7681):464–471

Gelvin SB (2017) Integration of Agrobacterium T-DNA into the plant genome. Annu Rev Genet 51(1):195–217

Gibson G (2011) Rare and common variants: twenty arguments. Nat Rev Genet 13(2):135–145

Gilbert LA, Larson MH, Morsut L, Liu Z, Brar GA, Torres SE, Stern-Ginossar N, Brandman O, Whitehead EH, Doudna JA, Lim WA (2013) CRISPR-mediated modular RNA-guided regulation of transcription in eukaryotes. Cell 154(2):442–451

Gilchrist E, Haughn G (2010) Reverse genetics techniques: engineering loss and gain of gene function in plants. Brief Funct Genomics 9(2):103–110

Golicz AA, Bayer PE, Barker GC, Edger PP, Kim H, Martinez PA, Chan CK, Severn-Ellis A, McCombie WR, Parkin IA, Paterson AH (2016) The pangenome of an agronomically important crop plant Brassica oleracea. Nat Commun 7(1):13390

Goodstein DM, Shu S, Howson R, Neupane R, Hayes RD, Fazo J, Mitros T, Dirks W, Hellsten U, Putnam N, Rokhsar DS (2012) Phytozome: a comparative platform for green plant genomics. Nucleic Acids Res 40(D1):D1178–D1186

Grohmann L, Keilwagen J, Duensing N, Dagand E, Hartung F, Wilhelm R, Bendiek J, Sprink T (2019) Detection and identification of genome editing in plants: challenges and opportunities. Front Plant Sci 10:236

Gu K, Yang B, Tian D, Wu L, Wang D, Sreekala C, Yang F, Chu Z, Wang GL, White FF, Yin Z (2005) R gene expression induced by a type-III effector triggers disease resistance in rice. Nature 435(7045):1122–1125

Guilinger JP, Pattanayak V, Reyon D, Tsai SQ, Sander JD, Joung JK, Liu DR (2014) Broad specificity profiling of TALENs results in engineered nucleases with improved DNA-cleavage specificity. Nat Methods 11(4):429–435

Guo AY, Zhu QH, Chen X, Luo JC (2007) GSDS: a gene structure display server. Hereditas 29(8): 1023–1026

Guo Z, Yang W, Chang Y, Ma X, Tu H, Xiong F, Jiang N, Feng H, Huang C, Yang P, Zhao H (2018) Genome-wide association studies of image traits reveal genetic architecture of drought resistance in rice. Mol Plant 11(6):789–805

Guo W, Schreiber M, Marosi VB, Bagnaresi P, Jørgensen ME, Braune KB, Chalmers K, Chapman B, Dang V, Dockter C, Fiebig A (2025) A barley pan-transcriptome reveals layers of genotype-dependent transcriptional complexity. Nat Genet 3:1–0

Gupta SK, Barg R, Arazi T (2021) Tomato agamous-like6 parthenocarpy is facilitated by ovule integument reprogramming involving the growth regulator KLUH. Plant Physiol 185(3): 969–984

Hale CR, Zhao P, Olson S, Duff MO, Graveley BR, Wells L, Terns RM, Terns MP (2009) RNA-guided RNA cleavage by a CRISPR RNA-Cas protein complex. Cell 139(5):945–956

Hamada H, Liu Y, Nagira Y, Miki R, Taoka N, Imai R (2018) Biolistic-delivery-based transient CRISPR/Cas9 expression enables in planta genome editing in wheat. Sci Rep 8(1):14422
Hameed A, Mehmood MA, Shahid M, Fatma S, Khan A, Ali S (2020) Prospects for potato genome editing to engineer resistance against viruses and cold-induced sweetening. GM Crops Food 11(4):185–205
Hao LI, Ruiying QIN, Xiaoshuang LIU, Shengxiang LIAO, Rongfang XU, Jianbo YANG, Pengcheng WEI (2019) CRISPR/Cas9-mediated adenine base editing in rice genome. Rice Sci 26(2):125–128
Hardison RC (2003) Comparative genomics. PLoS Biol 1(2):e58
Haun W, Coffman A, Clasen BM, Demorest ZL, Lowy A, Ray E, Retterath A, Stoddard T, Juillerat A, Cedrone F, Mathis L (2014) Improved soybean oil quality by targeted mutagenesis of the fatty acid desaturase 2 gene family. Plant Biotech J 12(7):934–940
He Q, Tang S, Zhi H, Chen J, Zhang J, Liang H, Alam O, Li H, Zhang H, Xing L, Li X (2023) A graph-based genome and pan-genome variation of the model plant Setaria. Nat Genet 55(7): 1232–1242
He X, Qi Z, Liu Z, Chang X, Zhang X, Li J, Wang M (2024) Pangenome analysis reveals transposon-driven genome evolution in cotton. BMC Biol 22(1):92
Heyer WD, Ehmsen KT, Liu J (2010) Regulation of homologous recombination in eukaryotes. Annu Rev Genet 44(1):113–139
Hiei Y, Ohta S, Komari T, Kumashiro T (1994) Efficient transformation of rice (Oryza sativa L.) mediated by Agrobacterium and sequence analysis of the boundaries of the T-DNA. Plant J 6(2):271–282
Hilton IB, D'ippolito AM, Vockley CM, Thakore PI, Crawford GE, Reddy TE, Gersbach CA (2015) Epigenome editing by a CRISPR-Cas9-based acetyltransferase activates genes from promoters and enhancers. Nat Biotechnol 33(5):510–517
Hirsch CN, Foerster JM, Johnson JM, Sekhon RS, Muttoni G, Vaillancourt B, Peñagaricano F, Lindquist E, Pedraza M, Barry K, Leon N, Kaeppler SM, Buell RC (2014) Insights into the maize pangenome and pan-transcriptome. Plant Cell 26(1):121–135
Horvath P, Barrangou R (2010) CRISPR/Cas, the immune system of bacteria and archaea. Science 327(5962):167–170
Hsu PD, Scott DA, Weinstein JA, Ran FA, Konermann S, Agarwala V, Li Y, Fine EJ, Wu X, Shalem O, Cradick TJ (2013) DNA targeting specificity of RNA-guided Cas9 nucleases. Nat Biotechnol 31(9):827–832
Hu Y, Han Z, Sun Y, Wang S, Wang T, Wang Y, Xu K, Zhang X, Xu X, Han Z, Wu T (2020) ERF4 affects fruit firmness through TPL4 by reducing ethylene production. Plant J 103(3):937–950
Hu C, Sheng O, Deng G, He W, Dong T, Yang Q, Dou T, Li C, Gao H, Liu S, Yi G (2021) CRISPR/Cas9-mediated genome editing of MaACO1 (aminocyclopropane-1-carboxylate oxidase 1) promotes the shelf life of banana fruit. Plant Biotechnol J 19(4):654
Huang H, Cui T, Zhang L, Yang Q, Yang Y, Xie K, Fan C, Zhou Y (2020) Modifications of fatty acid profile through targeted mutation at BnaFAD2 gene with CRISPR/Cas9-mediated gene editing in Brassica napus. Theoret Appl Genet 133:2401–2411
Hubbard T, Barker D, Birney E, Cameron G, Chen Y, Clark L, Cox T, Cuff J, Curwen V, Down T, Durbin R (2002) The Ensembl genome database project. Nucleic Acids Res 30(1):38–41
Hulo N, Bairoch A, Bulliard V, Cerutti L, De Castro E, Langendijk-Genevaux PS, Pagni M, Sigrist CJ (2006) The PROSITE database. Nucleic Acids Res 34(suppl_1):D227–D230
Hwang HH, Yu M, Lai EM (2017) Agrobacterium-mediated plant transformation: biology and applications. 15:e0186
Illouz-Eliaz N, Nissan I, Nir I, Ramon U, Shohat H, Weiss D (2020) Mutations in the tomato gibberellin receptors suppress xylem proliferation and reduce water loss under water-deficit conditions. J Exp Bot 71(12):3603–3612
Ishino Y, Shinagawa H, Makino K, Amemura M, Nakata A (1987) Nucleotide sequence of the iap gene, responsible for alkaline phosphatase isozyme conversion in Escherichia coli, and identification of the gene product. J Bacteriol 169(12):5429–5433

Jaiswal V, Gupta S, Gahlaut V, Muthamilarasan M, Bandyopadhyay T, Ramchiary N, Prasad M (2019) Genome-wide association study of major agronomic traits in Foxtail Millet (Setaria italica L.) using ddRAD sequencing. Sci Rep 9(1):5020

Jangra S, Chaudhary V, Yadav RC, Yadav NR (2021) High-throughput phenotyping: a platform to accelerate crop improvement. Phenomics 1:31–53

Jansen R, Embden JDV, Gaastra W, Schouls LM (2002) Identification of genes that are associated with DNA repeats in prokaryotes. Mol Microbiol 43(6):1565–1575

Jia H, Zhang Y, Orbović V, Xu J, White FF, Jones JB, Wang N (2017) Genome editing of the disease susceptibility gene Cs LOB 1 in citrus confers resistance to citrus canker. Plant Biotechnol J 15(7):817–823

Jiang F, Taylor DW, Chen JS, Kornfeld JE, Zhou K, Thompson AJ, Nogales E, Doudna JA (2016) Structures of a CRISPR-Cas9 R-loop complex primed for DNA cleavage. Science 351(6275): 867–871

Jin M, Liu H, He C, Fu J, Xiao Y, Wang Y, Xie W, Wang G, Yan J (2016) Maize pan-transcriptome provides novel insights into genome complexity and quantitative trait variation. Sci Rep 6: 18936

Jinek M, Chylinski K, Fonfara I, Hauer M, Doudna JA, Charpentier E (2012) A programmable dual-RNA–guided DNA endonuclease in adaptive bacterial immunity. Science 337(6096): 816–821

Jinek M, Jiang F, Taylor DW, Sternberg SH, Kaya E, Ma E, Anders C, Hauer M, Zhou K, Lin S, Kaplan M (2014) Structures of Cas9 endonucleases reveal RNA-mediated conformational activation. Science 343(6176):1247997

Jones P, Binns D, Chang HY, Fraser M, Li W, McAnulla C, McWilliam H, Maslen J, Mitchell A, Nuka G, Pesseat S (2014) InterProScan 5: genome-scale protein function classification. Bioinformatics 30(9):1236–1240

Jung YJ, Lee HJ, Kim JH, Kim DH, Kim HK, Cho YG, Bae S, Kang KK (2019) CRISPR/Cas9-targeted mutagenesis of F3′ H, DFR and LDOX, genes related to anthocyanin biosynthesis in black rice (Oryza sativa L.). Plant Biotechnol Rep 13:521–531

Kannababu N, Nanjundappa S, Narayanan N, Vetriventhan M, Venkateswarlu R, Das IK, Srikanth A, Viswanath A, Singh S, Malipatil R, Satyavathi TC, Thirunavukkarasu N (2025) Role of functional genes for seed vigor related traits through genome-wide association mapping in finger millet (Eleusine coracana L. Gaertn.). Sci Rep 15(1):5569

Kaur N, Alok A, Kumar P, Kaur N, Awasthi P, Chaturvedi S, Pandey P, Pandey A, Pandey AK, Tiwari S (2020) CRISPR/Cas9 directed editing of lycopene epsilon-cyclase modulates metabolic flux for β-carotene biosynthesis in banana fruit. Metab Eng 59:76–86

Kay S, Bonas U (2009) How Xanthomonas type III effectors manipulate the host plant. Curr Opin Microbiol 12(1):37–43

Khan MSS, Basnet R, Islam SA, Shu Q (2019) Mutational analysis of OsPLDα1 reveals its involvement in phytic acid biosynthesis in rice grains. J Agric Food Chem 67(41):11436–11443

Khan AA, Raess M, de Angelis MH (2021) Moving forward with forward genetics: a summary of the INFRAFRONTIER forward genetics panel discussion. F1000Res 10:456

Khan AW, Garg V, Sun S, Gupta S, Dudchenko O, Roorkiwal M, Chitikineni A, Bayer PE, Shi C, Upadhyaya HD, Bohra A (2024) Cicer super-pangenome provides insights into species evolution and agronomic trait loci for crop improvement in chickpea. Nat Genet 56(6):1225–1234

Khanday I, Skinner D, Yang B, Mercier R, Sundaresan V (2019) A male-expressed rice embryogenic trigger redirected for asexual propagation through seeds. Nature 565(7737):91–95

Kim YG, Cha J, Chandrasegaran S (1996) Hybrid restriction enzymes: zinc finger fusions to Fok I cleavage domain. Proc Natl Acad Sci USA 93(3):1156–1160

Kim HJ, Lee HJ, Kim H, Cho SW, Kim JS (2009) Targeted genome editing in human cells with zinc finger nucleases constructed via modular assembly. Genome Res 19(7):1279–1288

Kim H, Kim ST, Ryu J, Kang BC, Kim JS, Kim SG (2017) CRISPR/Cpf1-mediated DNA-free plant genome editing. Nat Commun 8(1):14406

Komor AC, Kim YB, Packer MS, Zuris JA, Liu DR (2016) Programmable editing of a target base in genomic DNA without double-stranded DNA cleavage. Nature 533(7603):420–424

Korte A, Farlow A (2013) The advantages and limitations of trait analysis with GWAS: a review. Plant Methods 9:29

Kumagai Y, Liu Y, Hamada H, Luo W, Zhu J, Kuroki M, Nagira Y, Taoka N, Katoh E, Imai R (2022) Introduction of a second "Green Revolution" mutation into wheat via in planta CRISPR/Cas9 delivery. Plant Physiol 188(4):1838–1842

Kumar J, Choudhary AK, Solanki RK, Pratap A (2011) Towards marker-assisted selection in pulses: a review. Plant Breed 130:297–313

Kumar S, Stecher G, Tamura K (2016) MEGA7: molecular evolutionary genetics analysis version 7.0 for bigger datasets. Mol Biol Evol 33(7):1870–1874

Kumar J, Kumar A, Sen Gupta D, Kumar S, DePauw RM (2022) Reverse genetic approaches for breeding nutrient-rich and climate-resilient cereal and food legume crops. Heredity 128(6): 473–496

Kuska M, Wahabzada M, Leucker M, Dehne HW, Kersting K, Oerke EC, Steiner U, Mahlein AK (2015) Hyperspectral phenotyping on the microscopic scale: towards automated characterization of plant-pathogen interactions. Plant Methods 11:1–5

Le H, Nguyen NH, Ta DT, Le TN, Bui TP, Le NT, Nguyen CX, Rolletschek H, Stacey G, Stacey MG, Pham NB (2020) CRISPR/Cas9-mediated knockout of galactinol synthase-encoding genes reduces raffinose family oligosaccharide levels in soybean seeds. Front Plant Sci 11:612942

Lee TH, Tang H, Wang X, Paterson AH (2012) PGDD: a database of gene and genome duplication in plants. Nucleic Acids Res 41(D1):D1152–D1158

Lee J, Chin JH, Ahn SN, Koh H (2015) Brief history and prespectives on plant breeding. In: Current technologies in plant molecular breeding. Springer, pp 1–14

Lemmon ZH, Reem NT, Dalrymple J, Soyk S, Swartwood KE, Rodriguez-Leal D, Van Eck J, Lippman ZB (2018) Rapid improvement of domestication traits in an orphan crop by genome editing. Nat Plants 4(10):766–770

Li T, Huang S, Zhao X, Wright DA, Carpenter S, Spalding MH, Weeks DP, Yang B (2011) Modularly assembled designer TAL effector nucleases for targeted gene knockout and gene replacement in eukaryotes. Nucleic Acids Res 39(14):6315–6325

Li T, Liu B, Spalding MH, Weeks DP, Yang B (2012) High-efficiency TALEN-based gene editing produces disease-resistant rice. Nat Biotechnol 30(5):390–392

Li T, Yang X, Yu Y, Si X, Zhai X, Zhang H, Dong W, Gao C, Xu C (2018a) Domestication of wild tomato is accelerated by genome editing. Nat Biotechnol 36(12):1160–1163

Li X, Wang Y, Chen S, Tian H, Fu D, Zhu B, Luo Y, Zhu H (2018b) Lycopene is enriched in tomato fruit by CRISPR/Cas9-mediated multiplex genome editing. Front Plant Sci 9:559

Li B, Chen L, Sun W, Wu D, Wang M, Yu Y, Chen G, Yang W, Lin Z, Zhang X, Duan L (2020) Phenomics-based GWAS analysis reveals the genetic architecture for drought resistance in cotton. Plant Biotechnol J 18(12):2533–2544

Li S, Zhang Y, Liu Y, Zhang P, Wang X, Chen B, Ding L, Nie Y, Li F, Ma Z, Kang Z (2024) The E3 ligase TaGW2 mediates transcription factor TaARR12 degradation to promote drought resistance in wheat. Plant Cell 36(3):605–625

Lian Z, Nguyen CD, Liu L, Wang G, Chen J, Wang S, Yi G, Wilson S, Ozias-Akins P, Gong H, Huo H (2022) Application of developmental regulators to improve in planta or in vitro transformation in plants. Plant Biotechnol J 20(8):1622–1635

Lieber MR (2010) The mechanism of double-strand DNA break repair by the nonhomologous DNA end-joining pathway. Annu Rev Biochem 79(1):181–211

Liu G, Li J, Godwin ID (2019) Genome editing by CRISPR/Cas9 in sorghum through biolistic bombardment. Methods Protoc 1931:169–183

Liu Y, Du H, Li P, Shen Y, Peng H, Liu S, Zhou GA, Zhang H, Liu Z, Shi M, Huang X (2020) Pan-genome of wild and cultivated soybeans. Cell 182(1):162–176

Liu X, Yang Y, Hou S, Men Y, Han Y (2022) The integration of genome-wide association study and homology analysis to explore the genomic regions and candidate genes for panicle-related traits in foxtail millet. Int J Mol Sci 23(23):14735
Liu D, Myers EA, Xuan S, Prichard LE, Donahue LI, Ellison EE, Starker CG, Voytas DF (2024a) Heritable, multinucleotide deletions in plants using viral delivery of a repair exonuclease and guide RNAs. Plant Physiol 194(4):2229–2239
Liu H, Zhang X, Shang Y, Zhao S, Li Y, Zhou X, Huo X, Qiao P, Wang X, Dai K, Li H, Guo J, Shi W (2024b) Genome-wide association study reveals genetic loci for ten trace elements in foxtail millet (Setaria italica). Theor Appl Genet 137(8):186
Loots GG, Ovcharenko I (2004) rVISTA 2.0: evolutionary analysis of transcription factor binding sites. Nucleic Acids Res 32(suppl_2):W217–W221
Loots GG, Locksley RM, Blankespoor CM, Wang ZE, Miller W, Rubin EM, Frazer KA (2000) Identification of a coordinate regulator of interleukins 4, 13, and 5 by cross-species sequence comparisons. Science 288(5463):136–140
Lor VS, Starker CG, Voytas DF, Weiss D, Olszewski NE (2014) Targeted mutagenesis of the tomato PROCERA gene using transcription activator-like effector nucleases. Plant Physiol 166(3):1288–1291
Lu J, Li S, Deng S, Wang M, Wu Y, Li M, Dong J, Lu S, Su C, Li G, Lang Z (2024) A method of genetic transformation and gene editing of succulents without tissue culture. Plant Biotechnol J 22(7):1981–1988
Lv H, Wang X, Dong X, Gao M, Dong D, Li C, Jing S, Guo YD, Zhang N (2024) CRISPR/Cas9 edited SlGT30 improved both drought resistance and fruit yield through endoreduplication. Plant Cell Environ 48(4):2581–2595
Ma L, Zhang D, Miao Q, Yang J, Xuan Y, Hu Y (2017) Essential role of sugar transporter OsSWEET11 during the early stage of rice grain filling. Plant Cell Physiol 58(5):863–873
Ma T, Ye Z, Wang L (2019a) Genome wide approaches to identify protein-DNA interactions. Curr Med Chem 26(42):7641–7654
Ma Y, Liu M, Stiller J, Liu C (2019b) A pan-transcriptome analysis shows that disease resistance genes have undergone more selection pressure during barley domestication. BMC Genomics 20(1):12
Ma W, Kang X, Liu P, Zhang Y, Lin X, Li B, Chen Z (2021) The analysis of transcription factor CsHB1 effects on caffeine accumulation in tea callus through CRISPR/Cas9 mediated gene editing. Process Biochem 101:304–311
Maeder ML, Thibodeau-Beganny S, Osiak A, Wright DA, Anthony RM, Eichtinger M, Jiang T, Foley JE, Winfrey RJ, Townsend JA, Unger-Wallace E (2008) Rapid "open-source" engineering of customized zinc-finger nucleases for highly efficient gene modification. Mol Cell 31(2): 294–301
Maher MF, Nasti RA, Vollbrecht M, Starker CG, Clark MD, Voytas DF (2020) Plant gene editing through de novo induction of meristems. Nat Biotechnol 38(1):84–89
Mak ANS, Bradley P, Bogdanove AJ, Stoddard BL (2013) TAL effectors: function, structure, engineering and applications. Curr Opin Struct Biol 23(1):93–99
Makarova KS, Haft DH, Barrangou R, Brouns SJ, Charpentier E, Horvath P, Moineau S, Mojica FJ, Wolf YI, Yakunin AF, Van Der Oost J (2011) Evolution and classification of the CRISPR–Cas systems. Nat Rev Microbiol 9(6):467–477
Makarova KS, Wolf YI, Alkhnbashi OS, Costa F, Shah SA, Saunders SJ, Barrangou R, Brouns SJ, Charpentier E, Haft DH, Horvath P (2015) An updated evolutionary classification of CRISPR–Cas systems. Nat Rev Microbiol 13(11):722–736
Makarova KS, Wolf YI, Iranzo J, Shmakov SA, Alkhnbashi OS, Brouns SJ, Charpentier E, Cheng D, Haft DH, Horvath P, Moineau S (2020) Evolutionary classification of CRISPR–Cas systems: a burst of class 2 and derived variants. Nat Rev Microbiol 18(2):67–83
Mali P, Aach J, Stranges PB, Esvelt KM, Moosburner M, Kosuri S, Yang L, Church GM (2013) CAS9 transcriptional activators for target specificity screening and paired nickases for cooperative genome engineering. Nat Biotechnol 31(9):833–838

Marchler-Bauer A, Panchenko AR, Shoemaker BA, Thiessen PA, Geer LY, Bryant SH (2002) CDD: a database of conserved domain alignments with links to domain three-dimensional structure. Nucleic Acids Res 30(1):281–283

Martignago D, Rico-Medina A, Blasco-Escámez D, Fontanet-Manzaneque JB, Caño-Delgado AI (2020) Drought resistance by engineering plant tissue-specific responses. Front Plant Sci 10: 1676

McHughen A (2016) A critical assessment of regulatory triggers for products of biotechnology: product vs. process. GM Crops Food 7(3-4):125–158

Medina CA, Samac DA, Yu LX (2021) Pan-transcriptome identifying master genes and regulation network in response to drought and salt stresses in Alfalfa (Medicago sativa L.). Sci Rep 11(1): 17203

Menz J, Modrzejewski D, Hartung F, Wilhelm R, Sprink T (2020) Genome edited crops touch the market: a view on the global development and regulatory environment. Front Plant Sci 11: 586027

Miller J, McLachlan AD, Klug A (1985) Repetitive zinc-binding domains in the protein transcription factor IIIA from Xenopus oocytes. EMBO J 4(6):1609–1614

Miller JC, Holmes MC, Wang J, Guschin DY, Lee YL, Rupniewski I, Beausejour CM, Waite AJ, Wang NS, Kim KA, Gregory PD (2007) An improved zinc-finger nuclease architecture for highly specific genome editing. Nat Biotechnol 25(7):778–785

Miller JC, Tan S, Qiao G, Barlow KA, Wang J, Xia DF, Meng X, Paschon DE, Leung E, Hinkley SJ, Dulay GP (2011) A TALE nuclease architecture for efficient genome editing. Nat Biotechnol 29(2):143–148

Mojica FJ, Díez-Villaseñor C, Soria E, Juez G (2000) MicroCorrespondence. Mol Microbiol 36(1):244–246

Mojica FJ, Díez-Villaseñor CS, García-Martínez J, Soria E (2005) Intervening sequences of regularly spaced prokaryotic repeats derive from foreign genetic elements. J Mol Evol 60: 174–182

Mojica FJ, Díez-Villaseñor C, García-Martínez J, Almendros C (2009) Short motif sequences determine the targets of the prokaryotic CRISPR defence system. Microbiol 155(3):733–740

Montenegro JD, Golicz AA, Bayer PE, Hurgobin B, Lee H, Chan CK, Visendi P, Lai K, Doležel J, Batley J, Edwards D (2017) The pangenome of hexaploid bread wheat. Plant J 90(5):1007–1013

Moscou MJ, Bogdanove AJ (2009) A simple cipher governs DNA recognition by TAL effectors. Science 326(5959):1501–1501

Nekrasov V, Wang C, Win J, Lanz C, Weigel D, Kamoun S (2017) Rapid generation of a transgene-free powdery mildew resistant tomato by genome deletion. Sci Rep 7(1):482

Nelson OE, Rines HW (1962) The enzymatic deficiency in the waxy mutant of maize. Biochem Biophys Res Commun 9(4):297–300

Neumann K, Zhao Y, Chu J, Keilwagen J, Reif JC, Kilian B, Graner A (2017) Genetic architecture and temporal patterns of biomass accumulation in spring barley revealed by image analysis. BMC Plant Biol 17:1–2

Newburger DE, Bulyk ML (2009) UniPROBE: an online database of protein binding microarray data on protein–DNA interactions. Nucleic Acids Res 37(suppl_1):D77–D82

Nguyen CC, Van Vu T, Shelake RM, Nguyen NT, Khanh TD, Kim WY, Kim JY (2024) Generation of parthenocarpic tomato plants in multiple elite cultivars using the CRISPR/Cas9 system. Mol Breed 44(2):13

Nishimasu H, Ran FA, Hsu PD, Konermann S, Shehata SI, Dohmae N, Ishitani R, Zhang F, Nureki O (2014) Crystal structure of Cas9 in complex with guide RNA and target DNA. Cell 156(5): 935–949

Nonaka S, Arai C, Takayama M, Matsukura C, Ezura H (2017) Efficient increase of γ-aminobutyric acid (GABA) content in tomato fruits by targeted mutagenesis. Sci Rep 7(1):7057

Nuñez JK, Kranzusch PJ, Noeske J, Wright AV, Davies CW, Doudna JA (2014) Cas1–Cas2 complex formation mediates spacer acquisition during CRISPR–Cas adaptive immunity. Nat Struct Mol Biol 21(6):528–534

Ortigosa A, Gimenez-Ibanez S, Leonhardt N, Solano R (2019) Design of a bacterial speck resistant tomato by CRISPR/Cas9-mediated editing of Sl JAZ 2. Plant Biotechnol J 17(3):665–673

Ou L, Li D, Lv J, Chen W, Zhang Z, Li X, Yang B, Zhou S, Yang S, Li W, Gao H (2018) Pan-genome of cultivated pepper (Capsicum) and its use in gene presence–absence variation analyses. New Phytol 220(2):360–363

Pabo CO, Peisach E, Grant RA (2001) Design and selection of novel Cys2His2 zinc finger proteins. Annu Rev Biochem 70(1):313–340

Pâques F, Duchateau P (2007) Meganucleases and DNA double-strand break-induced recombination: perspectives for gene therapy. Curr Gene Ther 7(1):49–66

Pavletich NP, Pabo CO (1991) Zinc finger-DNA recognition: crystal structure of a Zif268-DNA complex at 2.1 Å. Science 252(5007):809–817

Peng B, Yu M, Zhang B, Xu J, Ma R (2020) Differences in PpAAT1 activity in high-and low-aroma peach varieties affect γ-decalactone production. Plant Physiol 182(4):2065–2080

Pennacchio LA, Rubin EM (2001) Genomic strategies to identify mammalian regulatory sequences. Nat Rev Genet 2(2):100–109

Petek M, Zagorščak M, Ramšak Ž, Sanders S, Tomaž Š, Tseng E, Zouine M, Coll A, Gruden K (2020) Cultivar-specific transcriptome and pan-transcriptome reconstruction of tetraploid potato. Sci Data 7(1):249

Peters JL, Cnudde F, Gerats T (2003) Forward genetics and map-based cloning approaches. Trends Plant Sci 8(10):484–491

Pham AT, Maurer A, Pillen K, Brien C, Dowling K, Berger B, Eglinton JK, March TJ (2019) Genome-wide association of barley plant growth under drought stress using a nested association mapping population. BMC Plant Biol 19:1–6

Podevin N, Davies HV, Hartung F, Nogué F, Casacuberta JM (2013) Site-directed nucleases: a paradigm shift in predictable, knowledge-based plant breeding. Trends Biotechnol 31(6): 375–383

Porteus MH, Carroll D (2005) Gene targeting using zinc finger nucleases. Nat Biotechnol 23(8): 967–973

Pourcel C, Salvignol G, Vergnaud G (2005) CRISPR elements in Yersinia pestis acquire new repeats by preferential uptake of bacteriophage DNA, and provide additional tools for evolutionary studies. Microbiol 151(3):653–663

Pronozin A, Bragina MK, Salina EA (2021) Crop pangenomes. Vavilov J Genet Breeding 25(1):57

Puchta H (2005) The repair of double-strand breaks in plants: mechanisms and consequences for genome evolution. J Exp Bot 56(409):1–14

Pujar M, Gangaprasad S, Govindaraj M, Gangurde SS, Kanatti A, Kudapa H (2020) Genome-wide association study uncovers genomic regions associated with grain iron, zinc and protein content in pearl millet. Sci Rep 10(1):19473

Qi LS, Larson MH, Gilbert LA, Doudna JA, Weissman JS, Arkin AP, Lim WA (2013) Repurposing CRISPR as an RNA-guided platform for sequence-specific control of gene expression. Cell 152(5):1173–1183

Qian L, Jin H, Yang Q, Zhu L, Yu X, Fu X, Zhao M, Yuan F (2022) A sequence variation in GmBADH2 enhances soybean aroma and is a functional marker for improving soybean flavor. Int J Mol Sci 23(8):4116

Qin P, Lu H, Du H, Wang H, Chen W, Chen Z, He Q, Ou S, Zhang H, Li X, Li X (2021) Pan-genome analysis of 33 genetically diverse rice accessions reveals hidden genomic variations. Cell 184(13):3542–3558

Quesada MA, Blanco-Portales R, Posé S, García-Gago JA, Jiménez-Bermudez S, Munoz-Serrano-A, Caballero JL, Pliego-Alfaro F, Mercado JA, Munoz-Blanco J (2009) Antisense down-regulation of the FaPG1 gene reveals an unexpected central role for polygalacturonase in strawberry fruit softening. Plant Physiol 150(2):1022–1032

Ramalingam S, Kandavelou K, Rajenderan R, Chandrasegaran S (2011) Creating designed zinc-finger nucleases with minimal cytotoxicity. J Mol Biol 405(3):630–641

Ran FA, Hsu PD, Wright J, Agarwala V, Scott DA, Zhang F (2013) Genome engineering using the CRISPR-Cas9 system. Nat Protoc 8(11):2281–2308

Rapacz M, Wójcik-Jagła M, Fiust A, Kalaji HM, Kościelniak J (2019) Genome-wide associations of chlorophyll fluorescence OJIP transient parameters connected with soil drought response in barley. Front Plant Sci 10:78

Rasheed A, Xia X, Ogbonnaya F, Mahmood T, Zhang Z, Mujeeb-Kazi A, He Z (2014) Genome-wide association for grain morphology in synthetic hexaploid wheats using digital imaging analysis. BMC Plant Biol 14:1–21

Rebolledo MC, Peña AL, Duitama J, Cruz DF, Dingkuhn M, Grenier C, Tohme J (2016) Combining image analysis, genome wide association studies and different field trials to reveal stable genetic regions related to panicle architecture and the number of spikelets per panicle in rice. Front Plant Sci 7:1384

Ren Y, Huang Z, Jiang H, Wang Z, Wu F, Xiong Y, Yao J (2021) A heat stress responsive NAC transcription factor heterodimer plays key roles in rice grain filling. J Exp Bot 72(8):2947–2964

Reyon D, Khayter C, Regan MR, Joung JK, Sander JD (2012) Engineering designer transcription activator–like effector nucleases (TALENs) by REAL or REAL-fast assembly. Curr Protoc Mol Biol 100(1):12–15

Robertsen CD, Hjortshøj RL, Janss LL (2019) Genomic selection in cereal breeding. Agronomy 9: 95

Safder I, Shao G, Sheng Z, Hu P, Tang S (2022) Genome-wide identification studies–A primer to explore new genes in plant species. Plant Biol 24(1):9–22

Sahito JH, Zhang H, Gishkori ZGN, Ma C, Wang Z, Ding D, Zhang X, Tang J (2024) Advancements and prospects of genome-wide association studies (GWAS) in Maize. Int J Mol Sci 25(3): 1918

Saini DK, Chopra Y, Singh J, Sandhu KS, Kumar A, Bazzer S, Srivastava P (2021) Comprehensive evaluation of mapping complex traits in wheat using genome-wide association studies. Mol Breed 42(1):1

Sánchez-León S, Gil-Humanes J, Ozuna CV, Giménez MJ, Sousa C, Voytas DF, Barro F (2018) Low-gluten, nontransgenic wheat engineered with CRISPR/Cas9. Plant Biotechnol J 16(4): 902–910

Sandelin A, Alkema W, Engström P, Wasserman WW, Lenhard B (2004a) JASPAR: an open-access database for eukaryotic transcription factor binding profiles. Nucleic Acids Res 32 (suppl_1):D91–D94

Sandelin A, Wasserman WW, Lenhard B (2004b) ConSite: web-based prediction of regulatory elements using cross-species comparison. Nucleic Acids Res 32(suppl_2):W249–W252

Sander JD, Cade L, Khayter C, Reyon D, Peterson RT, Joung JK, Yeh JR (2011) Targeted gene disruption in somatic zebrafish cells using engineered TALENs. Nat Biotech 29(8):697–698

Santosh Kumar VV, Verma RK, Yadav SK, Yadav P, Watts A, Rao MV, Chinnusamy V (2020) CRISPR-Cas9 mediated genome editing of drought and salt tolerance (OsDST) gene in indica mega rice cultivar MTU1010. Physiol Mol Biol Plants 26:1099–1110

Sashidhar N, Harloff HJ, Potgieter L, Jung C (2020) Gene editing of three BnITPK genes in tetraploid oilseed rape leads to significant reduction of phytic acid in seeds. Plant Biotechnol J 18(11):2241–50

Schornack S, Moscou MJ, Ward ER, Horvath DM (2013) Engineering plant disease resistance based on TAL effectors. Annu Rev Phytopathol 51(1):383–406

Schultz J, Copley RR, Doerks T, Ponting CP, Bork P (2000) SMART: a web-based tool for the study of genetically mobile domains. Nucleic Acids Res 28(1):231–234

Segal DJ, Beerli RR, Blancafort P, Dreier B, Effertz K, Huber A, Koksch B, Lund CV, Magnenat L, Valente D, Barbas CF (2003) Evaluation of a modular strategy for the construction of novel polydactyl zinc finger DNA-binding proteins. Biochemist 42(7):2137–2148

Segura V, Vilhjalmsson BJ, Platt A, Korte A, Seren U, Long Q, Nordborg M (2012) An efficient multi-locus mixed-model approach for genome-wide association studies in structured populations. Nat Genet 44(7):825–830

Shan Q, Wang Y, Li J, Zhang Y, Chen K, Liang Z, Zhang K, Liu J, Xi JJ, Qiu JL, Gao C (2013) Targeted genome modification of crop plants using a CRISPR-Cas system. Nat Biotechnol 31: 686–688

Shan Q, Zhang Y, Chen K, Zhang K, Gao C (2015) Creation of fragrant rice by targeted knockout of the OsBADH 2 gene using TALEN technology. Plant Biotechnol J 13(6):791–800

Shang L, Li X, He H, Yuan Q, Song Y, Wei Z, Lin H, Hu M, Zhao F, Zhang C, Li Y (2022) A super pan-genomic landscape of rice. Cell Res 32(10):878–896

Shen L, Wang C, Fu Y, Wang J, Liu Q, Zhang X, Yan C, Qian Q, Wang K (2018) QTL editing confers opposing yield performance in different rice varieties. J Integr Plant Biol 60(2):89–93

Shi J, Gao H, Wang H, Lafitte HR, Archibald RL, Yang M, Hakimi SM, Mo H, Habben JE (2017) ARGOS 8 variants generated by CRISPR-Cas9 improve maize grain yield under field drought stress conditions. Plant Biotechnol J 15(2):207–216

Shlush BI, Samach A, Melamed-Bessudo C, Ben-Tov D, Dahan-Meir T, Filler-Hayut S, Levy AA (2020) CRISPR/Cas9 induced somatic recombination at the CRTISO locus in tomato. Genes 12(1):59

Shikha K, Shahi JP, Vinayan MT, Zaidi PH, Singh AK, Sinha B (2021) Genome-wide association mapping in maize: status and prospects. 3 Biotech 11(5):244

Shipman EN, Yu J, Zhou J, Albornoz K, Beckles DM (2021) Can gene editing reduce postharvest waste and loss of fruit, vegetables, and ornamentals? Hortic Res 8:1

Shmakov S, Smargon A, Scott D, Cox D, Pyzocha N, Yan W, Abudayyeh OO, Gootenberg JS, Makarova KS, Wolf YI, Severinov K (2017) Diversity and evolution of class 2 CRISPR–Cas systems. Nat Rev Microbiol 15(3):169–82

Shukla VK, Doyon Y, Miller JC, DeKelver RC, Moehle EA, Worden SE, Mitchell JC, Arnold NL, Gopalan S, Meng X, Choi VM (2009) Precise genome modification in the crop species Zea mays using zinc-finger nucleases. Nature 459(7245):437–441

Singh D, Wang X, Kumar U, Gao L, Noor M, Imtiaz M, Singh RP, Poland J (2019) High-throughput phenotyping enabled genetic dissection of crop lodging in wheat. Front Plant Sci 10:394

Smith JJ (2001) Studies on zinc finger chimeric nucleases to refine and improve them for in vivo applications. The Johns Hopkins University

Song JH, Shin G, Kim HJ, Lee SB, Moon JY, Jeong JC, Choi HK, Kim IA, Song HJ, Kim CY, Chung YS (2022) Mutation of GmIPK1 gene using CRISPR/Cas9 reduced phytic acid content in soybean seeds. Int J Mol Sci 23(18):10583

Sood S, Joshi DC, Rajashekara H, Tiwari A, Bhinda MS, Kumar A, Kant L, Pattanayak A (2023) Deciphering the genomic regions governing major agronomic traits and blast resistance using genome wide association mapping in finger millet. Gene 854:147115

Srivastava RK, Singh RB, Pujarula VL, Bollam S, Pusuluri M, Chellapilla TS, Yadav RS, Gupta R (2020) Genome-Wide association studies and genomic selection in pearl millet: advances and prospects. Front Genet 10:1389

Sternberg SH, LaFrance B, Kaplan M, Doudna JA (2015) Conformational control of DNA target cleavage by CRISPR–Cas9. Nature 527(7576):110–113

Stoddard BL (2005) Homing endonuclease structure and function. Q Rev Biophys 38(1):49–95

Stoddard BL (2011) Homing endonucleases: from microbial genetic invaders to reagents for targeted DNA modification. Structure 19(1):7–15

Stormo GD (2000) DNA binding sites: representation and discovery. Bioinformatics 16:16–23

Stormo GD, Zhao Y (2010) Determining the specificity of protein-DNA interactions. Nat Rev Genet 11:751–760

Strauss SH, Sax JK (2016) Ending event-based regulation of GMO crops. Nat Biotechnol 34(5): 474–477

Sun N, Zhao H (2013) Transcription activator-like effector nucleases (TALENs): a highly efficient and versatile tool for genome editing. Biotechnol Bioeng 110(7):1811–1821

Sun N, Abil Z, Zhao H (2012) Recent advances in targeted genome engineering in mammalian systems. Biotechnol J 7(9):1074–1087

Sun C, Hu Z, Zheng T, Lu K, Zhao Y, Wang W, Shi J, Wang C, Lu J, Zhang D, Li Z (2017a) RPAN: rice pan-genome browser for~ 3000 rice genomes. Nucleic Acids Res 45(2): 597–605
Sun Y, Jiao G, Liu Z, Zhang X, Li J, Guo X, Du W, Du J, Francis F, Zhao Y, Xia L (2017b) Generation of high-amylose rice through CRISPR/Cas9-mediated targeted mutagenesis of starch branching enzymes. Front Plant Sci 8:298
Sun D, Cen H, Weng H, Wan L, Abdalla A, El-Manawy AI, Zhu Y, Zhao N, Fu H, Tang J, Li X (2019) Using hyperspectral analysis as a potential high throughput phenotyping tool in GWAS for protein content of rice quality. Plant Methods 15:1–6
Svitashev S, Young JK, Schwartz C, Gao H, Falco SC, Cigan AM (2015) Targeted mutagenesis, precise gene editing, and site-specific gene insertion in maize using Cas9 and guide RNA. Plant Physiol 169(2):931–945
Swarup S, Yang Y, Kingsley MT, Gabriel DW (1992) An Xanthomonas citri pathogenicity gene, pthA, pleiotropically encodes gratuitous avirulence on nonhosts. Mol Plant-Microbe Interact 5(3):204–213
Sweeney DW, Sun J, Taagen E, Sorrells ME (2019) Genomic selection in wheat. In: Applications of genetic and genomic research in cereals. Woodhead Publishing, pp 273–302
Szczepek M, Brondani V, Büchel J, Serrano L, Segal DJ, Cathomen T (2007) Structure-based redesign of the dimerization interface reduces the toxicity of zinc-finger nucleases. Nat Biotechnol 25(7):786–793
Szklarczyk D, Franceschini A, Wyder S, Forslund K, Heller D, Huerta-Cepas J, Simonovic M, Roth A, Santos A, Tsafou KP, Kuhn M (2015) STRING v10: protein–protein interaction networks, integrated over the tree of life. Nucleic Acids Res 43(D1):D447–D452
Tang X, Zheng X, Qi Y, Zhang D, Cheng Y, Tang A, Voytas DF, Zhang Y (2016) A single transcript CRISPR-Cas9 system for efficient genome editing in plants. Mol Plant 9(7): 1088–1091
Tang Y, Huang J, Ji H, Pan L, Hu C, Qiu X, Zhu H, Sui J, Wang J, Qiao L (2022) Identification of AhFatB genes through genome-wide analysis and knockout of AhFatB reduces the content of saturated fatty acids in peanut (Arichis hypogaea L.). Plant Sci 319:111247
Tanksley SD (2004) The genetic, developmental, and molecular bases of fruit size and shape variation in tomato. Plant Cell 16 Suppl:S181–S189
Tao Y, Luo H, Xu J, Cruickshank A, Zhao X, Teng F, Hathorn A, Wu X, Liu Y, Shatte T, Jordan D (2021) Extensive variation within the pan-genome of cultivated and wild sorghum. Nat Plants 7(6):766–773
Tesson L, Usal C, Ménoret S, Leung E, Niles BJ, Remy S, Santiago Y, Vincent AI, Meng X, Zhang L, Gregory PD (2011) Knockout rats generated by embryo microinjection of TALENs. Nat Biotechnol 29(8):695–696
Thornsberry JM, Goodman MM, Doebley J, Kresovich S, Nielsen D, Buckler ES (2001) Dwarf8 polymorphisms associate with variation in flowering time. Nat Genet 28(3):286–289
Tsai SQ, Wyvekens N, Khayter C, Foden JA, Thapar V, Reyon D, Goodwin MJ, Aryee MJ, Joung JK (2014) Dimeric CRISPR RNA-guided FokI nucleases for highly specific genome editing. Nat Biotechnol 32(6):569–576
Tsai SQ, Zheng Z, Nguyen NT, Liebers M, Topkar VV, Thapar V, Wyvekens N, Khayter C, Iafrate AJ, Le LP, Aryee MJ (2015) GUIDE-seq enables genome-wide profiling of off-target cleavage by CRISPR-Cas nucleases. Nat Biotechnol 33(2):187–197
Ueta R, Abe C, Watanabe T, Sugano SS, Ishihara R, Ezura H, Osakabe Y, Osakabe K (2017) Rapid breeding of parthenocarpic tomato plants using CRISPR/Cas9. Sci Rep 7(1):507
Urnov FD, Rebar EJ, Holmes MC, Zhang HS, Gregory PD (2010) Genome editing with engineered zinc finger nucleases. Nat Rev Genet 11(9):636–646
Varshney RK, Sinha P, Singh VK, Kumar A, Zhang Q, Bennetzen JL (2020) 5Gs for crop genetic improvement. Curr Opin Plant Biol 56:190–196

Varshney RK, Roorkiwal M, Sun S, Bajaj P, Chitikineni A, Thudi M, Singh NP, Du X, Upadhyaya HD, Khan AW, Wang Y (2021) A chickpea genetic variation map based on the sequencing of 3,366 genomes. Nature 599(7886):622–627

Vernet A, Meynard D, Lian Q, Mieulet D, Gibert O, Bissah M, Rivallan R, Autran D, Leblanc O, Meunier AC, Frouin J (2022) High-frequency synthetic apomixis in hybrid rice. Nat Commun 13(1):7963

Vetriventhan M, Upadhyaya HD, Deshpande S, Johnson MS, Wallace JG, Victor A, Naresh D, Rayaprolu L, Singh K, Mayes S (2024) Genome-wide assessment of population structure and association mapping for agronomic and grain nutritional traits in proso millet (Panicum miliaceum L.). Sci Rep 14(1):21920

Voytas DF (2013) Plant genome engineering with sequence-specific nucleases. Annu Rev Plant Biol 64:327–350

Voytas DF, Gao C (2014) Precision genome engineering and agriculture: opportunities and regulatory challenges. PLoS Biol 12(6):e1001877

Vu TV, Sivankalyani V, Kim EJ, Doan DT, Tran MT, Kim J, Sung YW, Park M, Kang YJ, Kim JY (2020) Highly efficient homology-directed repair using CRISPR/Cpf1-geminiviral replicon in tomato. Plant Biotechnol J 18(10):2133–2143

Wang Y, Tang H, DeBarry JD, Tan X, Li J, Wang X, Lee TH, Jin H, Marler B, Guo H, Kissinger JC (2012) MCScanX: a toolkit for detection and evolutionary analysis of gene synteny and collinearity. Nucleic Acids Res 40(7):e49

Wang Y, Cheng X, Shan Q, Zhang Y, Liu J, Gao C, Qiu JL (2014) Simultaneous editing of three homoeoalleles in hexaploid bread wheat confers heritable resistance to powdery mildew. Nat Biotechnol 32(9):947–951

Wang C, Shen L, Fu Y, Yan C, Wang K (2015) A simple CRISPR/Cas9 system for multiplex genome editing in rice. J Genet Genomics 42(12):703–706

Wang W, Pan Q, He F, Akhunova A, Chao S, Trick H, Akhunov E (2018a) Transgenerational CRISPR-Cas9 activity facilitates multiplex gene editing in allopolyploid wheat. CRISPR J 1(1): 65–74

Wang X, Xu Y, Hu Z, Xu C (2018b) Genomic selection methods for crop improvement: current status and prospects. Crop J 6:330–340

Wang W, Pan Q, Tian B, He F, Chen Y, Bai G, Akhunova A, Trick HN, Akhunov E (2019a) Gene editing of the wheat homologs of TONNEAU 1-recruiting motif encoding gene affects grain shape and weight in wheat. Plant J 100(2):251–264

Wang X, Zhang R, Song W, Han L, Liu X, Sun X, Luo M, Chen K, Zhang Y, Yang H, Yang G (2019b) Dynamic plant height QTL revealed in maize through remote sensing phenotyping using a high-throughput unmanned aerial vehicle (UAV). Sci Rep 9(1):3458

Wang Q, Tang J, Han B, Huang X (2020a) Advances in genome-wide association studies of complex traits in rice. Theor Appl Genet 133(5):1415–1425

Wang S, Yang Y, Guo M, Zhong C, Yan C, Sun S (2020b) Targeted mutagenesis of amino acid transporter genes for rice quality improvement using the CRISPR/Cas9 system. Crop J 8(3): 457–464

Wang Y, Fuentes RR, van Rengs WM, Effgen S, Zaidan MW, Franzen R, Susanto T, Fernandes JB, Mercier R, Underwood CJ (2024) Harnessing clonal gametes in hybrid crops to engineer polyploid genomes. Nat Genet 56(6):1075–1079

Ware D, Jaiswal P, Ni J, Pan X, Chang K, Clark K, Teytelman L, Schmidt S, Zhao W, Cartinhour S, McCouch S (2002) Gramene: a resource for comparative grass genomics. Nucleic Acids Res 30(1):103–105

Wei W, Li S, Wang Y, Wang B, Fan G, Zeng Q, Zhao F, Xu C, Zhang X, Tang T, Feng X, Shi J, Shi G, Zhang W, Song G, Li H, Wang F, Zhang Y, Li X, Wang D, Zhang W, Pei J, Wang X, Zhao Z (2021) Metabolome-based genome-wide association study provides genetic insights into the natural variation of foxtail millet. Front Plant Sci 12:665530

Wei X, Liu C, Chen X, Lu H, Wang J, Yang S, Wang K (2023) Synthetic apomixis with normal hybrid rice seed production. Mol Plant 16(3):489–492

White FF, Yang B (2009) Host and pathogen factors controlling the rice-X anthomonas oryzae interaction. Plant Physiol 150(4):1677–1686
Wiles MV, Qin W, Cheng AW, Wang H (2015) CRISPR–Cas9-mediated genome editing and guide RNA design. Mamm Genome 26:501–510
Wolfe SA, Nekludova L, Pabo CO (2000) DNA recognition by Cys2His2 zinc finger proteins. Annu Rev Biophys Biomol Struct 29(1):183–212
Woo JW, Kim J, Kwon SI, Corvalán C, Cho SW, Kim H, Kim SG, Kim ST, Choe S, Kim JS (2015) DNA-free genome editing in plants with preassembled CRISPR-Cas9 ribonucleoproteins. Nat Biotechnol 33(11):1162–1164
Wu D, Guo Z, Ye J, Feng H, Liu J, Chen G, Zheng J, Yan D, Yang X, Xiong X, Liu Q (2019) Combining high-throughput micro-CT-RGB phenotyping and genome-wide association study to dissect the genetic architecture of tiller growth in rice. J Exp Bot 70(2):545–561
Xavier A, Hall B, Hearst AA, Cherkauer KA, Rainey KM (2017) Genetic architecture of phenomic-enabled canopy coverage in Glycine max. Genetics 206(2):1081–1089
Xiao Q, Bai X, Zhang C, He Y (2022) Advanced high-throughput plant phenotyping techniques for genome-wide association studies: A review. J Adv Res 35:215–30
Xie E, Li Y, Tang D, Lv Y, Shen Y, Cheng Z (2019) A strategy for generating rice apomixis by gene editing. J Integr Plant Biol 61(8):911–916
Xu R, Yang Y, Qin R, Li H, Qiu C, Li L, Wei P, Yang J (2016) Rapid improvement of grain weight via highly efficient CRISPR/Cas9-mediated multiplex genome editing in rice. J Genet Genom 43(8):529–532
Xu X, Zhiguo E, Zhang D, Yun Q, Zhou Y, Niu B, Chen C (2021a) OsYUC11-mediated auxin biosynthesis is essential for endosperm development of rice. Plant Physiol 185(3):934–950
Xu Y, Lin Q, Li X, Wang F, Chen Z, Wang J, Li W, Fan F, Tao Y, Jiang Y, Wei X (2021b) Fine-tuning the amylose content of rice by precise base editing of the Wx gene. Plant Biotechnol J 19(1):11–13
Yang B, Sugio A, White FF (2006) Os8N3 is a host disease-susceptibility gene for bacterial blight of rice. Proc Natl Acad Sci USA 103(27):10503–10508
Yang W, Guo Z, Huang C, Duan L, Chen G, Jiang N, Fang W, Feng H, Xie W, Lian X, Wang G (2014) Combining high-throughput phenotyping and genome-wide association studies to reveal natural genetic variation in rice. Nat Commun 5:5087
Yang X, Wang J, Xia X, Zhang Z, He J, Nong B, Luo T, Feng R, Wu Y, Pan Y, Xiong F (2021) OsTTG1, a WD40 repeat gene, regulates anthocyanin biosynthesis in rice. Plant J 107(1): 198–214
Yasir M, Kanwal HH, Hussain Q, Riaz MW, Sajjad M, Rong J, Jiang Y (2022) Status and prospects of genome-wide association studies in cotton. Front Plant Sci 13:1019347
Yates S, Mikaberidze A, Krattinger SG, Abrouk M, Hund A, Yu K, Studer B, Fouche S, Meile L, Pereira D, Karisto P (2019) Precision phenotyping reveals novel loci for quantitative resistance to septoria tritici blotch. Plant Phenomics 2019:3285904
Youssef D, Nihou A, Partier A, Tassy C, Paul W, Rogowsky PM, Beckert M, Barret P (2018) Induction of targeted deletions in transgenic bread wheat (Triticum aestivum L.) using customized meganuclease. Plant Mol Biol Report 36:71–81
Yu QH, Wang B, Li N, Tang Y, Yang S, Yang T, Xu J, Guo C, Yan P, Wang Q, Asmutola P (2017) CRISPR/Cas9-induced targeted mutagenesis and gene replacement to generate long-shelf life tomato lines. Sci Rep 7(1):11874
Yuan M, Zhu J, Gong L, He L, Lee C, Han S, Chen C, He G (2019) Mutagenesis of FAD2 genes in peanut with CRISPR/Cas9 based gene editing. BMC Biotechnol 19:1–7
Yuste-Lisbona FJ, Fernández-Lozano A, Pineda B, Bretones S, Ortíz-Atienza A, García-Sogo B, Müller NA, Angosto T, Capel J, Moreno V, Jiménez-Gómez JM (2020) ENO regulates tomato fruit size through the floral meristem development network. Proc Natl Acad Sci USA 117(14): 8187–8195

Zetsche B, Gootenberg JS, Abudayyeh OO, Slaymaker IM, Makarova KS, Essletzbichler P, Volz SE, Joung J, Van Der Oost J, Regev A, Koonin EV (2015) Cpf1 is a single RNA-guided endonuclease of a class 2 CRISPR-Cas system. Cell 163(3):759–771

Zhai Y, Yu K, Cai S, Hu L, Amoo O, Xu L, Yang Y, Ma B, Jiao Y, Zhang C, Khan MH (2020) Targeted mutagenesis of BnTT8 homologs controls yellow seed coat development for effective oil production in Brassica napus L. Plant Biotechnol J 18(5):1153–1168

Zhang Z, Ersoz E, Lai CQ, Todhunter RJ, Tiwari HK, Gore MA, Bradbury PJ, Yu J, Arnett DK, Ordovas JM (2010) Mixed linear model approach adapted for genome-wide association studies. Nat Genet 42(4):355–360

Zhang Y, Zhang F, Li X, Baller JA, Qi Y, Starker CG, Bogdanove AJ, Voytas DF (2013) Transcription activator-like effector nucleases enable efficient plant genome engineering. Plant Physiol 161(1):20–27

Zhang Y, Liang Z, Zong Y, Wang Y, Liu J, Chen K, Qiu JL, Gao C (2016) Efficient and transgene-free genome editing in wheat through transient expression of CRISPR/Cas9 DNA or RNA. Nat Commun 7(1):12617

Zhang Y, Bai Y, Wu G, Zou S, Chen Y, Gao C, Tang D (2017) Simultaneous modification of three homoeologs of Ta EDR 1 by genome editing enhances powdery mildew resistance in wheat. Plant J 91(4):714–724

Zhang J, Zhang H, Botella JR, Zhu JK (2018) Generation of new glutinous rice by CRISPR/Cas9-targeted mutagenesis of the Waxy gene in elite rice varieties. J Integr Plant Biol 60(5):369–375

Zhang A, Liu Y, Wang F, Li T, Chen Z, Kong D, Bi J, Zhang F, Luo X, Wang J, Tang J (2019) Enhanced rice salinity tolerance via CRISPR/Cas9-targeted mutagenesis of the OsRR22 gene. Mol Breed 39(3):47

Zhang D, Tang S, Xie P, Yang D, Wu Y, Cheng S, Du K, Xin P, Chu J, Yu F, Xie Q (2022) Creation of fragrant sorghum by CRISPR/Cas9. J Integr Plant Biol 64(5):961–964

Zhao DS, Li QF, Zhang CQ, Zhang C, Yang QQ, Pan LX, Ren XY, Lu J, Gu MH, Liu QQ (2018a) GS9 acts as a transcriptional activator to regulate rice grain shape and appearance quality. Nat Commun 9(1):1240

Zhao Q, Feng Q, Lu H, Li Y, Wang A, Tian Q, Zhan Q, Lu Y, Zhang L, Huang T, Wang Y (2018b) Pan-genome analysis highlights the extent of genomic variation in cultivated and wild rice. Nat Genet 50(2):278–284

Zhou J, Xin X, He Y, Chen H, Li Q, Tang X, Zhong Z, Deng K, Zheng X, Akher SA, Cai G (2019) Multiplex QTL editing of grain-related genes improves yield in elite rice varieties. Plant Cell Rep 38:475–485

Zhu J, Song N, Sun S, Yang W, Zhao H, Song W, Lai J (2016) Efficiency and inheritance of targeted mutagenesis in maize using CRISPR-Cas9. J Genet Genom 43(1):25–36

Zia MAB, Yousaf MF, Asim A, Naeem M (2022) An overview of genome-wide association mapping studies in Poaceae species (model crops: wheat and rice). Mol Biol Rep 49(12): 12077–12090

Zsögön A, Čermák T, Naves ER, Notini MM, Edel KH, Weinl S, Freschi L, Voytas DF, Kudla J, Peres LE (2018) De novo domestication of wild tomato using genome editing. Nat Biotechnol 36(12):1211–1216

MIX
Papier aus verantwortungsvollen Quellen
Paper from responsible sources
FSC® C105338

If you have any concerns about our products,
you can contact us on
ProductSafety@springernature.com

In case Publisher is established outside the EU,
the EU authorized representative is:
Springer Nature Customer Service Center GmbH
Europaplatz 3, 69115 Heidelberg, Germany

Printed by Libri Plureos GmbH
in Hamburg, Germany